T0262263

# New Frontiers in Biochemistry

# New Frontiers in Biochemistry

Edited by **Oliver Stone**

New York

Published by Callisto Reference,
106 Park Avenue, Suite 200,
New York, NY 10016, USA
www.callistoreference.com

**New Frontiers in Biochemistry**
Edited by Oliver Stone

International Standard Book Number: 978-1-63239-473-6 (Hardback)

# Contents

# Preface

Over the years, biochemistry has become significant in classifying living processes, so much so that many scientists in the field of life sciences are involved in biochemical research. This book presents an analysis of the research area of proteins, enzymes, cellular mechanisms and chemical compounds that are used in appropriate methods. It includes the basic issues and some of the current advancements in biochemistry. Emphasis is given on both theoretical and experimental facets of modern biochemistry. This book caters to students, researchers, biologists, chemists, chemical engineers and professionals who are keen to know more about biochemistry, molecular biology and other related fields. The chapters within the book have been contributed by renowned international scientists with expertise in protein biochemistry, enzymology, molecular biology and genetics; many of whom are active in biochemical and biomedical research. It will provide information for scientists about the complexities of some biochemical procedures; and will stimulate both professionals and students to devote a part of their future research in understanding related mechanisms and methods of biochemistry.

This book is a comprehensive compilation of works of different researchers from varied parts of the world. It includes valuable experiences of the researchers with the sole objective of providing the readers (learners) with a proper knowledge of the concerned field. This book will be beneficial in evoking inspiration and enhancing the knowledge of the interested readers.

In the end, I would like to extend my heartiest thanks to the authors who worked with great determination on their chapters. I also appreciate the publisher's support in the course of the book. I would also like to deeply acknowledge my family who stood by me as a source of inspiration during the project.

**Editor**

# Part 1

# Proteins and Hormones

# Peptides and Peptidomimetics as Tools to Probe Protein-Protein Interactions – Disruption of HIV-1 gp41 Fusion Core and Fusion Inhibitor Design

Lifeng Cai, Weiguo Shi and Keliang Liu
*Beijing Institute of Pharmacology & Toxicology, Beijing*
*China*

## 1. Introduction

Protein-protein interactions play important roles in many critical processes in the life sciences, such as signal transduction, lipid membrane fusion, receptor recognition, *etc.*, and many of them are important targets for drug development and design (Wilson 2009; Tavassoli 2011). Unlike an enzyme-substrate interaction which usually has a deep binding pocket in the protein for substrate binding, protein-protein interactions usually involve a large interacting interface; as a result, it is a big challenge for small molecule drugs to efficiently competitively occupy the interface and disrupt protein-protein interactions that modulate these life processes. Proteins are natural ligands that can modulate protein-protein interactions; however, they are not ideal therapeutic agents because of their expensive production costs and the fact that they are not able to be administered orally. In protein-protein interactions, energy is not always equally distributed throughout the binding interface; a couple of focused areas may account for the main protein-protein interaction energy, called a hot spot, which can be the target for a small molecule protein-protein interaction inhibitor (PPII).

Peptides, with suitable molecular size, provide a bridge between protein and small molecule drugs. Similar to proteins, many peptides are natural ligands that modulate protein-protein interactions in important life processes; they are used as drug leads and/or modified to increase potency and selectivity. Compared with small molecules, peptides are more efficient PPIIs due to their relatively large size, and can be useful tools to probe protein-protein interactions for PPII design.

HIV-1 gp41 mediated virus-cell membrane fusion is critical for HIV-1 infection and *in vivo* propagation (Eckert & Kim 2001; Caffrey 2011), and the mechanism is shared by many other viruses using a class 1 fusion protein as membrane fusion machinery, including some life threatening pathogens such as influenza virus, respiratory syncytial virus (RSV), Ebola virus, and severe acute respiratory syndrome (SARS) virus (Harrison 2008). A critical step in HIV-1 infection is a protein-protein interaction between the gp41 N- and C-terminal heptad repeats (NHR and CHR), that form a coiled-coil six-helical bundle (6-HB), providing energy for virus-cell membrane fusion (Fig. 1). Peptides derived from CHR or NHR can interact with their counterparts in gp41 to prevent fusogenic 6-HB formation and inhibit HIV-1-cell membrane fusion, thus preventing HIV-1 infection and replication. T20 (Fuzeon, enfuvirtide), a 36-mer peptide from HIV-1 gp41 CHR, was approved by the USA FDA in 2003 as the first fusion inhibitor for salvage therapy in HIV/AIDS patients unresponsive to common antiretroviral therapy. Its application has been limited by i) the high cost of peptide synthesis, ii) rapid *in vivo* proteolysis, and iii) poor efficacy against emerging T20-resistant strains. These drawbacks have called for a new generation of fusion inhibitors with improved antiviral and pharmacokinetic profiles.

In this chapter, we will focus on the development of HIV-1 fusion inhibitors, concentrating on C-peptide fusion inhibitors and their peptidomimetics, which have been used as probes and tools to elucidate gp41 NHR-CHR interactions for future fusion inhibitor design and improve, and in the long run, the development of small molecule inhibitors that can disrupt this important protein-protein interaction.

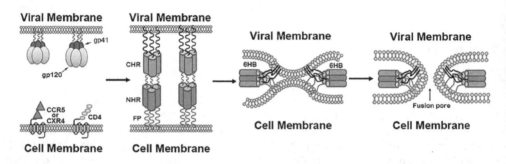

Fig. 1. HIV-1 gp41 mediated virus-cell membrane fusion.

## 2. Peptides as a model to build the HIV-1 gp41 fusion core

HIV-1 uses an envelope protein (ENV) mediated virus-cell membrane fusion to enter host cells for infection (Eckert & Kim 2001). HIV-1 ENV is composed of noncovalently associated gp120/gp41 trimers that form spikes and decorate the viral surface, in which the metastable transmembrane subunit gp41 is sequestered by the cell surface subunit gp120. During HIV-1 infection, gp120 first interacts with the T-cell receptor CD4, ensuring the viruses approach the target cells; then, the coreceptor binding sites in gp120 are sequentially exposed and gp120-coreceptor (CCR5 or CXR4) binding follows (Fig. 1). The resulting dissociation of the gp120-gp41 complex and the release of the unstable gp41 subunit trigger virus-cell membrane fusion. First, gp41 inserts into the target cell membrane using its fusion peptide, resulting in a pre-hairpin intermediate (PHI) in which its C-terminus anchors to the viral membrane and its N-terminus inserts into the host cell membrane, bridging the viral and cellular membranes (Fig. 1). The gp41 PHI automatically undergoes structure rearrangement with its NHR and CHR folding towards each other to form the fusogenic 6-HB. The energetic 6-HB formation drives the juxtaposition of the viral and cellular membrane, and finally results in virus-cell membrane fusion (Fig. 1). Agents that target the presumed gp41 PHI to prevent fusogenic 6-HB formation can terminate the virus-cell membrane fusion processes and be used as fusion inhibitors for antiretroviral therapy (Cai & Jiang 2010).

The discovery of potent anti-HIV peptides from HIV-1 gp41 NHR and CHR sequences suggests that gp41 is a target for fusion inhibitors (Wild et al. 1992; Jiang et al. 1993; Wild et al. 1994); these exogenous HIV-1 gp41 peptides interact with their counterparts in the gp41 6-HB, forming an unproductive complex that prevents gp41 fusion core formation. During the membrane fusion process, HIV-1 gp41 progressively undergoes a conformational change, and the gp41 PHI target exists for only a couple of minutes and then rapidly folds into a 6-HB; therefore, gp41 and its ectodomain are not suitable targets for a fusion inhibitor. Efforts to obtain a whole structure of the gp41 ectodomain also have been unsuccessful. So, the identification of a stable target in the PHI or gp41 fusion core is necessary for understanding the mechanism of gp41 mediated virus-cell membrane fusion for fusion inhibitor design and development.

The HIV-1 gp41 fusogenic 6-HB core has been constructed using synthesized peptides from the related gp41 wild-type sequences. Typical resolved crystal structures of the 6-HB fusogenic core include the N36/C34 complex (Chan et al. 1997), the IQNgp41/C43 complex (Weissenhorn et al. 1997), and the N34(L6)C28 trimer (Tan et al. 1997). These crystal structures provide atomic resolution of the interactions between NHR and CHR, verifying that NHR and CHR can be both a target and ligand from which a pharmacophore model can be deduced for fusion inhibitor design and optimization.

The crystal structures show that a parallel coiled-coil trimerized NHR forms the interior core, which is antiparallel packed with three CHR helices, to form a 6-HB (Fig. 2a,2b) (Chan et al. 1997). In the NHR interior core, the N-peptide uses its amino acid residues at the $a$ and $d$ positions of the heptads for self trimerization to stabilize the core; while the $e$ and $g$ residues of two adjacent helices form three hydrophobic grooves along the whole NHR trimer, which serve as targets that interact with the $a$ and $d$ residues of the C-peptides. Each groove contains a particularly deep cavity: Val-570, Lys-574, and Gln-577 from the left N36 (gp41$_{546-581}$) helix form the left side; Leu-568, Trp- 571, and Gly-572 from the right N36 helix form the right side; and Thr-569, Ile-573, and Leu-576 form the floor, resulting in a pocket of ~16 Å long, 7 Å wide, and 5–6 Å deep (Fig. 2d). With the exception of Ile-573, all of the residues forming the cavity are identical between HIV-1 and SIV. The NHR deep pocket accommodates three hydrophobic residues from the abutting C34 (gp41$_{628-661}$) helix: Ile-635, Trp-631, and Trp-628 constitute a WWI motif (Fig. 2c). The interaction between the NHR pocket and the WWI motif is predominately hydrophobic. A salt bridge between Lys-574 of NHR and Asp-632 of CHR immediately to the left of the cavity is also important for the NHR-CHR interaction (Chan et al. 1997). In addition to be the main binding sites for the C-peptide, the deep NHR pocket is also an attractive target for small molecule fusion inhibitors. Besides the deep pocket, the rest of the groove along the NHR helices also makes extensive contact with CHR, providing additional energy to stabilize the 6-HB. The N36/C34 complex shows striking structural similarity to the low-pH-induced conformation of the influenza HA2 subunit (TBHA2) and the TM subunit of Mo-MLV, both of which have been proposed to be in a fusogenic conformation, suggesting a common mechanism of virus-cell membrane fusion among enveloped viruses (Chan et al. 1997).

During 6-HB formation, NHR and CHR are mutual target and ligand, so either can be the target for fusion inhibitor design. In a 6-HB, NHRs form a trimerized interior core that contains three grooves, and each with a deep pocket, which is more like a target, especially for small molecule fusion inhibitors. An electrostatic potential map of the N36 coiled-coil trimer shows that its surface is largely uncharged; and the grooves that are the sites for C34 interaction are aligned with predominantly hydrophobic residues that would be expected to lead to aggregation upon exposure to solvent. In contrast, the N36/C34 complex shows a much more highly charged surface due to acidic residues on the outside of the C34 helices, resulting in greater solubility of the heterodimeric complex (Chan et al. 1997). As a result, N-peptides are prone to aggregate in the absence of C-peptides under physiological conditions. This also accounts for a much weaker inhibitory potency for N-peptides compared to C-peptides, since they must form a stable discrete trimerized inner core to efficiently interact with the CHR. Thus, construction of a stable and soluble discrete trimerized gp41 NHR core as a target is important for fusion inhibitor design and development.

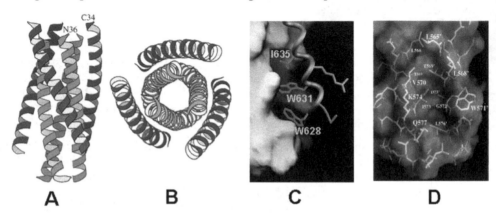

**A**  **B**  **C**  **D**

Fig. 2. Crystal structures of the HIV-1 gp41 fusion core. (A) 6-HB structure of the gp41 N36/C34 fusion core; (B) the top to bottom view of the N36/C34 6-HB structure (Chan et al. 1997); (C) the deep pocket in the NHR groove interacts with the WWI motif of CHR (Chan et al. 1998); (D) the NHR deep pocket.

The key for constructing an efficient NHR target is to promote trimerization of N-peptides without changing their native binding sites and conformation. Addition of physicochemical restraints in N-peptides has been shown to be an efficient way to construct a stable and discrete NHR trimer. Typical NHR constructs include: IQN17 (3) and IZN17 (4) (Eckert & Kim 2001), 5-helix (Root et al. 2001; Frey et al. 2006), and Env2.0 (5) and Env5.0 (6) (Cai & Gochin 2007; Cai et al. 2009). These stable NHR-trimers can be efficient targets for fusion inhibitor discovery and development. Through forming discrete and stable trimers, they are also highly potent HIV-1 fusion inhibitors by themselves. The sequences of the N-peptide targets are shown in Fig. 3.

IQN17/IZN17 (Fig. 4): A trimeric coiled-coil GCN4 isoleucine zipper was used to construct the first HIV-1 gp41 fusion core for an x-ray crystallographic study (Weissenhorn et al. 1997). IQN17 was constructed by fusing a modified GCN4-pI$_Q$I peptide sequence to the 17-mer N-peptide gp41$_{565-581}$ (N17) that comprises the gp41 hydrophobic pocket (Eckert et al. 1999). The resulting peptide, IQN17, is a fully helical discrete trimer in solution, as determined by circular dichroism (CD) and sedimentation equilibrium experiments. The crystal structure of the IQN17/D10-p1 complex, a cyclic D-peptide fusion inhibitor, showed that the overall architecture of the HIV-1 gp41 hydrophobic pocket in the complex is almost identical to that in the wild-type HIV-1 gp41 N36/C34 structure, with a $C_\alpha$ root mean square deviation (rmsd) of 0.65 Å. In follow-up studies, a new version, IZN17, was designed using the same strategy. IZN17 is more thermally stable than IQN17, with a $T_m > 100$ °C, compared with ~100 °C for IQN17; the enhancement of thermal stability was further confirmed by measuring the $T_m$ in 2 M guanidine chloride, with a $T_m$ of 66 °C and 74 °C for IQN17 and QZN17, respectively. IZN17 is also more soluble than IQN17 under physiological conditions (Eckert & Kim 2001). Both IQN17 and IZN17 were used as targets in a mirror-image phage display experiment to identify D-peptide fusion inhibitors (Eckert et al. 1999; Welch et al. 2007; Welch et al. 2010).

| | | |
|---|---|---|
| DP107 | (1) | NNLLRAIEAQQHLLQLTVWGIKQLQARILAVERYLKDQ |
| N36 | (2) | SGIVQQQNNLLRAIEAQQHLLQLTVWGIKQLQARIL |
| IQN17 | (3) | RMKQIEDKIEEIESKQKKIENEIARIKKLLQLTVWGIKQLQARIL |
| IZN17 | (4) | IKKEIEAIKKEQEAIKKKIEAIEKLLQLTVWGIKQLQARIL |
| Env2.0 | (5) | Bpy-GQAVEAQQHLLQLTVWGIKQLQARILAVEKK |
| Env5.0 | (6) | Bpy-GQAVSGIVQQQNNLLRAIEAQQHLLQLTVWGIKQLQARILAVEKK |

Fig. 3. NHR target sequences. The sequences and groups responsible for physicochemical constraint are shown in grey.

5-Helix (Fig. 4): 5-Helix was designed using the 6-HB as a motif (Root et al. 2001). In 5-helix, five of the six helices that make up the 6-HB core structure are connected by short peptide linkers. The 5-helix protein lacks a third C-peptide helix, and this vacancy is expected to create a high-affinity binding site for the gp41 CHR. Under physiological conditions, 5-helix is soluble and a well folded protein that adopts >95% helical content, as expected from the design, and is extremely stable. In addition, denaturation was not observed, even at 96 °C or in 8 M guanidine chloride. 5-Helix interacts strongly and specifically with C-peptides, inducing a helical conformation in the bound C-peptide as judged by CD. 5-Helix was successfully used as the target in a fluorescence polarization assay to identify small molecule fusion inhibitors (Frey et al. 2006).

Env2.0/Env5.0 (Fig. 4): A trivalent coordination metal complex was used to fortify the gp41 NHR trimer (Gochin et al. 2003). 5-Carboxy-2,2'-bipyridine (BPY) was attached to an N-peptide that contains a deep pocket. Addition of a metal ion such as $Fe^{2+}$ or $Ni^{2+}$ resulted in the formation of a tris-BPY metal complex, which stabilizes the coiled-coil structure. The resulting magenta $Fe^{2+}(BPY)_3$ complex solution was due to a $Fe^{2+}$–BPY charge transfer band at 545 nm and confirmed $Fe^{2+}$-BPY binding. The apo-Env2.0 displayed 40% α-helical structure that increased to 89% upon the addition of $Fe^{2+}$ ions, as measured by CD. The integrity of the binding grooves in $Fe^{2+}(Env2.0)_3$ was confirmed by its efficient binding with a matched C-peptide, as shown by CD and NMR (Gochin et al. 2006). The 545 nm absorbance agrees well with the emission maxima of the fluorophores fluorescein and Lucifer yellow. Fluorescence quenching by fluorescence resonance energy transfer (FRET) should occur if the fluorophore is brought close to the $Fe^{2+}$–BPY center. This enables direct determination of binding by using a fluorophore labeled C-peptide as the probe. Compounds which are able to bind to the NHR target and displace the probe can be measured with a competitive inhibition assay by following the recovery of probe fluorescence intensity (Cai & Gochin 2007). The BPY-metal complex FRET strategy is generally applicable to different interacting peptide pairs, as long as the two peptide sequences are matched. This has been confirmed by the development of a longer gp41 N-peptide/C-peptide pair, Env5.0 and CP5; the peptide pair showed nanomolar binding affinity and can be used to screen more potent fusion inhibitors. Env5.0 contains the whole groove, and it has been used to identify ligands that interact with the range of the groove outside of the deep pocket by designing suitable probes (Cai et al. 2009). In addition, Env2.0 has been successfully used as a target for a screening assay to identify small molecule fusion inhibitors (Cai & Gochin 2007; Zhou et al. 2010).

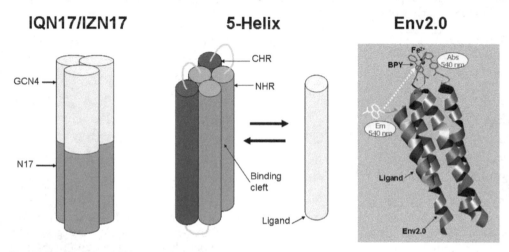

Fig. 4. Designed soluble and discrete NHR target.

In summary, peptides have been used to construct the HIV-1 gp41 fusion core, which is a 6-HB. Crystal structure analysis showed that the gp41 NHR trimer forms an interior core, which contains three hydrophobic grooves as the binding site for C-peptide. A deep pocket in the groove is a hot spot for the NHR-CHR interaction, and can be a target for small molecule fusion inhibitors. The NHR target can be constructed by adding physicochemical constraints in the N-peptides to promote the formation of a soluble and discrete NHR trimer, which can be used for screening HIV-1 fusion inhibitors targeting gp41 NHR or the deep pocket.

## 3. Peptide fusion inhibitors target the gp41 NHR core

Peptides, especially C-peptides (sequence see Fig. 5), can efficiently block the gp41 NHR-CHR interaction to inhibit HIV-cell membrane fusion and infection. They act in a dominant-negative manner by binding to the transiently exposed coiled-coil N-peptide region in the PHI (Eckert & Kim 2001). The wild-type C-peptide sequences have been shown to have low nanomolar $IC_{50}$ values for HIV-1 ENV mediated membrane fusion and viral infection. Peptide engineering has been employed on wild-type C-peptide sequences to obtain structure activity relationship (SAR) data for the peptide fusion inhibitors, resulting in peptides with an improved anti-HIV profile and a better understanding of the mechanism of gp41 mediated virus-cell membrane fusion (Otaka et al. 2002; Dwyer et al. 2007). The insight gained from these works was finally tested by the artificial design of peptide fusion inhibitors with few sequence homologies to natural peptides or protein sequences (Qi et al. 2008; Shi et al. 2008).

## 3.1 Peptides from the wild-type gp41 sequence

The first highly potent HIV-1 fusion inhibitors were independently discovered by two groups in the early 1990s, including SJ-2176 (gp41$_{630-659}$) (Jiang et al. 1993; Jiang et al. 1993) and DP178 (gp41$_{638-673}$, later named T20) (Wild et al. 1994), which were both derived from the gp41 CHR wild-type sequence. Due to their stronger anti-HIV activity compared with N-peptides, most of the exploited fusion inhibitors were C-peptides, among them, T20 (7) and C34 (gp41$_{628-661}$, 8) were extensively studied. C-peptide fusion inhibitors are usually unstructured in solution by themselves, and form α-helical structures in a 6-HB after interaction with NHR.

T20, originally named DP178, was developed into the first HIV-1 fusion inhibitor with the brand name Enfuvirtide (Lazzarin et al. 2003; Walmsley et al. 2003; Su et al. 2004). It has low nanomolar antiretroviral activity. Under physiological conditions, it is unstructured and cannot form a stable 6-HB with N-peptide; however, it is highly soluble, making it a good drug candidate. Its mechanism of action has been controversial until now, since it cannot form a 6-HB with N-peptide, which is an established interaction model of HIV-1 peptide fusion inhibitors that has been supported by x-ray crystallography (Chan et al. 1997). T20 does not contain the WWI motif necessary to bind with the primary NHR deep pocket. This may account for its relatively weak binding with NHR and the resulting loss of activity against emerging drug resistant HIV-1 isolates. The 8-residue C-terminus of T20 contains three Trp residues and is highly hydrophobic, which enables T20 to bind with the lipid membrane; thus, this 8-residue motif is called the lipid binding domain (LBD). The hydrophobic residues in the LBD are critical for T20 to maintain high anti-HIV activity, although the LBD elicits no anti-HIV activity by itself. It seems that T20 may interact with both the gp41 NHR groove and the lipid membrane to interfere with 6-HB formation, thus inhibiting HIV-1 infection (Liu et al. 2005; Liu et al. 2007).

C34 forms a stable 6-HB with NHR, thus preventing productive 6-HB formation, a mechanism well supported by x-ray crystallography (Chan et al. 1997). It also displays stronger antiretroviral activity than T20, while its poor solubility under physiological conditions hinders it as a promising drug candidate (Otaka et al. 2002). Like all other wild-type gp41 C-peptides, C34 is unstructured under physiological conditions, while it adopts a nearly full α-helical structure when interacting with N-peptide to form a 6-HB. C34 contains the WWI motif, so it can interact with the primary binding pocket in NHR to form a stable complex with N-peptides. N-PAGE has shown that C34 can form a stable 6-HB in the presence of N36 or N46; and thermal denaturation has shown that the N36/C34 complex displays typical two-state denaturation behavior with a $T_m$ value of ~61 °C (Pan et al. 2009). In a 6-HB, C34 uses residues $a$ and $d$ to interact with NHR. The $a$ and $d$ residues in the N-terminal half of C34 are uniformly hydrophobic and elicit a predominantly hydrophobic interaction with residues $e$ and $g$ in NHR and bury these residues in the 6-HB; and the $a$ and $d$ residues in the C-terminal half of C34 form a hydrophilic layer spanning four α-helical turns, which is assumed to match the similar hydrophilic layer in the related NHR sequence. Thus, C34 is widely used as a tool to study the mechanism of HIV fusion inhibitors, as well as the lead or template for next generation fusion inhibitor design, and will be discussed in Section 3.2.

CP32 (gp41$_{621-652}$, 14) is another identified highly potent wild-type gp41 C-peptide fusion inhibitor that targets NHR sequences other than T20 and C34 (He et al. 2008). It contains a 7-residue motif upstream of the C34 sequence. Interestingly, the CP32 sequence matches the

T21 sequence, the first identified peptide HIV-1 fusion inhibitor from gp41 under the name DP107 (Wild et al. 1992); the match is expected based on the N36/C34 complex and the anti-parallel interactions between gp41 NHR and CHR. CP32 contains the WWI motif, so it can interact with the NHR deep pocket to form a stable 6-HB. The CP32/T21 complex is ~100% α-helical with a $T_m$ of 82 °C, which is more stable than the N36/C34 complex. The discrete 6-HB conformation of the CP32/T21 complex was supported by N-PAGE, size exclusion chromatography, as well as analytical ultracentrifugation. CP32 showed an $IC_{50}$ value of 4.2 nM against HIV-1 ENV mediated cell-cell fusion and an $IC_{50}$ value of 4.6 nm against HIV-$1_{NL4-3wt}$ infection of MT-2 cells. Although it has a similar potency as C34 against wild-type HIV-1 isolates, CP32 is ~20-fold and >500-fold more potent than C34 and T20, respectively, against the drug-resistant HIV-$1_{NL4-3-V38SE/N42S}$ isolate, possibly due to the fact that it targets a different sequence in gp41 NHR.

There are also longer C-peptides fusion inhibitors, such as C43 and C52, which include both partial or complete sequences of C34 and T20; however, none of these longer peptides have improved anti-HIV potency compared with T20 and C34 (Deng et al. 2007). In the PHI, a long groove may expand throughout the gp41 NHR and beyond, and it may be targeted by its CHR counterpart. The C-terminal half of the gp41 ectodomain may make contact with the N-terminal half at a certain time during fusion processes, so the C-peptide sequence may expand to the whole C-terminal half of the gp41 ectodomain and interact with the PHI to inhibit gp41 mediated virus-cell membrane fusion. The energy along the gp41 NHR-CHR interface is not evenly distributed; the WWI motif and the LBD serve as hot spots in the gp41 NHR-CHR interaction. A highly potent C-peptide fusion inhibitor must contain at least the WWI motif or the LBD; in addition, a suitable length of total peptide sequence is required to provide additional interactions in the NHR groove to stabilize the C-peptide-NHR interaction. Though they form stable α-helical structures in the 6-HB, C-peptides and N-peptides from the wild-type gp41 sequence are usually unstructured in solution; thus, they are prone to proteolysis. The viral strains resistant to T20 also required the development of a highly potent fusion inhibitor to overcome drug resistance. The use of protein/peptide engineering to improve the physicochemical properties of the wild-type C-peptide sequence and to increase the stability of the C-peptide-NHR complex is discussed below.

## 3.2 Engineered peptides

New generations of peptide fusion inhibitors have been developed by engineering C34-related sequences in order to increase the *in vivo* stability and NHR binding affinity, and to overcome T20 resistance. It is well accepted that increasing the helicity of the peptide fusion inhibitor will increase its antiretroviral potency by increasing its binding affinity with NHR and the *in vivo* stability (Otaka et al. 2002). In a 6-HB, C-peptides interact with NHR with their *a* and *d* residues, which are considered to be critical for molecular recognition between CHR and NHR; while amino acid residues at the *b*, *c*, *f*, and *g* positions are exposed to solution and are not considered to be critical for the gp41 NHR and CHR interaction (Chan et al. 1997). However, the solvent exposed residues have a global effect on the solubility, stability, and other physicochemical properties of the C-peptides, so they affect the *in vivo* activity and the druggability of peptide fusion inhibitors. Salt bridges and helical enhancers have been engineered by replacing the solvent exposed residues with the desired residues in order to get more potent HIV-1 fusion inhibitors.

T1249 (13) was developed by Trimeris as a second generation peptide HIV-1 fusion inhibitor after T20 (Miralles et al. 2003; Eggink et al. 2008; Pan et al. 2009). It was designed to include both hot spots, the WWI motif of C34 and the LBD of T20. To keep the peptides a suitable length, the seven residues following the WWI motif were considered to be not critical for the NHR interaction and were deleted; thus, the WQEWEQKI motif remained. It also contained the conserved amino acid residues from SIV and HIV-2 that are essential for fighting contains the T20 resistant virus. In addition, alanine substitutions and salt bridges were added to increase the α-helicity, resulting in a 39-mer highly mutated peptide based on the wild-type HIV-1 gp41 sequence. T1249 showed enhanced antiretroviral activity against the T20 resistant virus, and ~50% α-helicity compared to the unstructured character of the wild-type gp41 peptide. It entered into phase II clinical trials, but it was terminated due to side effects (Miralles et al. 2003).

T1144 (9) and T2635 (31) are third generation peptide fusion inhibitors developed by Trimeris (Pan et al. 2011). They fully exploited the strategy used in the development of T1249, however, they are based on the $gp41_{626-663}$ sequence (Dwyer et al. 2007). T1144 and T2635 showed strong activity against both native and highly T20 resistant HIV-1 strains. They form stable α-helices in solution with a helical content of 97% and 75% for T1144 and T2635, respectively. In addition, they both form a very stable 6-HB with NHR under physiological conditions. Analytical ultracentrifugation showed that these highly helical peptides form trimers in solution, which may make them more resistant to proteolysis and increase their *in vivo* stability. Ultra stable C-peptides from the same CHR sequence were also obtained, which showed ~100% helical content in solution and formed ultra stable 6-HBs with N-peptide with $T_m$ values >100 °C, even in 8 M urea solutions. However, these peptides showed very weak antiretroviral activity. This indicated that it required a suitable degree of stability and α-helical content for the C-peptide to efficiently inhibit 6-HB formation to stop the HIV-1-cell fusion process.

Sifuvirtide (SFT, 12) was developed by FusoGen and was based on the C34-related sequence $gp41_{627-662}$ (He et al. 2008; Liu et al. 2011). It was derived from the HIV-1 subtype E sequence and was engineered to mutate the exposed residues to salt bridges to increase the helical content and solubility. Like C34, Sifuvirtide is featureless under physiological conditions, while it forms a nearly full α-helical 6-HB with N36, with a $T_m$ of 72 °C, 10 °C higher than that of N36/C34. As expected from its sequence origin, Sifuvirtide does not interact with the lipid membrane. Sifuvirtide showed low nanomolar inhibitory activity against HIV-1 ENV mediated cell-cell fusion and HIV-1 infection, including T20-resistant HIV-1 isolates. It showed an *in vivo* half-life of 20 h in a single dose administration in 12 healthy volunteers, which is much more stable than T20 and suitable for a once daily administration. It has finished phase IIb clinical trials in China and has shown promising antiretroviral profiles against both T20 resistant and T20 sensitive HIV-1 strains (Wang et al. 2009). The same strategy was applied to CP32 and resulted in CP32M with an improved anti-HIV profile (He et al. 2008).

SC35EK (10), also based on C34, was developed by Fujii's group. Most of the *b, c, f,* and *g* residues were substituted with glutamic acid and lysine residues in order to form EE-KK double salt bridges to fortify the α-helical structure (Otaka et al. 2002). SC35EK showed a little bit more potency than C34 in a multinuclear activation of galactosidase indicator (MAGI) assay ($IC_{50}$ from 0.68 to 0.39 nM), while the salt bridge greatly enhanced its solubility and made it a suitable drug candidate. Its structure is still largely random in solution, while the $T_m$ of its 6-HB formed with N36 increased from 57 °C to 77 °C, which is 20 °C higher than C34. SC35EK was further shortened to SC29EK (**11**), with similar potency (Naito et al. 2009). The same strategy was applied to T20, the resulting T20EK (**16**) showed eight times more potency than T20 and can efficiently inhibit T20 resistant HIV-1 strains (Oishi et al. 2008).

In summary, the C-peptide fusion inhibitor could be engineered to improve the anti-HIV profile. Exposed residues in the 6-HB were substituted to build salt-bridges to significantly stabilize the C-peptide-NHR complex. This type of substitution can improve the solubility of the peptide fusion inhibitor to improve its druggability. The substitution also improved the pharmacokinetic profile, resulting in a longer *in vivo* half life. The helicity of isolated C-peptides were greatly increased by replacing the $a$ and $d$ residues in the hydrophilic layer, resulting in thermally stable C-peptide fusion inhibitors with high α-helical content; they formed an extremely stable complex with NHR. Some of these structured C-peptides showed high anti-HIV potency, especially against highly drug-resistant HIV-1 isolates; while too thermally stable C-peptides of this type caused abolishment of their inhibitory activities.

| T20 | (7) | YTSLIHSLIEESQNQQEKNEQELLELDKWASLWNWF |
|---|---|---|
| C34 | (8) | WMEWDREINNYTSLIHSLIEESQNQQEKNEQELL |
| T1144 | (9) | TTWEAWDRAIAEYAARIEALLRALQEQQEKNEAALREL |
| SC35EK | (10) | WEEWDKKIEEYTKKIEELIKKSEEQQKKNEEELKK |
| SC29KE | (11) | WEEWDKKIEEYTKKIEELIKKSEEQQKKN |
| Sifuvirtide | (12) | SWETWEREIENYTRQIYRILEESQEQQDRNERDLLE |
| T1249 | (13) | WQEWEQKITALLEQAQIQQEKNEYELQKLDKWASLWEWF |
| CP32 | (14) | QIWNNMTWMEWDREINNYTSLIHSLLEESQNQ |
| CP32M | (15) | VEWNEMTWMEWEREIENYTKLIYKILEESQEQ |
| T20EK | (16) | YTSLIEELIKKSEEQQKKNEEELKKLEEWAKKWNWF |

Fig. 5. C-peptide fusion inhibitors

### 3.3 Artificially designed peptides

Artificial design was employed to design unknown peptide sequences with few homologies to natural peptide sequences (Qi et al. 2008; Shi et al. 2008). Based on the crystal structures of the HIV-1 gp41 fusion core, C-peptide uses it hydrophobic $a$ and $d$ residues to interact with the NHR. An EEYTKKI heptad unit (HR) was designed, with the heptad repeat 'bcdefga', as the building block. The $d$ and $a$ positions in the HR were hydrophobic Tyr and Ile residues, respectively, which were expected to form a hydrophobic face to interact with the hydrophobic NHR grooves. The residues at the $b$ and $c$ positions in the HR were negatively charged Glu, which were expected to form an intrahelical salt bridge with positively charged Lys at the $f$ and $g$ positions to stabilize the helical structure; these highly polar residues also form a highly hydrophilic face that increases the solubility of the peptides.

A 35-mer 5HR (**17**) (Fig. 6), which contains five copies of the HR, based on the length of most highly potent HIV-1 fusion inhibitors, was used as a template to build peptides to disrupt the HIV-1 gp41 NHR-CHR interaction. The interaction between 5HR and N46 (gp41$_{536-581}$) was modeled by using a PyMOL program based on the crystal structure of the N36/C34 6-HB, and compared with that of C34. The binding between the residues of 5HR and N46 was less complementary than that of the residues between C34 and N46. 5HR showed weak anti-HIV-1 activity (IC$_{50}$ = 156 ± 8 µg/mL), as measured by a dye transfer HIV-1-mediated cell-cell fusion assay. The WWI motif and LBD were used to replace the HR unit at the N- or C-terminus of 5HR, respectively, or both, based on the SAR of the C-peptide fusion inhibitors, resulting in PBD-4HR (**18**), 4HR-LBD (**19**), and PBD-3HR-LBD (**20**). Inserting a LBD or WWI motif in the 5HR sequences resulted in 2-fold and 6-fold increased potency, with an IC$_{50}$ value of 74 ± 4 and 26 ± 0.4 µg/mL for 4HR-LBD and PBD-4HR, respectively. The increasing potency was synergistic and PBD-3HR-LBD had an IC$_{50}$ value of 4.8 ± 0.3 µg/mL, a striking 33-fold increase over 5HR. As expected from the design, peptides containing PBD, e.g. PBD-4HR and PBD-3HR-LBD, could form a stable 6-HB with the N-peptide N46 and effectively blocked gp41 core formation, as measured by CD spectroscopy and N-PAGE; peptides containing the LBD, including 4HRLBD and PBD-3HR-LBD, were bound tightly to lipid vehicles, with an association constant of 6.80 × 10$^4$ and 1.27 × 10$^5$ M$^{-1}$, respectively, as determined by isothermal titration calorimetry (ITC). These results suggest that the HR sequence can be efficiently docked into the NHR groove and act as a structural domain; and the interaction can be greatly increased by including the WWI motif and LBD in the sequence. Thus, 4HR-LBD, PBD-4HR, and PBD-3HR-LBD are artificial fusion inhibitors that mimic T20, C34, and T1249 – the three typical highly potent HIV-1 fusion inhibitors target different sites of gp41 NHR, respectively.

The anti-HIV-1 activities of 4HR-LBD and PBD-4HR are lower than those of T20 and C34, which may be due to less sequence complementarity between the artificially designed HR and HIV-1 gp41 NHR. The resulting less tight binding suggests that a specific interaction should be uncovered and be addressed for the design of highly potent fusion inhibitors targeting specific viruses.

5HR (**17**) EEYTKKIEEYTKKIEEYTKKIEEYTKKIEEYTKKI

4HR–LBD (**18**) EEYTKKIEEYTKKIEEYTKKIEEYTKKIWASLWNWF

PBD–4HR (**19**) WMEWDREIEEYTKKIEEYTKKIEEYTKKIEEYTKKI

PBD–4HR–LBD (**20**) WMEWDREIEEYTKKIEEYTKKIEEYTKKIWASLWNWF

Fig. 6. Artificially designed peptide fusion inhibitors

In summary, C-peptide fusion inhibitors interact with gp41 NHR to prevent fusogenic 6-HB formation, and thus terminal HIV-1 ENV mediated virus-cell membrane fusion. A WWI motif in C-peptide that interacts with the NHR deep pocket is critical to the C-peptide-NHR interaction, and an extended interaction between C-peptide and the rest of the groove in the NHR trimer provides additional energy to stabilize the 6-HB. Artificial peptide design, based on the knowledge learned from SAR studies of the C-peptides, provides an alternative for peptide fusion inhibitor design; it also provides a stringent test for the knowledge gained and sets a new starting point for fully understanding the fundamentals of virus-cell membrane fusion in order to guide future fusion inhibitor design against HIV and other viruses with class I fusion proteins.

## 4. Peptidomimetics as probes and inhibitors to study the gp41 NHR-CHR interaction

Several SAR studies of highly potent peptide fusion inhibitors have provided an efficient way to disrupt the HIV-1 gp41 NHR-CHR interaction for anti-HIV therapy; they have also deepened our understanding of the gp41 NHR-CHR interaction. Peptide drugs have their intrinsic weaknesses, however, such as high-cost, and unsuitability for oral administration due to *in vivo* proteolysis. Peptidomimetics that use unnatural building blocks may overcome the *in vivo* instability of peptide drugs, leading to orally bioavailable drugs. Peptidomimetics are more like small molecules than peptide drugs, so highly potent peptidomimetic fusion inhibitor studies can be useful for guiding small molecule fusion inhibitor design. Peptidomimetic fusion inhibitors that target gp41 NHR, including D-peptides, foldamers, and covalently linked restrained α-helical peptides (sequences or structures see Fig. 7), are discussed in this section.

### 4.1 D-peptides

As enantiomers of natural L-peptides, D-peptides are not degraded by proteases and have the potential for oral bioavailability. D-Peptides that target a specific protein or peptide target can be discovered by mirror-image phage display (Eckert et al. 1999). The target is synthesized chemically with D-amino acids, resulting in a product that is the mirror image of the natural L-amino acid form, which is used to screen phage that expresses a peptide library of phage coat proteins, to select phage clones with L-peptide sequences that specifically bind to the D-target. The mirror images of the phage-expressed L-peptide sequences are chemically synthesized with D-amino acids. By symmetry, these D-peptides should bind to the natural L-amino acid target.

Cyclic D-peptide HIV-1 fusion inhibitors targeting IQN17 have been identified by mirror-image phage display (Eckert et al. 1999). The phage-expressed peptide library contained ten random amino acid residues flanked by either a cysteine or a serine on both sides. Of the 12 identified IQN17-specific phage clones, nine were pocket specific binders, and eight contained the consensus sequence CXXXXXEWXWLC. The corresponding D-peptides were synthesized and were oxidized to form disulfide bonds. Lysines were added to improve the solubility. An intramolecular disulfide bond was critical for pocket binding and viral inhibition by these D-peptides, since cysteines were selected from an initial phage library containing either Cys or Ser at these positions. Replacing the Cys with Ala in the most potent derivative D10-p5-2K (**22**, $IC_{50}$ of 3.6 μM) caused complete loss of inhibitory activity in a gp41 mediated cell/cell fusion assay.

A IQN17/D10-p1 (**21**) co-crystal was obtained and resolved to 1.5 Å resolution by x-ray crystallography. Structural superposition showed that the overall architecture of the gp41 NHR deep pocket in the IQN17/D10-p1 complex is almost identical to that in the wild-type N36/C34 structure (Chan et al. 1997), with a $C_\alpha$ rmsd of 0.65 Å. D10-p1 forms a circular structure and binds only to the gp41 region of IQN17. Ala-2 to Ala-5 and Ala-11 to Ala-16 form short left-handed α-helices, and the middle region is unstructured. The overall

Peptides and Peptidomimetics as Tools to Probe Protein-Protein Interactions – Disruption of HIV-1 gp41 Fusion Core and Fusion Inhibitor Design

15

positions of the D10-p1 and C34 helices closely overlap, but most of the side chains are significantly different, corresponding to the opposite handedness of the inhibitors. Of the 16 residues in D10-p1, only six interact directly with the gp41 pocket of IQN17, including Trp-10, Trp-12, and Leu-13 in the conserved EWXWL sequence, and Gly-1, Ala-2, and Ala-16 in the invariant original flanking phage sequence. The side chains of Trp-10, Trp-12, Leu-13, and Ala-16 are deeply buried in the hydrophobic pocket of IQN17. A hydrogen bond is formed between a pocket residue Gln-577 and Trp-12 in D10-p1. The packing difference between the Trp-12 and Leu-13 side chains in D10-p1 and Trp-631 and Ile-635 in C34 results in slight changes in the shape of the pocket. Overall, however, the hydrophobic pocket maintains its integrity between the N36/C34 and IQN17/D10-p1 structures. NHR chemical shift differences showed that, for all of the identified D-peptides, Trp-10, Trp-12, and Leu-13 are buried in the IQN17 pocket, validating the pocket as a target for drug development.

In follow-up work, the consensus residues in the sequence (CX5EWXWLC) reported above were fixed so that a constrained library was constructed in which the other six positions were randomized (Welch et al. 2007). The mirror-image phase display using IQN17 as the target identified, incidentally, the potent 8-mer D-peptide 2K-PIE1 (23) in the 10-mer template phage library. The x-ray crystal structure showed that 2K-PIE1 interacts in a similar manner as D10-p1 to IQN17, and 2K-PIE1 forms a more compact structure with IQN17. So, a comprehensive $1.5 \times 10^8$ member 8-mer phage library of the form CX4WXWLC ($3.4 \times 10^7$ possible sequences) was generated, and was screened using IZN17 as the target (Eckert & Kim 2001). The resulting PIE7 (24) was the most potent inhibitor (IC$_{50}$ = 620 nM) and is 15-fold more potent than the best first-generation D-peptide (D10-p5). Comparison of the crystal structures of 2K-PIE1 and PIE7 complexed with IQN17 reveals several interesting differences. First, an intramolecular polar contact between the hydroxyl of D-Ser7 and the carbonyl of D-Gly3 in 2K-PIE1 is lost in PIE7 but is replaced with a new interaction between the side chain carboxylate of D-Asp6 and the amide of D-Gly3. Second, new hydrophobic interactions are created in PIE7 between the ring carbons of D-Tyr7 and the pocket residue Trp-571. Third, the carbonyl of D-Lys2 of PIE7, although somewhat flexible in orientation, forms a direct hydrogen bond with the $\varepsilon$ nitrogen of Trp-571 in some of the structures. Fourth, in some of the structures the hydroxyl of D-Tyr7 in PIE7 forms a new water-mediated hydrogen bond with the pocket residue Gln-575, and this interaction cannot be formed in the 2K-PIE1 structure. Dimerized or trimerized PIE7 was constructed via PEG cross-linkers. The resulting (PIE7)$_2$ and (PIE7)$_3$ have IC$_{50}$ values of 1.9 nM and 250 pM against HXB2, respectively. In contrast, PIE7 inhibits both JRFL, a primary R5-tropic strain, (IC$_{50}$ = 24 µM) and BaL (IC$_{50}$ = 2.2 µM) entry, although ~40- and 4-fold less potently than HXB2 entry, respectively; the PIE7 trimer is a moderately potent inhibitor of this strain (IC$_{50}$ = 220 nM) and an extremely potent inhibitor against BaL (IC$_{50}$ = 650 pM).

Structure-guided phage display was used to optimize the flanking residues for further improvement of PIE7 (Welch et al. 2010). The crystal structure shows significant contacts between the presumed inert flanking residues (Gly-Ala on the N-terminus and Ala-Ala on the C-terminus) and the NHR deep pocket. A new phage library was designed using XXCDYPEWQWLCXX as the template. PIE12 (25) was identified as the most potent (40-fold more potent than PIE7 against the JRFL strain). The x-ray crystal structure showed similarity between PIE12/IZN17 and PIE7/IZN17 structures with a RMSD of 0.6 to 1.2 Å on all C$_\alpha$ atoms. In PIE12/IZN17, new N-terminal flank residues (His1 and Pro2) form favorable ring stacking interactions with the pocket (IQN17-Trp571), the substitution of Leu for Ala in the C-terminal flank sequence causes it to be buried an additional 50 Å into the hydrophobic surface area of the pocket, and the new interactions with the flanking sequence do not perturb the pocket-binding structure of the core PIE7 residues. These differences may account for the improved activity of PIE12 over PIE7. CD thermal denaturation showed that the PIE12-trimer forms the more stable complex with IZN17 with a $T_m$ of 81 °C in 2 M Guanidine chloride (Gua.HCl), 8 °C higher than that of the PIE7-trimer complex. The anti-HIV-1 breadths of the PIE7-trimer, PIE12-trimer, and PIE12 were tested by a pseudovirion assay against a panel of 23 pseudotyped viruses representing clades A to D, several CRFs, and enfuvirtide-resistant strains. Both PIE7 and PIE12-trimers potently inhibited all strains tested, though PIE12-trimer was generally a superior inhibitor.

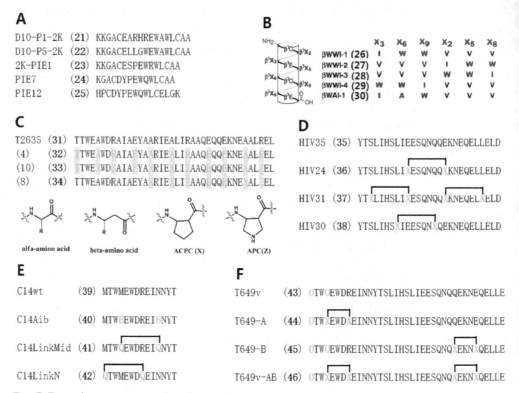

Fig. 7. Peptidomimetics used to disrupt the HIV-1 gp41 NHR-CHR interaction. (A) D-peptides; (B) β-foldamers; (C) α/β-foldamers; (D)/(E) linked peptides; (F) stapled peptides.

Viral passage studies were conducted to select for resistant strains. A strain bearing E560K/V570I mutations, which conferred a 400-fold resistance to PIE7-dimer, was selected with 20 weeks of propagation. These mutations dramatically weaken the binding of D-peptides to the gp41 pocket but not the C-peptide inhibitor C37. Despite this loss of affinity, the escape mutations had a minimal effect on the potencies of PIE12-dimer and PIE12-trimer. PIE12-dimer and PIE12-trimer resistant virus were identified after 40 and 65 weeks of propagation, respectively, using a much slower escalation strategy; only a Q577R single substitution was identified. Interestingly, this substitution is present in nearly all group O isolates but is rare among group M isolates. Examination of the PIE12 crystal structure shows that Q577 makes hydrogen bonds with Glu7 and Trp10 in PIE12, which may explain the disruptive effects of this mutation.

## 4.2 Foldamers

A foldamer is a discrete chain molecule or oligomer that adopts a secondary structure stabilized by noncovalent interactions. Foldamers use unnatural building blocks instead of natural amino acids or nucleotides; as a result, they are more resistant to enzymatic degradation and show enhanced *in vivo* stability. They can mimic the ability of proteins, nucleic acids, and polysaccharides to fold into well-defined conformations, such as helices and $\beta$-sheets.

Short $\beta^3$-foldamers have been designed that mimic the WWI motif of C-peptide fusion inhibitors (Stephens et al. 2005). A $\beta$-amino acid contains an additional methylene unit between the amine and carboxylic acid (Fig. 7c), and the amide bonds in $\beta$-peptides can resist *in vivo* proteolysis. A set of $\beta^3$-decapeptides, $\beta$-WWI-1-4 (**26-29**), in which the WWI motif is presented on one face of a short 1,4-helix (Fig. 7b), were designed. Each $\beta$-peptide was fluorescently labeled at the N-terminus and was used in direct fluorescence polarization experiments to determine its binding affinity to IZN17. $\beta$-WWI-1-4-Flu bound IZN17 well, with equilibrium affinities of $0.75 \pm 0.1$, $1.0 \pm 0.3$, $2.4 \pm 0.7$, and $1.5 \pm 0.4$ µM, respectively. A WWI-1 analog $\beta$-WAI-1-Flu (**30**), containing Ala in place of the central Trp of the WWI motif, bound IZN17 with lower affinity ($K_d > 20$ µM), suggesting the WWI motif is critical for pocket binding. The binding affinities are consistent with the cell-cell fusion assay results; $\beta$-WWI-1-4 inhibited cell-cell fusion with $EC_{50}$ values of $27 \pm 2.5$, $15 \pm 1.6$, $13 \pm 1.9$, and $5.3 \pm 0.5$ µM, respectively, whereas $\beta$-WAI-1 was inactive under the same conditions.

In a follow-up study (Bautista et al. 2009), the second Trp in $\beta$-WWI-4 (**29**) was replaced with unnatural residues to probe steric and electronic effects on the NHR deep pocket binding. Most of the new $\beta$-peptides ($EC_{50}$ 8.2-19 µM) are more potent than $\beta$WWI-1 (**26**) ($EC_{50}$ = 56 µM) at promoting the survival of HIV-infected cells. However, high cytotoxicities, with a selective index ($CC_{50}/EC_{50}$) <10, render these short $\beta$-peptides unsuitable as drug leads.

An $\alpha/\beta$ foldamer with partial $\beta$-amino acid replacement was used to modify a highly potent C-peptide fusion inhibitor to increase its *in vivo* stability (Horne et al. 2009). A two-stage design strategy was employed to modify T2635 (**31**), a highly potent third generation peptide fusion inhibitor (Dwyer et al. 2007). A fluorescent polarization binding assay using 5-helix as the target, a cell-cell fusion assay, and a protease K assay were used to assess the peptide and designed foldamer. In the first stage, one amino acid residue in each $\alpha$-helix turn at the same position was replaced by a $\beta$-amino acid. The optimized $\alpha/\beta$ foldamer (**32**), containing systematic $\beta$-amino acid substitutions at positions $c$ and $f$, showed weak binding affinity ($K_i$ = 3800 nM) and cell-cell fusion inhibitory activity ($IC_{50}$ = 390 nM), compared with those of T2635 ($K_i < 0.2$ nM and $IC_{50}$ = 9 nM). However, **32** showed a 20-fold increase in half-life (14 min) in the protease K assay, compared with that for T2635. The $\beta$-amino acid, with one additional methylene unit, may make the backbone of the $\alpha/\beta$ foldamer too flexible to adapt a suitable conformation and results in the loss of activity. Accordingly, in the second stage of design, $\beta$-amino acids at certain positions were replaced with cyclic $\beta$-

amino acids to restore the rigidity of the backbone. The resulting $\alpha/\beta$ foldamer (33) had an $IC_{50}$ value of 5 nM in the cell-cell fusion assay, similar to that of T2635; its binding affinity to 5-helix was 9 nM, similar to its cell-cell fusion inhibitory potency, despite its significantly weaker binding affinity to 5-helix than T2635. The $\alpha/\beta$ foldamer (33) showed a half-life of 200 min in the protease K assay, a 280-fold increase from T2635. The anti-HIV-1 infection activity of the $\alpha/\beta$ foldamer was also similar to T2635, as measured by HIV-1 infection of TZM-bl (JC53BL) cells using both R5 and X4 HIV-1 strains.

X-ray crystallography was used to characterize the structures of N36/T2635 and N36/33 complexes. The N36/T2635 6-HB structure is almost identical to that of the wild-type N36/C34 6-HB, with a rmsd of 0.73 Å for the $C_\alpha$ atoms. However, the N36/33 complex showed large structure distortion in the N-terminus (4.2 Å $C_\alpha$ rmsd for residues 2–15); the side chains of $Trp^3$ and $Trp^5$ were not resolved in electron density, suggesting a high degree of disorder, indicating that the N-terminal segment of 33 does not engage the NHR binding pocket in the complex. However, removal of the first ten residues of 33, where the WWI motif is located, causes the loss of binding to 5-helix ($K_i$ >10 µM), indicating that the N-terminal segment of 33 is essential for high-affinity 5-helix binding. The N36/34 complex, maintaining the intact WWI motif in the foldamer sequence (8), was crystallized and resolved to 2.8 Å resolution. Relative to 33, 34 tracks much more closely with T2635, with a 1.4 Å $C_\alpha$ rmsd for residues 2–33 between the two structures. The side chains of the WWI motif in the N-terminal segment of 34 show the expected packing into the binding pocket on the NHR core trimer. The above results suggest that the lack of direct contact between the N-terminal portion of 33 and the NHR trimer in the N36/33 complex may be an artifact of crystal packing.

### 4.3 Covalent-linked constrained peptides

Helical structure is critical for C-peptide fusion inhibitors to make proper contacts with the NHR binding sites to elicit potent inhibition. Constraining methods that add structural constraints into the peptide sequence by covalently cross-linking amino acid residues at suitable positions can promote the formation of the $\alpha$-helical conformation, even in short peptides. The covalent linker can be a longer linker between the $i$ and $i + 7$ residues, or a short linker called a stapler between the $i$ and $i + 4$ residues.

The first selected gp41 C-peptide was truncated T20 that lacks the LBD sequence, called HIV35 ($gp41_{638-665}$, 35) (Judice et al. 1997). A covalent cross-linker between the $i$ and $i + 7$ residues of the polypeptide chain locks the intervening residues into a $\alpha$-helical conformation. Residues at adjacent $f$ positions on the opposite face of the helix were selected for cross-linking to enforce the residues at positions $a$ and $d$ to adopt a suitable conformation for target binding. Analogs of HIV35 were prepared containing either one, HIV24 (36), or two, HIV31 (37), tethers to impart increasing helicity. A control peptide, HIV30 (38), was prepared in which a tether was introduced between successive $d$ residues to stabilize the helicity while blocking potential binding interactions across the $a$-$d$ face. HIV24 and HIV30 were partially $\alpha$-helical as measured by CD. By contrast, the doubly constrained analog HIV31 was mostly $\alpha$-helical. HIV35 showed very weak inhibitory activity against HIV-1 in primary infectivity assays by using peripheral blood mononuclear cells with the virus JRCSF, a nonsyncytium-inducing strain, and BZ167, a syncytium-inducing HIV-1 strain. Single restrained HIV24 is more potent than HIV35, partially restoring the inhibitory activity of T20. Doubly constrained HIV31 shows dramatically higher potency, and its activity was comparable with T20 in both HIV-1 infection assays.

In another report, a 14-residue C-peptide C14 (gp41$_{626-639}$, **39**) was selected (Sia et al. 2002). A cell–cell fusion assay was used to evaluate the biological activity of the peptides. Two strategies were employed, substitution with 2-aminoisobutyric acid (Aib) or a diaminoalkane crosslinker, to stabilize the helical conformation of C14. Six peptides were designed and produced, C14linkmid (**41**) was the most potent inhibitor against syncytia formation (IC$_{50}$ = 35 µM), followed by C14Aib (**40**) (IC$_{50}$ = 144 µM). C14linkmid and C14Aib bind to IQN17 with a $K_d$ of 1.2 µM, respectively, as measured by ITC. The efficacy of the cross-linker on the inhibitory activities depends on its position in the peptide sequence, N-terminal cross-linked C14linkN does not inhibit cell–cell fusion, whereas the middle cross-linked C14linkmid inhibits cell–cell fusion at micromolar concentrations. The cell–cell fusion inhibitory activities of the peptides generally correlated with their NHR binding affinities, although the cell–cell fusion activities were consistently ~10-fold less potent than the $K_d$ of NHR binding. Additional factors, other than binding affinity to the target, may be necessary for blocking viral entry. The crystal structure of the C14linkmid/IQN17 complex showed that C14linkmid binds to the gp41 hydrophobic pocket in essentially the same conformation as the pocket-binding region of C34, demonstrating that the crosslink imparts no detectable distortion on the backbone of the C14 peptide in the bound conformation.

Chemical staples have been used to fortify peptides to overcome the proteolytic shortcomings of highly potent peptide HIV fusion inhibitors as therapeutics. As an example, chemical staples were inserted at the N- or C-termini of T649v (**43**) by substituting (S)-2-(((9H-fluoren-9-yl)methoxy)carbonylamino)-2-methyl-hept-6-enoic acid at select ($i$ and $i$ + 4) positions, followed by ruthenium-catalyzed olefin metathesis (Bird et al. 2010). Sites for unnatural amino acid insertion were carefully selected to avoid disruption of the critical hydrophobic interface between NHR and CHR helices as delineated by the crystal structure of N36/C34. Three stapled peptides were designed by inserting single or double staples at selected positions. The activities of related peptides were measured using a luciferase-based HIV-1 infectivity assay, using viruses derived from HXBc2 and the neutralization-resistant primary R5 isolate, YU2. All of the peptides showed low nanomolar IC$_{50}$ values against HXBc2 strains, suggesting that chemical modification in the stapled peptides does not disrupt its NHR interaction. Moreover, all of the stapled peptides showed higher inhibitory activities against drug resistant HIV-1 isolates, such as YU2 and the HIV-1 HXBc2 virus bearing the T20-resistant V38A/N42T or V38E/N42S double mutations in gp41 NHR, with a rank order of SAH-gp41$_{626-662}$ (**46**) > **44** > **45** > T649v > enfuvirtide. SAH-gp41$_{626-662}$ (**46**) displayed medium to low nanomolar IC$_{50}$ values for all of the viruses tested, including T20 and the T649v-resistant YU2 isolate.

The pharmacokinetic properties of **44** were evaluated in a mouse model (Bird et al. 2010). The total body clearance of **44** (1.0 mL/min/kg) was 10-fold more slow than that of the unmodified T649v peptide (9.5 mL/min/kg). A proteolysis assay using both chymotrypsin and pepsin suggested that the striking protease resistance of stapled peptides is conferred by a combination of (1) decreased rate of proteolysis due to induction of α-helical structure and (2) complete blockage of peptidase cleavage at sites localized within or immediately adjacent to the ($i$, $i$ + 4)-crosslinked segment. In addition, a pilot study was undertaken to compare the oral absorption of T649v and **44** using a mouse model. Measurable concentrations of the full-length peptide were found in plasma samples from all **44** treated animals after oral dosing, and the concentration was dose dependent; no T649v was detected in plasma under the same conditions. The hydrocarbon double-stapling confers striking protease resistance of the peptide fusion inhibitor, which translates into markedly improved pharmacokinetic properties, including oral absorption, thus unlocking the therapeutic potential of natural bioactive polypeptides.

In summary, highly potent peptidomimetic HIV-1 fusion inhibitors have been discovered based on peptide fusion inhibitors, including: D-peptide fusion inhibitors discovered by mirror-image phage display using a D-amino acid form of the HIV-1 gp41 target; foldamers constructed from highly potent C-peptide fusion inhibitors by proper substitution of selected residues with $\beta$-amino acid residues; and structurally constrained peptides by covalently linking two residues at the same positions in a helical turn to promote $\alpha$-helical structure formation. More like small molecule drugs, these peptidomimetics are potentially orally bioavailable and also provide clues for small molecule fusion inhibitor design.

## 5. Small molecule helix mimetics

The ultimate goal for drug development is small molecule drugs; it is also the main challenge in PPII development. The NHR deep pocket is a hot spot for the NHR-CHR interaction; it has an internal volume of roughly 400 $\text{Å}^3$, and could be filled by a molecule with a molecular weight of approximately 500 Da, raising the possibility that it could be targeted by small molecule drugs (Chan et al. 1997). Several groups have identified small molecules that show low micromolar inhibitory potency against HIV-1 ENV mediated cell-cell fusion and virus infection (Debnath et al. 1999; Frey et al. 2006; Cai & Gochin 2007; Zhou et al. 2010), however no direct evidence supports that these small molecule fusion inhibitors bind to the deep pocket (Gochin & Cai 2009; Cai & Jiang 2010). Therefore, providing direct structural evidence that a small molecule can bind to the NHR deep pocket, so that a small molecule pharmacophore model can be deduced, is highly desired for small molecule HIV fusion inhibitor design and development.

### 5.1 Small molecule-peptide conjugates

To identify small molecule ligands that specifically bind to the gp41 NHR deep pocket, Harrison's group has synthesized a biased peptide conjugate library (Ferrer et al. 1999). It contained ~60,000 compounds and used three small molecule building blocks to replace the WWI motif in C-peptide and links to the same peptide sequence. The library was synthesized and screened against 5-helix (Weissenhorn et al. 1997) using an on-bead affinity-based assay. A small molecule moiety was identified, which sequentially contained cyclopentyl propionic acid–$\varepsilon$-glutamic acid–p-(N-carboxyethyl) aminomethyl benzoic acid (Fig. 8) (**47**). The moiety alone had no activity based on an HIV-1 ENV mediated cell-cell fusion assay. However, when conjugated to a 30-mer C-peptide C30 (gp41$_{636-665}$) without the PBD sequence, the resulting conjugate peptide showed an IC$_{50}$ value of 0.3 µM, which was 20-fold increase compared with the IC$_{50}$ value of 7 µM for C30. The conjugated peptide still had a much lower potency than a 38-mer (gp41$_{628-665}$) C-peptide containing the PBD that showed an IC$_{50}$ value of 3 nM. The conjugated peptide could form a stable complex with N-peptide, as shown by size exclusion chromatography and native N-PAGE. This indicated that the small molecule moiety could partially mimic the WWI motif of C-peptide to occupy the deep binding pocket of the NHR, while structure modification is needed to optimize the binding.

Crystallography was used to characterize the interaction between the conjugated peptide and gp41 NHR (Zhou et al. 2000). The full length of the non-peptide moiety is visible in electron density maps, but unexpectedly in two orientations, each with about 50% occupancy. The two binding modes share the same aminobenzoic acid position (F1) but diverge at the two more distal building blocks. Also, the electron density for the amino acid at the connection to the non-peptide moiety is poor, suggesting disorder in the peptide linkage to the non-peptide moiety.

## 5.2 Small molecules

A small molecule α-helical mimetic based on a substituted terphenyl scaffold was designed to inhibit the assembly of the 6-HB core (Ernst et al. 2002). Tris-functionalized 3,2′,2″-terphenyl derivatives can serve as effective mimics of the surface functionality projected along one face of an α-helix. Compound 1a (48) was designed to mimic the side chains of an $i, i+4, i+7$ hydrophobic surface in an α-helix, using the branched alkyl substituents isobutyl and isopropyl (to avoid complications from chirality in a sec-butyl group) to mimic the side chains of the most prevalent Leu and Ile in the $a$ and $d$ positions of a 3-4 heptad repeat. Terminal carboxylate groups were also added to mimic the anionic character of the C-peptide and to improve the aqueous solubility. The ability of 1a (48) to disrupt the gp41 core was studied by CD spectroscopy, using a N36/C34 6-HB model ($T_m$ = 66 °C). Titration of 1a into a 10 μM solution of N36/C34 resulted in a decrease of the CD signal at 222 and 208 nm, which corresponds to a reduction in the helicity of the 6-HB. A plot of $θ_{222}$ versus inhibitor concentration shows saturation at approximately three equivalents of 1a. The CD spectrum with excess 1a was similar to the theoretical addition of the individual N36 and C34 spectra at the same concentration; and the thermal denaturation curve of the gp41 core in the presence of 50 μM 1a shows a significant 18 °C drop in the $T_m$ value and closely resembles the melting transition of N36 alone at the same concentration. These data suggest that the 6-HB structure is completely disrupted by helix mimetic 1a. Both the hydrophobic and electrostatic features of 1a are important for its ability to disrupt the bundle. Analogs lacking the key alkyl side chains or carboxylic groups have little effect on the CD spectrum of the protein, even at high concentrations. Mimetic 1a effectively disrupts N36/C34 complexation with an $IC_{50}$ value of 13.18 ± 2.54 μg.mL$^{-1}$, as measured by an ELISA assay using NC-1 (Jiang et al. 1998). Compound 1a inhibits HIV-1 mediated cell-to-cell fusion with an $IC_{50}$ value of 15.70 ± 1.30 μg.mL$^{-1}$, using a dye-transfer cell fusion assay. In comparison, analogs lacking hydrophobic side chains or carboxylic groups had no inhibitory activity and proved to be cytotoxic at similar concentrations. Compound 1b (49), with larger hydrophobic groups than 1a, showed marginally enhanced activity than 1a.

Cai and Gochin identified a set of small molecule fusion inhibitors from a peptidomimetic library using a fluorescent biochemistry assay using Env2.0 as the target (Cai & Gochin 2007). Compounds 54 [3,5] and 55 [6,11] showed $K_i$ values of 1.51 ± 0.16 and 1.34 ± 0.19 μM, respectively, in a competitive binding assay using Env2.0 as the target; and an $IC_{50}$ value of ~8 μM in an HIV-1 gp41 mediated cell-cell fusion assay using a CCR5/CXR4 dual dependent target cell line (JI et al. 2006). These compounds contain two units that are covalently linked by an amide bond, and each unit contains two aromatic rings that may bind into the gp41 NHR hydrophobic pocket. A carboxyl group provides electrostatic interaction with K574 in the binding pocket and is critical for the activity of these small molecules; methylation of the carboxyl group resulted in loss of activities of the compounds in both the biochemical assay and cell-cell fusion assay. Three-unit compounds are prone to form aggregates under the assay conditions used and showed no activity, while single-unit compounds, such as M1 (**56**), display submillimolar inhibitory activity (Cai et al. 2009). Compound 1 (**57**), based on M1, was developed, which displayed an $IC_{50}$ value of $4.5 \pm 0.5$ and $3.2 \pm 0.5$ µM in a fluorescence biochemical assay and a cell-cell fusion assay, respectively (Zhou et al. 2010). Compound 1 (**57**) showed very low cytotoxcity ($IC_{50} > 500$ µM); with a relatively small size, it is a promising lead for fusion inhibitor design.

Others have reported well-characterized small molecule fusion inhibitors targeting gp41, including SDS-J1 (**50**) (Debnath et al. 1999), NB64 (**51**), NB2 (**52**) (Jiang et al. 2004), and 4M041 (**53**) (Frey et al. 2006). These fusion inhibitors were selected from an active compound library by visual screening, then identified by high-throughput screening, and finally verified by a cell-cell fusion assay or HIV-1 infection assay. They usually showed low micromolar $IC_{50}$ values for fusion inhibition; however, the following work to optimize the structures to obtain more potent fusion inhibitors were less fruitful, resulting in the identification of more small molecules with similar activity (Jiang et al. 2011). Also, their exact binding model with the gp41 NHR deep pocket still needs to be verified.

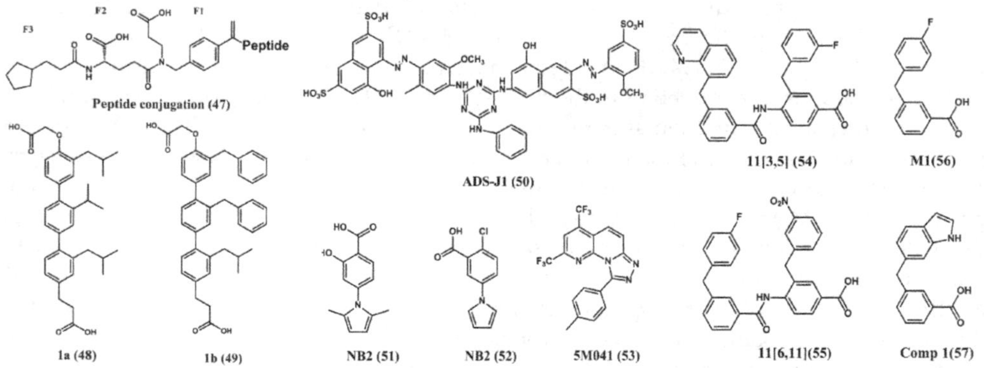

Fig. 8. Small molecule fusion inhibitors.

## 6. Conclusion

Peptides and peptidomimetics are efficient tools to study the HIV-1 gp41 NHR-CHR interaction, a key protein-protein interaction for HIV-1 gp41 mediated virus-cell membrane fusion, which enables HIV-1 enters and ultimately infects host cells. Peptides derived from wild-type HIV-1 gp41 sequences have been used to model the HIV-1 gp41 fusogenic core, a 6-HB formed by the NHR trimer as the inner core, and anti-parallel bind with three CHRs. Crystallographic structure analysis of the 6-HB has uncovered structure details for the gp41 NHR-CHR interaction. A deep pocket in the surface of NHR is a hot spot for the NHR-CHR interaction and a potential target for small molecule fusion inhibitors. N-peptides can be efficient targets for screening fusion inhibitors targeting the gp41 deep pocket by adding structural modulators to promote the trimerization of N-peptide.

Natural C-peptides can efficiently inhibit the gp41 NHR-CHR interaction by interacting with their counterpart in the gp41 6-HB; therefore, they can be used as fusion inhibitors against HIV-1 ENV mediated virus-cell fusion. They use residues at the $a$ and $d$ positions in heptad registration to bind the NHR hydrophobic grooves. The WWI motif of C-peptide provides a critical interaction with the NHR deep pocket, and an additional interaction between the C-peptide and NHR groove is required for a highly potent peptide fusion inhibitor of 30–40 residues. The $b$, $c$, $f$, and $g$ residues in the C-peptide that form the predominantly hydrophilic surface in the 6-HB can be modified for increasing the secondary structure and solubility of the C-peptide, in order to increase its anti-HIV potency. The knowledge gained has been tested by artificial design of highly potent peptide fusion inhibitors with few similarities from known peptide sequences.

Peptidomimetics using unnatural building blocks have been successfully employed to mimic the molecular structures involved in the gp41 NHR-CHR interaction, resulting in highly potent HIV-1 fusion inhibitors with extraordinary *in vivo* stability to overcome the weakness of peptide drugs with potential oral administration possibilities. The achievements of the high potency peptidomimetic fusion inhibitors can also be used to guide small molecule fusion inhibitor design to disrupt this important protein-protein interaction.

In summary, HIV-1 fusion inhibitor development provides a model for using peptides as tools to probe protein-protein interactions for small molecule PPII design and development. The methods and results described in this chapter not only provide clues for future HIV-1 fusion inhibitor design, but also can be used for other viruses using a familiar virus-cell membrane fusion mechanism, as well as to guide other PPII design and development.

## 7. Acknowledgment

This work was supported by National Key Technologies R&D Program for New Drugs Grant 2009ZX09301-002 and National Natural Science Foundation of China Grant 81072581.

## 8. References

Bautista, A. D., et al. (2009). Identification of a beta(3)-peptide HIV fusion inhibitor with improved potency in live cells. *Bioorganic & Medicinal Chemistry Letters,* Vol. 19, No. 14, pp. 3736-3738, ISSN 0960-894X.

Bird, G. H., et al. (2010). Hydrocarbon double-stapling remedies the proteolytic instability of a lengthy peptide therapeutic. *Proceedings of the National Academy of Sciences of the United States of America,* Vol. 107, No. 32, pp. 14093-14098, ISSN 0027-8424.

Caffrey, M. (2011). HIV envelope: challenges and opportunities for development of entry inhibitors. *Trends in Microbiology,* Vol. 19, No. 4, pp. 191-197, ISSN 0966-842X.

Cai, L. & S. Jiang (2010). Development of peptide and small-molecule HIV-1 fusion inhibitors that target gp41. *Chemmedchem,* Vol. 5, No. 11, pp. 1813-1824, ISSN 1860-7187.

Cai, L. F., E. Balogh & M. Gochin (2009). Stable Extended Human Immunodeficiency Virus Type 1 gp41 Coiled Coil as an Effective Target in an Assay for High-Affinity Fusion

Inhibitors. *Antimicrobial Agents and Chemotherapy*, Vol. 53, No. 6, pp. 2444-2449, ISSN 0066-4804.

Cai, L. F. & M. Gochin (2007). A novel fluorescence intensity screening assay identifies new low-molecular-weight inhibitors of the gp41 coiled-coil domain of human immunodeficiency virus type 1. *Antimicrobial Agents and Chemotherapy*, Vol. 51, No. 7, pp. 2388-2395, ISSN 0066-4804.

Chan, D. C., C. T. Chutkowski & P. S. Kim (1998). Evidence that a prominent cavity in the coiled coil of HIV type 1 gp41 is an attractive drug target. *Proceedings of the National Academy of Sciences of the United States of America*, Vol. 95, No. 26, pp. 15613-15617, ISSN 0027-8424.

Chan, D. C., D. Fass, J. M. Berger & P. S. Kim (1997). Core structure of gp41 from the HIV envelope glycoprotein. *Cell*, Vol. 89, No. 2, pp. 263-273, ISSN 0092-8674.

Debnath, A. K., L. Radigan & S. B. Jiang (1999). Structure-based identification of small molecule antiviral compounds targeted to the gp41 core structure of the human immunodeficiency virus type 1. *Journal of Medicinal Chemistry*, Vol. 42, No. 17, pp. 3203-3209, ISSN 0022-2623.

Deng, Y. Q., Q. Zheng, T. J. Ketas, J. P. Moore & M. Lu (2007). Protein design of a bacterially expressed HIV-1 gp41 fusion inhibitor. *Biochemistry*, Vol. 46, No. 14, pp. 4360-4369, ISSN 0006-2960.

Dwyer, J. J., et al. (2007). Design of helical, oligomeric HIV-1 fusion inhibitor peptides with potent activity against enfuvirtide-resistant virus. *Proceedings of the National Academy of Sciences of the United States of America*, Vol. 104, No. 31, pp. 12772-12777, ISSN 0027-8424.

Eckert, D. M. & P. S. Kim (2001). Design of potent inhibitors of HIV-1 entry from the gp41 N-peptide region. *Proceedings of the National Academy of Sciences of the United States of America*, Vol. 98, No. 20, pp. 11187-11192, ISSN 0027-8424.

Eckert, D. M. & P. S. Kim (2001). Mechanisms of viral membrane fusion and its inhibition. *Annual Review of Biochemistry*, Vol. 70, pp. 777-810, ISSN 0066-4154.

Eckert, D. M., V. N. Malashkevich, L. H. Hong, P. A. Carr & P. S. Kim (1999). Inhibiting HIV-1 entry: Discovery of D-peptide inhibitors that target the gp41 coiled-coil pocket. *Cell*, Vol. 99, No. 1, pp. 103-115, ISSN 0092-8674.

Eggink, D., et al. (2008). Selection of T1249-resistant human immunodeficiency virus type 1 variants. *Journal of Virology*, Vol. 82, No. 13, pp. 6678-6688, ISSN 0022-538X.

Ernst, J. T., et al. (2002). Design of a protein surface antagonist based on alpha-helix mimicry: Inhibition of gp41 assembly and viral fusion. *Angewandte Chemie-International Edition*, Vol. 41, No. 2, pp. 278-+, ISSN 1433-7851.

Ferrer, M., et al. (1999). Selection of gp41-mediated HIV-1 cell entry inhibitors from biased combinatorial libraries of non-natural binding elements. *Nature Structural Biology*, Vol. 6, No. 10, pp. 953-960, ISSN 1072-8368.

Frey, G., et al. (2006). Small molecules that bind the inner core of gp41 and inhibit HIV envelope-mediated fusion. *Proceedings of the National Academy of Sciences of the United States of America*, Vol. 103, No. 38, pp. 13938-13943, ISSN 0027-8424.

Peptides and Peptidomimetics as Tools to Probe Protein-Protein Interactions – Disruption of HIV-1 gp41 Fusion Core and Fusion Inhibitor Design

25

Gochin, M. & L. F. Cai (2009). The Role of Amphiphilicity and Negative Charge in Glycoprotein 41 Interactions in the Hydrophobic Pocket. *Journal of Medicinal Chemistry*, Vol. 52, No. 14, pp. 4338-4344, ISSN 0022-2623.

Gochin, M., R. K. Guy & M. A. Case (2003). A metallopeptide assembly of the HIV-1 gp41 coiled coil is an ideal receptor in fluorescence detection of ligand binding. *Angewandte Chemie-International Edition*, Vol. 42, No. 43, pp. 5325-5328, ISSN 1433-7851.

Gochin, M., R. Savage, S. Hinckley & L. F. Cai (2006). A fluorescence assay for rapid detection of ligand binding affinity to HIV-1 gp41. *Biological Chemistry*, Vol. 387, No. 4, pp. 477-483, ISSN 1431-6730.

Harrison, S. C. (2008). Viral membrane fusion. *Nature Structural & Molecular Biology*, Vol. 15, No. 7, pp. 690-698, ISSN 1545-9985.

He, Y. X., et al. (2008). Identification of a critical motif for the human immunodeficiency virus type 1 (HIV-1) gp41 core structure: Implications for designing novel anti-HIV fusion inhibitors. *Journal of Virology*, Vol. 82, No. 13, pp. 6349-6358, ISSN 0022-538X.

He, Y. X., et al. (2008). Potent HIV fusion inhibitors against Enfuvirtide-resistant HIV-1 strains. *Proceedings of the National Academy of Sciences of the United States of America*, Vol. 105, No. 42, pp. 16332-16337, ISSN 0027-8424.

He, Y. X., et al. (2008). Design and evaluation of sifuvirtide, a novel HIV-1 fusion inhibitor. *Journal of Biological Chemistry*, Vol. 283, No. 17, pp. 11126-11134, ISSN 0021-9258.

Horne, W. S., et al. (2009). Structural and biological mimicry of protein surface recognition by alpha/beta-peptide foldamers. *Proceedings of the National Academy of Sciences of the United States of America*, Vol. 106, No. 35, pp. 14751-14756, ISSN 0027-8424.

JI, C., J. ZHANG, N. CAMMACK & S. SANKURATRI (2006). Development of a Novel Dual CCR5-Dependent and CXCR4-Dependent Cell-Cell Fusion Assay System with Inducible gp160 Expression. *Journal of Biomolecular Screening*, Vol. 11, No. 1, pp. 65-74, ISSN 1087-0571.

Jiang, S., K. Lin & M. Lu (1998). A conformation-specific monoclonal antibody reacting with fusion-active gp41 from the human immunodeficiency virus type 1 envelope glycoprotein. *Journal of Virology*, Vol. 72, No. 12, pp. 10213-10217, ISSN 0022-538X.

Jiang, S., et al. (2011). Design, Synthesis, and Biological Activity of Novel 5-((Arylfuran/1H-pyrrol-2-yl)methylene)-2-thioxo-3-(3-(trifluoromethyl)phenyl)thiazolidin-4-ones as HIV-1 Fusion Inhibitors Targeting gp41. *Journal of Medicinal Chemistry*, Vol. 54, No. 2, pp. 572-579, ISSN 0022-2623.

Jiang, S. B., K. Lin, N. Strick & A. R. Neurath (1993). HIV-1 inhibition by a peptide. *Nature*, Vol. 365, No. 6442, pp. 113-113, ISSN 0028-0836.

Jiang, S. B., K. Lin, N. Strick & A. R. Neurath (1993). Inhibition of hiv-1 infection by a fusion domain binding peptide from the HIV-1 envelope glycoprotein-gp41. *Biochemical and Biophysical Research Communications*, Vol. 195, No. 2, pp. 533-538, ISSN 0006-291X.

Jiang, S. B., et al. (2004). N-substituted pyrrole derivatives as novel human immunodeficiency virus type 1 entry inhibitors that interfere with the gp41 six-helix bundle formation and block virus fusion. *Antimicrobial Agents and Chemotherapy*, Vol. 48, No. 11, pp. 4349-4359, ISSN 0066-4804.

Judice, J. K., et al. (1997). Inhibition of HIV type 1 infectivity by constrained alpha-helical peptides: Implications for the viral fusion mechanism. *Proceedings of the National Academy of Sciences of the United States of America*, Vol. 94, No. 25, pp. 13426-13430, ISSN 0027-8424.

Lazzarin, A., et al. (2003). Efficacy of enfuvirtide in patients infected with drug-resistant HIV-1 in Europe and Australia. *New England Journal of Medicine*, Vol. 348, No. 22, pp. 2186-2195, ISSN 0028-4793.

Liu, S. W., et al. (2007). HIV gp41 C-terminal heptad repeat contains multifunctional domains - Relation to mechanisms of action of anti-HIV peptides. *Journal of Biological Chemistry*, Vol. 282, No. 13, pp. 9612-9620, ISSN 0021-9258.

Liu, S. W., et al. (2005). Different from the HIV fusion inhibitor C34, the anti-HIV drug fuzeon (T-20) inhibits HIV-1 entry by targeting multiple sites in gp41 and gp120. *Journal of Biological Chemistry*, Vol. 280, No. 12, pp. 11259-11273, ISSN 0021-9258.

Liu, Z., et al. (2011). In Vitro Selection and Characterization of HIV-1 Variants with Increased Resistance to Sifuvirtide, a Novel HIV-1 Fusion Inhibitor. *Journal of Biological Chemistry*, Vol. 286, No. 5, pp. 3277-3287, ISSN 0021-9258.

Miralles, G. D., et al. (2003). Baseline and on-treatment gp41 genotype and susceptibility to enfuvirtide (ENF) and T-1249 in a 10-day study of T-1249 in patients failing an ENF-containing regimen (T1249-102). *Antiviral Therapy*, Vol. 8, No. 3, pp. 21, ISSN 1359-6535.

Naito, T., et al. (2009). SC29EK, a Peptide Fusion Inhibitor with Enhanced alpha-Helicity, Inhibits Replication of Human Immunodeficiency Virus Type 1 Mutants Resistant to Enfuvirtide. *Antimicrobial Agents and Chemotherapy*, Vol. 53, No. 3, pp. 1013-1018, ISSN 0066-4804.

Oishi, S., et al. (2008). Design of a novel HIV-1 fusion inhibitor that displays a minimal interface for binding affinity. *Journal of Medicinal Chemistry*, Vol. 51, No. 3, pp. 388-391, ISSN 0022-2623.

Otaka, A., et al. (2002). Remodeling of gp41-C34 peptide leads to highly effective inhibitors of the fusion of HIV-1 with target cells. *Angewandte Chemie-International Edition*, Vol. 41, No. 16, pp. 2938-2940, ISSN 1433-7851.

Pan, C., L. Cai, H. Lu, L. Lu & S. Jiang (2011). A novel chimeric protein-based HIV-1 fusion inhibitor targeting gp41 with high potency and stability. *Journal of Biological Chemistry*, Vol. 286, No. 32, pp. 28425-28434, ISSN 0021-9258.

Pan, C. G., L. F. Cai, H. Lu, Z. Qi & S. B. Jiang (2009). Combinations of the First and Next Generations of Human Immunodeficiency Virus (HIV) Fusion Inhibitors Exhibit a Highly Potent Synergistic Effect against Enfuvirtide-Sensitive and -Resistant HIV Type 1 Strains. *Journal of Virology*, Vol. 83, No. 16, pp. 7862-7872, ISSN 0022-538X.

Qi, Z., et al. (2008). Rationally Designed Anti-HIV Peptides Containing Multifunctional Domains as Molecule Probes for Studying the Mechanisms of Action of the First and Second Generation HIV Fusion Inhibitors. *Journal of Biological Chemistry*, Vol. 283, No. 44, pp. 30376-30384, ISSN 0021-9258.

Root, M. J., M. S. Kay & P. S. Kim (2001). Protein design of an HIV-1 entry inhibitor. *Science*, Vol. 291, No. 5505, pp. 884-888, ISSN 0036-8075.

Shi, W. G., et al. (2008). Novel anti-HIV peptides containing multiple copies of artificially designed heptad repeat motifs. *Biochemical and Biophysical Research Communications*, Vol. 374, No. 4, pp. 767-772, ISSN 0006-291X.

Sia, S. K., P. A. Carr, A. G. Cochran, V. N. Malashkevich & P. S. Kim (2002). Short constrained peptides that inhibit HIV-1 entry. *Proceedings of the National Academy of Sciences of the United States of America*, Vol. 99, No. 23, pp. 14664-14669, ISSN 0027-8424.

Stephens, O. M., et al. (2005). Inhibiting HIV Fusion with a $\beta$-Peptide Foldamer. *Journal of the American Chemical Society*, Vol. 127 pp. 13126-13127, ISSN 0002-7863.

Su, C., et al. (2004). Substitutions within HIV gp41 amino acids 36-45 are identified as the primary determinants for loss of in vitro susceptibility to enfuvirtide: results of data mining analyses of genotypic changes in gp41 in TORO 1 and TORO 2 that associate with changes in phenotypic susceptibility to enfuvirtide. *Antiviral Therapy*, Vol. 9, No. 4, pp. U120-U121, ISSN 1359-6535.

Tan, K. M., J. H. Liu, J. H. Wang, S. Shen & M. Lu (1997). Atomic structure of a thermostable subdomain of HIV-1 gp41. *Proceedings of the National Academy of Sciences of the United States of America*, Vol. 94, No. 23, pp. 12303-12308, ISSN 0027-8424.

Tavassoli, A. (2011). Targeting the protein-protein interactions of the HIV lifecycle. *Chemical Society Reviews*, Vol. 40, No. 3, pp. 1337-1346, ISSN 0306-0012.

Walmsley, S., et al. (2003). Enfuvirtide (T-20) cross-reactive glycoprotein 41 antibody does not impair the efficacy or safety of enfuvirtide. *Journal of Infectious Diseases*, Vol. 188, No. 12, pp. 1827-1833, ISSN 0022-1899.

Wang, R. R., et al. (2009). Sifuvirtide, a potent HIV fusion inhibitor peptide. *Biochemical and Biophysical Research Communications*, Vol. 382, No. 3, pp. 540-544, ISSN 0006-291X.

Weissenhorn, W., A. Dessen, S. C. Harrison, J. J. Skehel & D. C. Wiley (1997). Atomic structure of the ectodomain from HIV-1 gp41. *Nature*, Vol. 387, No. 6631, pp. 426-430, ISSN 0028-0836.

Welch, B. D., et al. (2010). Design of a Potent D-Peptide HIV-1 Entry Inhibitor with a Strong Barrier to Resistance. *Journal of Virology*, Vol. 84, No. 21, pp. 11235-11244, ISSN 0022-538X.

Welch, B. D., A. P. VanDemark, A. Heroux, C. P. Hill & M. S. Kay (2007). Potent D-peptide inhibitors of HIV-1 entry. *Proceedings of the National Academy of Sciences of the United States of America*, Vol. 104, No. 43, pp. 16828-16833, ISSN 0027-8424.

Wild, C., T. Oas, C. McDanal, D. Bolognesi & T. Matthews (1992). A synthetic peptide inhibitor of human-immunodeficiency-virus replication - correlation between solution structure and viral inhibition. *Proceedings of the National Academy of Sciences of the United States of America*, Vol. 89, No. 21, pp. 10537-10541, ISSN 0027-8424.

Wild, C. T., D. C. Shugars, T. K. Greenwell, C. B. McDanal & T. J. Matthews (1994). Peptides corresponding to a predictive alpha-helical domain of human-immunodeficiency-virus type-1 gp41 are potent inhibitors of virus-infection. *Proceedings of the National*

*Academy of Sciences of the United States of America,* Vol. 91, No. 21, pp. 9770-9774, ISSN 0027-8424.

Wilson, A. J. (2009). Inhibition of protein–protein interactions using designed molecules. *Chemical Society Reviews,* Vol. 38, pp. 3289-3300, ISSN 0306-0012.

Zhou, G. F., et al. (2000). The structure of an HIV-1 specific cell entry inhibitor in complex with the HIV-1 gp41 trimeric core. *Bioorganic & Medicinal Chemistry,* Vol. 8, No. 9, pp. 2219-2227, ISSN 0968-0896.

Zhou, G. Y., D. Wu, E. Hermel, E. Balogh & M. Gochin (2010). Design, synthesis, and evaluation of indole compounds as novel inhibitors targeting Gp41. *Bioorganic & Medicinal Chemistry Letters,* Vol. 20, No. 5, pp. 1500-1503, ISSN 0960-894X.

# 2

# Profilin, and Vascular Diseases

Mohammad T. Elnakish and Hamdy H. Hassanain

*Department of Anesthesiology, Dorothy M. Davis Heart & Lung Research Institute*
*Molecular, Cellular, and Developmental Biology Program*
*The Ohio State University, Columbus, OH*
*USA*

## 1. Introduction

Actin is a highly dynamic network. It is essential for several important activities, such as muscle contraction and transmembrane signaling (Luna & Hitt, 1992; Salmon, 1989). Actin consists of actin filaments and a variety of associated proteins (Schmidt & Hall, 1998). Many proteins associated with the actin cytoskeleton control actin assembly and disassembly. These proteins regulate actin assembly at multiple levels, including the organization of actin monomers into actin polymers (Schmidt & Hall, 1998).

One key actin-regulatory protein is profilin, which associates with polymerization of actin. Profilin is a ubiquitous small (12–15 kDa) actin-binding protein expressed in eukaryotes (Kwiatkowski & Bruns, 1988; Magdolen et al., 1988; Sonobe et al., 1986; Tseng et al., 1984; Valenta et al., 1991a, b; Widada et al., 1989) and some viruses (Machesky et al., 1994). Profilin plays an important role in the regulation of actin polymerization in a number of motility functions (Haarer & Brown, 1990). The ability of profilin to bind to many ligands suggests that profilin is involved in signal transduction and may link transmembrane signaling to the control of the microfilament system (Korenbaum et al., 1998; Pantaloni & Carlier, 1993).

Early biochemical studies indicated that profilin interacts with actin in a 1:1 ratio and participates in the addition of monomers at the free barbed end of the filament then disassociates at the barbed end (Pantaloni & Carlier, 1993). Latest work has suggested several more functions of profilin aside from its monomer-sequestering ability. Profilin promotes the exchange of adenine nucleotide bound to actin monomer and also effectively lowers the critical concentration of monomer actin for polymerization of actin (Borisy & Svitkina, 2000; Theriot & Mitchison, 1993). It also promotes nucleotide exchange on an actin monomer by lowering the affinity of the actin monomer for its bound nucleotide by 1000-fold (Goldschmidt-Clermont et al., 1991).

It became progressively clear that profilins are vital constituents of the cytoskeleton. Additionally, the role of profilins in several cytoskeleton-based processes of clinical relevance has been proven. Several studies showed abnormal profilin levels in some pathological conditions. For example, high levels of profilin expression have been reported in human gastric cancer (Tanaka et al., 1992). On the contrary, profilin-I has been described as a tumor suppressor in some other types of cancer such as breast cancer (Das et al., 2009;

Zuo et al., 2007). Another clinical problem in which profilins may be involved is the lateral spreading of some infectious diseases (Pistor et al., 1995; Smith et al., 1996; Zeile et al., 1996). Moreover, profilins got a clinical consideration in other unexpected milieu. In this regard, profilins have been reported as major allergens implicated in pollen and food allergies in approximately 20% of type I allergy patients (Ebner et al., 1995; Valenta et al., 1991c, 1992). Furthermore, we (Hassona et al., 2010, 2011; Moustafa-Bayoumi et al., 2007) and others (Caglayan et al., 2010; Romeo & Kazlauskas, 2008; Romeo et al., 2004, 2007) have shown that profilin-I is an unexpectedly novel molecule that plays a highly significant role in vascular problems that predict a higher risk for developing arteriosclerosis, hypertension, stroke, heart failure, and finally death. Therefore, the aim of this chapter is to shed light on the significance of profilin-I via understanding the molecular and cellular aspects of this molecule, and its role in the vascular diseases.

## 2. Profilin family

### 2.1 Gene expression, products & intracellular localization of profilins

So far, there are four profilin genes that have been identified in the mouse and humans. Normally, the isoforms are expressed by diverse genes; nevertheless, differentially spliced isoforms are known to be present as well. It has been reported that in human, bovine, mouse, and rat, profilin-II is alternatively spliced into profilin-IIA and -IIB (Di Nardo et al., 2000; Lambrechts et al., 2000). In humans, profilin-I is expressed in every cell, while other isoforms are expressed in specific tissues. For example, profilin-IIA and -IIB are found to be brain specific and they are essential for neuronal development (Witke et al., 2001). Profilin-II complexes with other proteins such as synapsin and dynamin-I, well- known proteins that implicated in membrane trafficking. In addition, in humans and mouse profilin-III has been shown to be expressed in the testis and kidney and entirely in developing spermatids (Braun et al., 2002). At the amino acid level profilin-III and -IV exhibited only 30% identity among themselves and with other mammalian profilins (Obermann et al., 2005). Profilin-IV plays a key role in acrosome production and sperm morphogenesis. The same study by Obermann et al. proposes that profilin-III and -IV are transcribed in the germ cells. Yet, the expression timing was different during the rat testis post-natal development and in the rat spermatogenetic cycle. In the human testis, there is a correlation between profilin-IV mRNA expression and the presence of germ cells. Profilin-III and -IV may control testicular actin cytoskeleton dynamics and be a factor in acrosome production and spermatid nuclear shaping (Obermann et al. 2005).

Additionally, in *Caenorhabditis elegans* three profilin isoforms, profilin-I, profilin-II, and profilin-III, have been reported, among them profilin-I is crucial; however, profilin-II and profilin-III are not (Polet et al., 2006). As evident by immunostaining expression patterns for the profilin isoforms was different. At the early stages of embryogenesis, profilin-I confines to the cytoplasm and to the cellular contacts, while at the later stages of embryogenesis it confines to the nerve ring. At the late stages of embryogenesis, it has been shown that profilin-III expresses exclusively in the muscle cell walls. On the other hand, during adulthood, profilin-I is expressed in the neurons, the vulva, and the somatic gonad, profilin-II in the intestinal wall, the spermatheca, and the pharynx, and profilin-III, as dots, in the muscle cells of the body wall (Polet et al., 2006). Furthermore, two profilin isoforms (I and II) have been identified in *Dictyostelium amoebae*; profilin-I is fundamental for growth and development, where profilin-II is not. Moreover, it has been reported that *Saccharomyces cerevisiae* and *S. pombe* have only a single profilin isoform (Ezezika et al., 2009; Magdolen et al., 1988).

Based on the small sizes of profilin (15 kDa) and the profilactin complex (57 kDa) one might expect that they can easily diffuse to the nucleus. Nonetheless, profilin ordinarily is excluded from the nucleus and can be found only in the cytoplasm. Either the most part of profilin is bound in the cytoplasm and only a small portion can diffuse freely or there is a particular export mechanism that can actively take the profilin out of the nucleus (Witke, 2004). Recently, Stuven et al., (2003) reported a profilin-specific exportin present in the mammalian cells. Exportin 6 identifies the actin-bound profilin only, as a cargo and moves it out of the nucleus. The reasons for the existence of this profilactin-specific exportin still unclear, but this finding proposes that the nuclear levels of profilin and actin should be strictly regulated (Witke, 2004). Conversely, there are numerous reports about a nuclear fraction of profilin. For example, it has been reported that profilin-I is linked with subnuclear structures such as ribonuclear particles and Cajal bodies, and anti-profilin antibodies interfere with splicing *in vitro*. This implies a role for profilin-I in pre-mRNA processing (Skare et al., 2003). Also, it has been proposed that in the nucleus profilin-I and profilin-II interact with the survival of motor neuron (SMN) protein, a nuclear factor that is mutated in spinal muscular atrophy (Giesemann et al., 1999). SMN is important for splicing regulation yet it is not known whether this requires profilin binding or not. Still, in cell culture, co-localization of profilin-I and profilin-II with SMN in nuclear gems has been established (Giesemann et al., 1999).

To date the nuclear localization of profilins is a mystifying finding. Only a role for profilin and actin in splicing, chromatin remodeling or transcriptional regulation can be speculated. A more detailed understanding of the dynamics and properties of nuclear profilin and actin is required. It is possible that in the nucleus these proteins are considered necessary momentarily during the cell cycle or, particularly, in cells experiencing transcriptional activity changes (Witke, 2004).

## 2.2 Structural aspects of profilin

All recognized profilins share common structural and biochemical properties, though the amino acid sequences of the analogous isoforms in distantly related species may demonstrate less than 25% homology (Schlüter et al., 1997). Numerous studies on profilins from different origins demonstrate that they have highly similar tertiary structures (Fedorov et al., 1994, 1997; Metzler et al., 1993; Schutt et al., 1993; Thorn et al., 1997; Vinson et al., 1993) (Figure. 1). The profilin polypeptide consists of 100-131amino acids (Krishnan & Moens, 2009) and it is folded into a central $\beta$-pleated sheet formed of 5-7 antiparallel $\beta$-strands (Schlüter et al., 1997). On one side, this core is flanked by N- and C-terminal $\alpha$-helices, with both termini next to each other, and on the opposed side by an extra $\alpha$-helix attached to either additional $\alpha$-helix or a small $\beta$-strand (Schlüter et al., 1997) (Figure. 1).

It has been reported that there are three groups of ligands characterize profilins: (1) G-actin and actin-related proteins (Machesky et al., 1994; Schutt et al., 1989; Tobacman et al., 1983) (2) polyphosphoinositides (Lassing & Lindberg, 1985, 1988) (3) poly-L-proline (PLP) with the exception of *Vaccinia* profilin (Kaiser et al., 1989; Lindberg et al., 1988; Tanaka & Shibata 1985), existing either as a peptide or as a sequence motif in particular proteins.

In this context, Gieselmann et al., (1995) showed that human profilin-I exhibits about five folds higher affinity for actin than profilin-II. Radiography analyses of the structures of human profilin isoforms imply that the substitution of profilin-I S29 by Y29 in profilin-II participates in the higher affinity of profilin-II for proline-rich sequences (Nodelman et al., 1999). In spite of the similarity in the 3D structures of human profilin-I and -II, the surface characteristics, such as exposure of hydrophobic patches (Figure 2), and biochemical properties of each isoform are different (Krishnan & Moens, 2009).

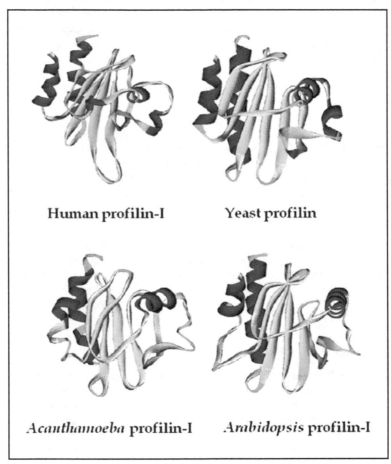

Human profilin-I    Yeast profilin

*Acanthamoeba* profilin-I    *Arabidopsis* profilin-I

Fig. 1. Profilin-I isoforms from different organisms showing a similar helix (red) and strand (cyan) structure (PDB database: 1PFL, 1KOK, 2PRF, and 3NUL) with the loops highlighted in green, adapted from Krishnan & Moens, (2009) with permission.

## 2.3 Profilin ligands

Despite its relatively small size, many profilin ligands have by now been recognized, such as actin and actin-related proteins, polyphosphoinositides, PLP, annexin-I, and the list still increasing (Schlüter et al., 1997). Recently, there are more than 50 described profilin-binding ligands from diverse origins. However, this represents only a part of the real number of profilin-binding partners. Figure 3 shows the identified profilin ligands in mammalian cells. These do not include only molecules of focal contacts that could link profilin directly to actin polymerization such as VASP (vasodilator-stimulated Phosphoprotein) or Mena (mouse homolog of Drosophila enabled) (Gertler et al., 1996; Parast & Otey, 2000; Reinhard et al., 1995) but also include other molecules such as nuclear-export receptors (Boettner et al., 2000; Camera et al., 2003), regulators of endocytosis and membrane trafficking (Witke et al., 1998), Rac and Rho effectors molecules (Alvarez-Martinez et al., 1996; Miki et al., 1998; Ramesh et al., 1997; Suetsugu et al., 1998; Watanabe et al., 1997; Witke et al., 1998; Yayoshi-Yamamoto et al., 2000) and synaptic scaffold proteins (Mammoto et al., 1998; Miyagi et al., 2002; Wang et al., 1999). While a small number of these interactions demonstrated a physiological relevance, the recognition of profilin-interacting proteins could explain the unpredicted roles of profilin in mammalian cells. The profilin-ligands binding might help in linking different pathways to cytoskeletal dynamics via a mechanism that still unknown. Instead, the profilin–ligand interaction might work independently of actin to control the ligands directly (Witke, 2004).

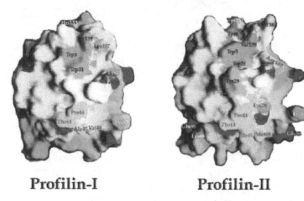

## Profilin-I                          Profilin-II

Fig. 2. Structure of human profilin-I and –II: differences in the surface-charge distribution might account for the ligand-binding specificity of profilin-I and -II. Colored regions highlight amino acid residues that are different in profilin-I and -II. Non-conserved residues are shown in blue; conserved residues are shown in brown, adapted from Witke, (2004) with permission.

Among this large number of profilin ligands we will focus on the binding of profilin to some of those ligands believed to be of relevant role in vascular problems such as actin and ligands in Rho/Rac pathway.

### 2.3.1 Profilin, actin & cytoskeleton

*In vitro*, Profilins can interact with and sequester actin monomers, in that way diminishing the concentration of free actin monomers that are accessible for filament elongation (Carlsson et al., 1977). They refill the pool of ATP-actin monomers via rising the nucleotide exchange rate by 1000-fold in comparison with that rate obtained from simple diffusion (Goldschmidt-Clermont et al., 1992). The profilin–ATP–actin complex can bind to the fast growing, barbed, or plus end of the actin filament and liberate the ATP–actin monomer, which is after that added to the filament (Figure 4). As a result, the elongating filament is made of ATP-actin. Down the filament, the ATP is slowly hydrolyzed via the actin intrinsic

ATPase activity. This produces ADP–actin in the older part of the filament. ADP–actin can be liberated gradually from the pointed or minus end of the filament by depolymerization or at faster rate by actin-depolymerizing proteins (Witke, 2004).

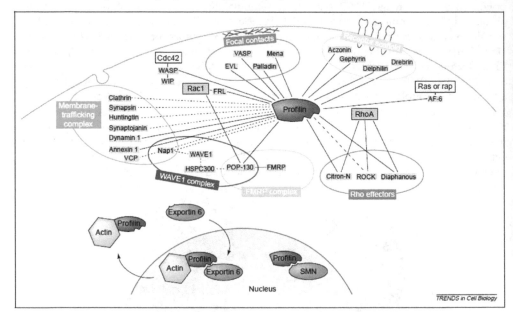

Fig. 3. Network of molecular interactions of profilin. Proteins that are known to interact with profilin are grouped according to their cellular location or the complexes in which they are found. Some of the profilin ligands are shared among different complexes (indicated by the intersecting fields), which suggests a crosstalk among signaling platforms, with profilin as the common denominator. Several links exist to small GTPases such as Rac1, RhoA, cdc42, Ras and Rap that are part of pathways that signal to the actin cytoskeleton. For simplicity, the term profilin is commonly used for profilin-I and profilin-II. Direct interactions between profilin and the ligands are indicated by unbroken lines, whereas potentially direct interactions are indicated by broken lines. Abbreviations: AF-6, All-1 fusion partner from chromosome 6; EVL, Ena VASP like; FMRP, fragile X mental retardation protein; FRL, formin-related gene in leukocytes; HSP, heat-shock protein; Mena, mouse homolog of Drosophila enabled; POP, partner of profilin; SMN, survival of motor neuron; VASP, vasodilator-stimulated phosphoprotein; VCP, valosine-containing protein; WASP, Wiskott–Aldrich syndrome protein; WAVE, WASP family verprolin-homologous protein; WIP, WASP-interacting protein, adapted from Witke, (2004) with permission.

It is worth noting that the presence of other G-actin binding proteins, such as thymosin β4 or any of the ADF family members can alter these processes (Pantaloni & Carlier, 1993). Additionally, capping the plus end of the filaments inhibits the addition of the profilin-actin complexes and consequently limits the activity of profilin to a simple sequestering effect (Pantaloni & Carlier, 1993; Perelroizen et al., 1996; Pring et al., 1992). Thus, the presence of other G-actin binding and/or capping proteins could regulate the profilin effect on cellular actin (Schlüter et al., 1997).

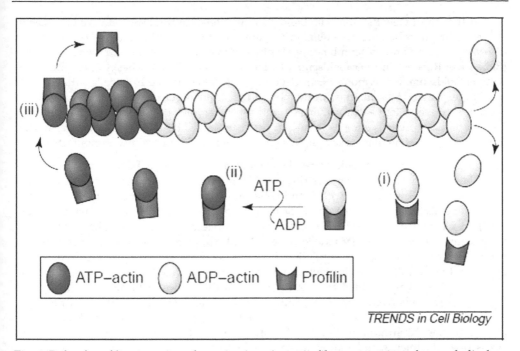

TRENDS in Cell Biology

Fig. 4. Role of profilins in actin polymerization. An actin filament consists of two α-helical protofilaments. *In vitro*, three major functions have been identified for profilins in the regulation of actin polymerization. (i) Profilins can bind to and sequester actin monomers, thereby decreasing the concentration of free actin monomers that are available for filament elongation. (ii) Profilins replenish the pool of ATP–actin monomers (red) by increasing the rate of nucleotide exchange on the bound actin monomer 1000-fold compared with the rate of exchange based on simple diffusion. (iii) The profilin–ATP–actin complex can interact with the fast growing end of the actin filament and release the ATP–actin monomer, which is then added to the filament. Consequently, the elongating filament consists of ATP–actin. Along the filament, the ATP is slowly hydrolyzed by the intrinsic ATPase activity of actin, which generates ADP–actin (orange) in the older part of the filament. ADP–actin can be released slowly from the end of the filament by depolymerization or at an accelerated rate by 'actin-depolymerizing proteins' (not shown), adapted from Witke, (2004) with permission.

Previously, it has been considered that the profilins effect on nucleotide exchange on actin directly regulates their ability to promote filament assembly at the plus end. Polymerization of filament is coupled with the actin-bound ATP hydrolysis and thus far, profilins are unique microfilament associated proteins that can work as nucleotide exchange factors. Polymerization of ATP-actin occurs more rapidly and at a lower critical concentration than ADP-actin (Pollard, 1986). Nonetheless, profilin isoforms I and III in *Arabidopsis* are unable to speed up the rate of nucleotide exchange on G-actin yet still reduce the critical concentration at the plus ends of filaments, similar to vertebrate profilin (Perelroizen et al., 1996). These data demonstrate that the major effect of profilins on actin polymerization cannot be linked with their capacity to work as nucleotide exchange factors.

*In vivo*, the global view that the main biological function of profilin was observed in its actin sequestering effect became debatable, principally due to the finding that the concentration of profilin in cells and its actin-binding affinity are inadequate to stabilize the G-actin pool (Babcock & Rubenstein 1993; Goldschmidt-Clermont et al., 1991; Machesky & Pollard, 1993; Sohn & Goldschmidt-Clermont, 1994). Generally, the data obtained from cells with different profilin levels are in harmony with the notion that in lower eukaryotes the central role of profilin is to sequester G-actin, whereas in higher eukaryotes this is mainly done via other G-actin binding proteins such as thymosin $\beta 4$ (Safer et al., 1991), and profilins are mostly implicated in the actin filament dynamics control (Sohn & Goldschmidt-Clermont, 1994).

Based on this notion, lower eukaryotes deficient in profilins should exhibit an increase in F-actin, however in higher eukaryotes this would not be the principal outcome. Compatible with this model, *S. pombe* cells with profilin overproduction showed undetectable amount of actin filaments, and are incapable of forming a contractile ring (Balasubramanian et al., 1994). In *S. cerevisiae* cells harmful effects due to actin overexpression, could be compensated by profilin overexpression (Magdolen et al., 1993). Conversely, several studies reported about filament-stabilizing or -regulating functions of profilin in higher eukaryotes. For example, the overall F-actin content and stability were elevated whereas; a considerable amount of F-actin was shifted from stress fibers to the cortical actin network in Chinese hamster ovary cells overexpressing profilin (Finkel et al., 1994). Likewise, actin filaments were stabilized against cytochalasin D and latrunculin in baby hamster kidney cells overexpressing birch profilin (Rothkegel et al., 1996). In addition, a shift in F-actin from stress fibers to thick peripheral actin filament bundles with a corresponding increase in cellular adhesion to fibronectin has been reported in cultured human endothelial cells overexpressing profilin (Moldovan et al., 1997).

Although these findings indicated a differential role of profilins between lower and higher eukaryotes, a few studies showed contradictory data to these reports (Cao et al., 1992; Edwards et al., 1994; Staiger et al., 1994). Consequently, a final conclusion on the validity of the assumption regarding differential functions of profilins in higher and lower eukaryotes needs to be confirmed with further experimentations (Schlüter et al., 1997).

### 2.3.2 Profilin & Rho/Rac pathway

Rho/Rac signaling pathway represents one of the well-known pathways in the regulation of actin cytoskeleton, as indicated by the Rac1-dependent membrane ruffling and RhoA-stimulated stress-fiber formation (Nobes & Hall, 1995). Although there is no any report about the direct interaction between profilins and Rho and/or Rac or any other small GTPases, many of the profilin ligands are well-recognized Rho/Rac effector molecules (Witke, 2004). In this regard, our recent data showed that profilin overexpression in vascular smooth muscle cells (VSMC) of transgenic mice results in vascular remodeling and hypertension. These were associated with increased Rho-GTPase activity and Rho-dependent coiled-coil kinase (ROCK) expression (Hassona et al., 2010; Moustafa-Bayoumi et al., 2007). As well, it has been reported that ROCK is a part of the profilin-II complex in the brain (Witke et al., 1998) and this binding is significant in the regulation of neurite outgrowth by ROCK (Da Silva et al., 2003). Furthermore, two other proteins that connect profilin to the Rac pathway were recognized in the profilin-II complex in the brain, Nck-associated protein (Nap 1) and partner of profilin (POP)-130 (Witke et al., 1998). GTP–Rac1 interacts with POP-130 and can detach the tetrameric WAVE 1 [Wiskott–Aldrich syndrome protein (WASP) family verprolin-homologous protein]1 complex, resulting in the activation of actin polymerization by WAVE1. Yet, the role of profilin binding to POP-130 is not apparent however it is possible that profilin might manage the complex formation between WAVE1 and POP-130 and between FMRP (fragile X mental retardation protein) and POP-130, in the same way as Rac1 (Witke, 2004).

Additional small-GTPase-binding molecules that can interact with profilin are the Rho-binding molecules, mouse homologs of the *Drosophila* gene diaphanous (mDia1, mDia2 and mDia3) which are known as potent nucleators of actin polymerization (Wallar & Alberts, 2003). Generally, the diaphanous protein exists in an inactive conformation due to folding back of its N terminal GTPase-binding domain onto its C-terminal Dia-autoregulatory domain resulting in association and autoinhibition. RhoA binding to the N terminus releases the autoinhibition and activates actin nucleation (Alberts, 2002). Profilin binding occurs through the proline-rich formin homology domain that present in the core of that diaphanous molecule (Watanabe et al., 1997). Yet, the significance of that binding is not clear. One interesting possibility is that diaphanous can move actin after it has been sequestered by profilin and activate actin polymerization (Li & Higgs, 2003). However, this is limited by the argument that the studies of profilin-diaphanous binding used truncated versions of diaphanous, rather than the full-length protein. *In vivo*, it has been suggested that large complex of diaphanous oligomers is present as well (Li & Higgs, 2003), which via diaphanous monomers can interact with profilin and/or profilactin molecules. Nevertheless, the structure and regulation of this enormous signaling platform for actin nucleation need to be understood (Witke, 2004).

### 2.3.3 Ligands binding sites

In this section we will discuss the binding sites of the main profilin ligands, actin, phatidylinositol 4,5-bisphosphate (PtdIns 4,5-P), and PLP. Initially, profilin binds with high affinity (micromolar range) to G-actin in a 1:1 stoichiometric complex (Schlüter et al., 1997). The amino acid motif LADYL in the C-terminal $\alpha$-helix was first proposed to be implicated in actin binding depending on (1) the presence of this motif in most of profilins, (2) the presence of homologous sequences in a range of actin-binding proteins such as DNase I, fragmin, gelsolin, severin, villin and the vitamin D-binding protein (Binette et al., 1990; Tellam et al., 1989; Vandekerckhove, 1989; Vinson et al., 1993). Nevertheless, this hypothesis was neglected due to (1) the absence of this sequence in mammalian profilins, (2) the ability of Saccharomyces profilin to interact with actin even after deletion of this motif (Haarer et al., 1993). Now, the LADYL-motif is believed to be a central element in the dense structure of these proteins (Ampe & Vandekerckhove, 1994; Fedorov et al., 1994; Haarer et al., 1993; McLaughlin & Weeds, 1995). Studies on bovine profilin-I and $\beta$-actin showed that the actin binding sites on profilin are localized in the $\alpha$-helix 3, the proximal part of $\alpha$-helix 4, and in the $\beta$-strands 4, 5 and 6 (Schutt et al., 1993) (Figure 5). These residues bind to subdomains 1 and 3 on the actin molecule; however, they do not exhibit a conserved sequence motif (Thorn et al., 1997). In the bovine complex, Phe375 appears to be a key residue that interacts with Ile73, His119, Gly121 and Asn124 on the profilin side (Schutt et al., 1993). Similarly, other studies on *Acanthamoeba* reported that actin-related proteins such as Arp2 interact with profilin using the same binding site (Kelleher et al., 1995; Machesky, 1997).

Fig. 5. Topographical relation of the main ligand binding domains as seen on the X-ray structure of bovine profilin (Schutt et al., 1993). The binding domains of actin and actin related proteins (blue; Schutt et al., 1993) and PtdIns 4,5-P2 (red; Sohn et al., 1995) overlap, while that for proline-cluster sequences (green; Metzler et al., 1994) is located at the opposite side of the profilin molecule, adapted from Schlüter et al., (1997) with permission.

On the other hand, studies on *Acanthamoeba* described a positively charged area opposing both termini, placed in the G-actin binding site as the binding motif of the second key ligand of profilin, PtdIns 4,5-$P_2$ (Fedorov et al., 1994) (Figure 5). This was supported by mutation studies on *Saccharomyces* profilin and human profilin-I. Point mutations in this region diminished the binding affinity of profilin to PtdIns 4,5-$P_2$ (Haarer et al., 1993; Sohn et al., 1995). In line with the observation that the binding sites of G-actin and PtdIns 4,5-$P_2$ on profilin overlap (Figure 5), it has been reported that these ligands compete with each other for binding to profilin (Lassing & Lindberg, 1985; 1988; Machesky et al., 1990). In addition, other reports showed that binding of PtdIns 4,5-P2 results in a conformational change in profilin and disrupts the profilin-actin complex (Raghunathan et al., 1992). Also, it has been revealed that profilin can bind a variety of phosphatidylinositol and the binding affinity of human profilin-I to phosphatidylinositol 3,4-bisphosphate (PtdIns 3,4-$P_2$), and phosphatidylinositol 3,4,5-trisphosphate (PtdIns 3,4,5-$P_3$) is higher than its affinity to PtdIns 4,5-$P_2$ (Lu et al., 1996). Furthermore, phosphoinositide (PI) 3-kinase activity may be regulated by profilin through direct binding to the p85 subunit of this enzyme (Singh et al., 1996). PI 3-kinase has no effect on the binding of actin to profilin (Singh et al., 1996), signifying that the binding sites of actin and p85 on profilin are different.

Conversely, the profilin-PtdIns 4,5-P2 complex can be hydrolyzed only via phospholipase C$\gamma$1 (PLC$\gamma$1). Phosphorylation and activation of this lipase as a result of transmembrane signaling (Goldschmidt-Clermont et al., 1990; 1991) leads to the conclusions that (1) profilins are implicated in the metabolism of phosphoinositide and (2) hydrolysis of PtdIns 4,5-P2 causes profilin to move out from the membrane to the cytosol where it can bind to actin or other ligands. These conclusions propose that profilin–phosphoinositide binding plays a vital role *in vivo* (Janmey et al., 1995; Ostrander et al., 1995).

Mutation (Björkegren et al., 1993; Haarer et al., 1993) and NMR (Archer et al., 1994; Metzler et al., 1994) analyses described the binding site of profilin to the third main ligand, PLP as a hydrophobic patch including the NH- and COOH-terminal α-helices and the upper face of the antiparallel β-sheet, opposing to the actin/PtdIns 4,5-P2 binding region (Figure 5). The binding of PLP to profilins has no effect on the interaction with G-actin or PtdIns 4,5-P2 (Archer et al., 1994; Kaiser et al., 1989), indicating that PLP has a distinct binding site (Figure 4). Expediently, this specific PLP-profilin binding is used for profilins purification (Kaiser et al., 1989; Lindberg et al., 1988). For effective profilin binding, it has been proposed that 6 continuous prolines would be sufficient (Metzler et al., 1994). Nevertheless, other reports demonstrated that at least 8–10 prolines are required for efficient binding (Domke et al., 1997; Machesky & Pollard, 1993; Perelroizen et al., 1994; Petrella et al., 1996).These proline stretches may be interrupted by single glycine residues (Domke et al., 1997; Lambrechts et al., 1997) and may be capable of simultaneous binding of two profilins (Lambrechts et al., 1997), depending on the ability of profilin to oligomerize (Babich et al., 1996).

The first recognized ligand for PLP was VASP, a focal adhesion molecule that was reported to interact directly with F-actin (Jockusch et al., 1995; Reinhard et al., 1995), and it also described as a substrate of both cGMP- and cAMP-dependent protein kinases in platelets (Halbrügge et al., 1990). VASP has a central proline-rich domain with a single copy and a 3-fold tandem repeat of a remarkable $(G)P_5$ motif (Haffner et al., 1995). This motif is both required and sufficient for profilin binding (Domke et al., 1997; Lambrechts et al., 1997; Reinhard et al., 1995). Another PLP-binding ligand similar to VASP is a VASP-related mouse protein, Mena (Gertler et al., 1996). Additional PLP-biding ligands are the formin-related proteins, *S. cerevisiae* Bni1p and Bnr1p, *S. pombe* Cdc12p, *Drosophila* cappuccino and p140mDia, the mammalian homologue of the *Drosophila* protein diaphanous (Chang et al., 1997; Evangelista et al., 1997; Imamura et al., 1997; Manseau et al., 1996; Watanabe et al., 1997). These proteins have a proline-rich domain with numerous proline stretches consisting of 5–13 residues and a C-terminal consensus sequence of approximately 100 amino acids (Castrillon & Wasserman, 1994). Due to the high specific binding of Bni1p, Bnr1p and p140mDia to the GTP-bound form of Rho family members (Kohno et al., 1996; Imamura et al., 1997; Watanabe et al., 1997); they perhaps represent significant connectors between signal transduction, profilin and the cytoskeleton. Furthermore, adenylyl cyclase-associated protein (CAP) has been described as PLP-binding ligand. CAP has a $G(P)_6$ $G(P)_5$ motif and it can bind to profilin (Domke et al., 1997; Lambrechts et al., 1997). Nevertheless, other studies demonstrated that CAP exists in a folded configuration (Lambrechts et al., 1997) and hence its binding to profilin may be firmly regulated.

### 2.3.4 Regulation of profilin-ligands binding

The important factors that could help in understanding the process of profilin-ligand binding regulation include the structural requirements for the binding of profilin to the ligand, the binding specificity of ligands to different profilins and the mechanisms of ligand release. Initially, the structural requirements for the profilin-ligand binding are not completely understood. In spite of binding of profilin to an extremely diverse group of ligands either directly or as part of a larger complex, the binding sites on both profilin and ligands appear to be well conserved. The majority of ligands are believed to interact with the PLP domain of profilin that contains the N- and C-terminal helices. The only exception, so far, to this model is gephyrin, which appears to bind to a special profilin domain (Giesemann et al., 2003). All profilin ligands are characterized by the presence of stretched or nearly stretched proline-rich domains that are required for profilin binding. Still, a contiguous prolines stretch is insufficient. Depending on the data obtained from *in vitro* studies using synthetic PLP peptides of different length high-affinity binding requires a decamer as a minimum, (Perelroizen et al., 1994) however this cannot be extended to cover proteins or to be used for recognizing or evaluating the ability of profilin to bind to a ligand. A lot of profilin ligands contain in their proline-rich domains proline repeats of no more than three or four successive prolines. Further amino acids, mostly glycines, appear to be capable of replacing proline, and an efficient profilin-binding domain appears to include numerous repeats that have the consensus sequence ZPPX (where Z=P, G or A; and X= any hydrophobic amino acid) (Witke et al., 1998).

The second important factor in regulating the binding of profilins to their ligands is the binding specificity of ligands to the different profilins. Previous reports showed that the interaction of ligands with profilin-I and profilin-II occurs in a highly specific manner (Witke et al., 1998) and it looks likely that it is not only the PLP-binding domains but also other complex binding parameters have to be considered. Comparative studies on the structures of mammalian profilin-I and profilin-II indicated that they are approximately superimposable (Nodelman et al., 1999). Nevertheless, the distribution of surface charges in profilin-I and profilin-II is significantly different and this perhaps participates in the ligand-binding specificity (Figure 2). Eventually, identifying the structural features of different profilin complexes will be helpful to understand the basis of specificity.

Finally, the profilin- ligands binding should be a dynamic process and the mechanisms of ligand release under physiological conditions have to be determined. For example, actin can be released from profilactin complex via $PtdIns(4,5)P_2$, and an analogous mechanism might be used for ligand binding regulation. For instance, $PtdIns(4,5)P_2$ can regulate the interaction between dynamin 1 and profilin-II, but not the Mena–profilin or VASP–profilin complexes. Regulation of Mena, VASP, and other ligands binding might be achieved in different ways such as profilin or ligand phosphorylation (Witke, 2004).

## 2.4 Role of profilin in signal transduction

Profilins bind to several ligands, and a lot of these ligands are part of various complexes or interact with each other as well (Figure 3). This results in an intimate crosstalk among these complexes that can substitute and distribute components and, thus, could assimilate signals from other signaling pathways such as small-GTPase and phosphoinositide pathways. In these signaling platforms profilins appear to be a common denominator (Witke, 2004). Figure 6 is a schematic representation demonstrating various interactions between profilin, the microfilament system and different signaling pathways.

Profilins are linked to the phosphatidylinositol cycle and in turns to the receptor tyrosine kinase pathway through their binding to PtdIns $4,5-P_2$. Profilin-bound PtdIns $4,5-P_2$ is resistant to hydrolysis by phospholipase $C\gamma1$ (Goldschmidt-Clermont et al., 1990). However, this resistance can be overcome after activating phospholipase via receptor tyrosine kinases-dependant phosphorylation (Goldschmidt-Clermont et al., 1991). This activation process results in PtdIns $4,5$-$P_2$ hydrolysis with subsequent formation of other two second messengers, diacylglycerol and inositol 1,4,5-trisphosphate. Additionally, profilin releases from the membrane, which might initiate fast, local actin polymerization. Conversely, activated phospholipase $C\gamma1$ cannot hydrolyze other PI 3-kinase activity products such as PtdIns $3,4$-$P_2$ and PtdIns $3,4,5$-$P_3$ ,that bind to profilin with a higher affinity than PtdIns $4,5$-$P_2$. Consequently, it has been revealed that PtdIns $3,4$-$P_2$ and PtdIns $3,4,5$-$P_3$ may regulate phospholipase $C\gamma1$-controlled turnover of PtdIns $4,5$-$P_2$ (Lu et al., 1996).

In addition, the profilin ligands of the formin-related proteins such as p140mDia connect the GTPase-related signaling cascade, which is also coupled with the PtdIns $4,5$-$P_2$ signaling pathway to the microfilament system. The small GTPases of the Rho family are active members that are involved in regulating the cytoskeleton-based processes such as cell morphology, adhesion and cytokinesis (Tapon & Hall, 1997). Most likely, these formin-related proteins are down-stream effectors of Rho in this cascade (Evangelista et al., 1997; Watanabe et al., 1997).

Also, the microfilament system is linked to the adenylyl cyclase-related pathway via substrates of the cAMP/cGMP-dependent protein kinases such as VASP/Mena family (Butt et al., 1994; Gertler et al., 1996) and the putative profilin ligand CAP, which is an adenylyl cyclase activator (Fedor-Chaiken et al., 1990; Field et al., 1990; Toda et al., 1985). This linking can be executed through either direct binding of CAP and VASP proteins to actin (Freeman et al., 1995; Gieselmann & Mann, 1992; Gottwald et al., 1996; Hubberstey et al., 1996; Reinhard et al., 1992) or recruiting profilin and profilin-actin complexes to areas of dynamic actin remodelling via the interaction of VASP proteins with cell contact proteins such as zyxin and vinculin (Brindle et al., 1996; Gertler et al., 1996; Reinhard et al., 1995, 1996).

Furthermore, annexin I could be involved in this crosstalk depending on previous reports that described the sensitivity of annexin I-profilin binding to PtdIns $4,5$-$P_2$ and actin (Alvarez-Martinez et al., 1996). On top of that the annexins activity is controlled by the free $Ca^{2+}$ level, which is adjusted via PtdIns $4,5$-$P_2$ hydrolysis upon the action of the activated phospholipase $C\gamma$ 1 (Figure 6). In addition to annexin I, $Ca^{2+}$ level will affect various $Ca^{2+}$-actin-binding and –severing proteins which slice the actin filaments and create new plus ends to which profilin-actin complexes can be added (Schlüter et al., 1997) (Figure 6).

Interestingly, in mesangial cells extracellular profilin was shown to bind specifically to a putative receptor and stimulates AP-1, a key element in signal transduction that is involved in the regulation of the transcription of several genes and cell growth (Tamura et al, 2000).

With the current large number of profilins ligands the future challenge is to determine their role in this complicated signaling crosstalk. One possibility is that profilins may act as regulators for the composition of the complexes and facilitate entrance or exit of certain ligands. Additionally, they might act as direct regulators for the ligands activities. Identification of all profilins molecular interactions, their ligands, and recognizing the structure of these complexes will be helpful to understand the mechanisms by which profilins can control this diverse signaling complexes (Witke, 2004).

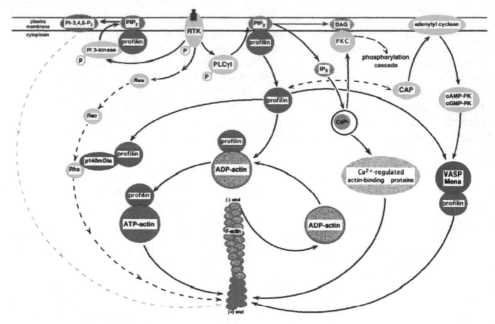

Fig. 6. The involvement of profilin (red) in different signaling routes. This schematic drawing depicts only the main connections established so far. Molecules and second messengers of the polyphosphoinositide signaling pathway are indicated in yellow, protein members of signaling routes are marked green, proline-cluster proteins identified as profilin ligands are marked purple, the actin cycle is seen in blue, $Ca^{2+}$ in intracellular stores and $Ca^{2+}$ regulated microfilament proteins are marked grey. For simplicity, the solid arrows indicate either direct interactions between components, as shown by biochemical assays, or point to pathways. Broken arrows indicate suspected or indirect interactions. Abbreviations: PI-3,4,5-$P_3$: phosphatidylinositol 3,4,5-trisphosphate; $PIP_2$: phosphatidylinositol 4,5-bisphosphate; RTK: receptor tyrosine kinase; DAG: diacylglycerol; PLC$\gamma$ 1: phospholipase C$\gamma$ 1; cAMP/cGMP- PK: cAMP/cGMP dependent protein kinase; $IP_3$: inositol 1,4,5-trisphosphate, adapted from Schlüter et al., (1997) with permission.

## 3. Profilin & vascular diseases

### 3.1 Role of profilin in vascular smooth muscle & endothelial cells

#### 3.1.1 Profilin & vascular smooth muscle cells migration & proliferation

Migration of smooth muscle cell takes place throughout vascular development, as a result of vascular injury, and throughout atherogenesis. Throughout vascular development, platelet-derived growth factor promotes migration of pericyte or other precursors of smooth muscle that is required for the formation of correct vessel wall (Hellstrom et al., 1999). Clinically, vascular injury takes place after angioplasty, vascular stent implantation, or organ transplantation. In vascular injury in animals, thickening of intima and media has been attributed to VSM proliferation and migration from media to intima (Clowes et al., 1989; Majesky & Schwartz, 1990; Reidy, 1992). Throughout atherogenesis, VSMCs migrate to occupy the intima, either from the media (Murry et al., 1997) or from the circulation via $CD34^+$ hematopoietic progenitor cells migration, resulting in smooth muscle progenitor cells (Yeh et al., 2003). Figure 7 shows the inner lining of a normal artery.

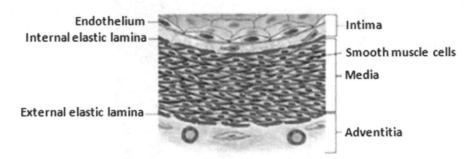

Fig. 7. Inner lining of normal artery

VSMC migration requires the extension of lamellipodia toward the stimulus via actin polymerization, trailing edge detachment via focal contacts degradation, and force generation via myosin II in the cellular body to drive the cell forward (Gerthoffer, 2007). Initiation of new filaments nucleation is achieved by actin-related proteins such as ARP2/3 complex interacting with the minus end and elimination at the plus end of capping proteins that are $PIP_2$-sensitive. Extension of new actin filaments is improved by formin-related proteins such as mDia1 and mDia2 that operate along with profilin on the plus end. Activation of The formins mDia1 and mDia2 is achieved by RhoA and Cdc42, respectively. Profilin released from the binding sites of membrane phospholipid enhances nucleotide exchange on G-actin monomers and promotes actin polymerization. Stimulation of filament branching is accomplished via activating WAVE complex and WASP by Rac and Cdc42, respectively. WAVE and WASP increase nucleation and branching through activating actin-related proteins such as ARP2/3 complex. Severing of Actin-filament by gelsolin is stimulated by $Ca^{2+}$, and nucleation is favored via liberating gelsolin from plus ends of F-actin by PtdIns 4,5-P2. Stimulation of actin depolymerization is executed by cofilin at the minus end. Cofilin acts to limit the filaments length and to induce the existing filaments turnover. These operations have been reported to be sufficient for force generation to expand the leading edge of the cell toward the stimulus (Mogilner & Oster, 2003; Prass et al., 2006). Consistent with these findings our recent data confirmed the significant role of profilin-I in VSMC migration. Migration assays performed on VSMC isolated from the aorta of transgenic mice that overexpress the cDNA of profilin-I or profilin-I-dominant negative mutant (88R/L) and nontransgenic controls showed that the rate of cell migration of profilin-I VSMCs is significantly higher than that of the control and 88R/L. Conversely, 88R/L mice exhibited a significantly lower rates compared to nontransgenic controls (Figure 8) (Hassanain HH, unpublished).

On the other hand, it has been shown that profilin plays a vital role in the proliferation and differentiation of normal cell. Disruption in the profilin results in embryonic lethality due to gross impairment in growth, motility, and cytokinesis in single cells (Haugwitz et al., 1994; Witke, 2004; Witke et al., 2001). Also, profilin-1 was demonstrated to exert cellular responses such as DNA synthesis and increasing the binding activity of AP-1 DNA in mesangial cells via activating putative cell surface receptors (Tamura et al., 2000).

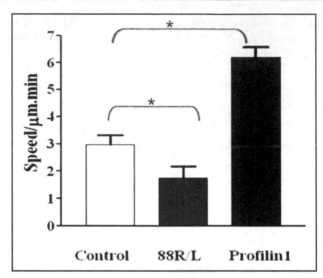

Fig. 8. The MetaMorph image analysis of the mean speed of the individual cells (μm/min) of profilin-I, 88 R/L and control VSMCs. The differences in mean were determined by ANOVA. *P < 0.05, compared with corresponding control, is considered to be significant (Hassanain HH, unpublished).

In line with the established role of profilin in cellular migration and proliferation, it has been shown that recombinant profilin-I stimulates DNA-synthesis and migration of both rat and human VSMCs in a concentration-dependent manner (Caglayan et al., 2010). The same study indicated that profilin-induced VSMCs migration is dependent on PI3K activity (Caglayan et al., 2010). Moreover, Cheng et al., (2011) found that profilin-I plays a key role in Angiotensin (Ang) II-induced VSMCs proliferation. They also suggested that Ang-II increases profilin-I expression and promotes VSMCs proliferation via activating AT1 receptor/JAK2/STAT3 pathway (Cheng et al., 2011). On the contrary, other studies described the involvement of  phospho-extracellular signal-regulated kinase1/2 (P-ERK1/2) and phospho-c-Jun NH2-terminal kinase (P-JNK) in Ang-II-induced profilin-I expression (Zhong et al., 2011), and that PI3-kinase, Src, and, to a lesser extent, P-ERK1/2 are required for profilin-I-dependent VSMCs proliferation (Caglayan et al., 2010). Consequently, Cheng et al., (2011) proposed that the interaction of these signaling pathways mediating the role of profilin-I in VSMCs proliferation requires further investigation. Consistent with these data, we observed that the treatment of mouse aortic VSMCs with Ang-II (100 nM/10 min) resulted in increased profilin-I expression (Hassanain HH, unpublished)

### 3.1.2 Profilin & vascular smooth muscle contraction

Regulation of smooth muscle contraction has been thought to be only dependent on the 20-kDa regulatory light chain of myosin (MLC20) that in turn modulates cross-bridge cycling of actomyosin. Numerous studies showed that contractile stimulation promotes actin polymerization in vascular and airway smooth muscle tissues (Cipolla & Osol, 1998; Jones et al., 1999; D. Mehta & Gunst, 1999) and in cultured smooth muscle cells (An et al., 2002; Barany, et al., 2001; Hirshman & Emala, 1999). In addition, inhibition of actin polymerization by specific inhibitors such as latrunculin decreases the contractile stimuli- activated force development in smooth muscle (Cipolla & Osol, 1998; D. Mehta & Gunst, 1999; Youn et al., 1998). However, this does not affect contractile stimulation-induced MLC20 phosphorylation (34), suggesting that actin polymerization plays a central role during smooth muscle contraction. Tang & Tan, (2003) investigated the effect of profilin, the main actin-regulatory protein on the regulation of smooth muscle contraction. They demonstrated that profilin downregulation with antisense repressed force generation, without affecting MLC20 phosphorylation, signifying that profilin is crucial for smooth muscle contraction and that it does not regulate the activation of contractile protein. Yet, profilin downregulation repressed increases in the F-actin/G-actin ratio in return to agonist stimulation, showing that profilin is essential for actin dynamics during contractile stimulation of smooth muscle (Tang & Tan, 2003). In harmony with these finding our results showed higher expression of stress fibers and membrane ruffling in vascular smooth muscle cells from profilin-I transgenic mice compared with nontransgenic control and 88R/L. The 88R/L cells, however, showed lower expression of stress fiber formation and ruffling than the nontransgenic controls (Figure 9A) (Moustafa-Bayoumi et al., 2007). In addition, we confirmed these findings by assessing the ratio of F-actin/G-actin in the aortic smooth muscle cells from profilin-I. Our results showed a significant increase in F/G actin ratio in the aortic smooth muscle cells from profilin-I mice compared with the nontransgenic controls (Figure 9B) (Moustafa-Bayoumi et al., 2007). Furthermore, we showed that profilin-I plays a significant role in increased contractility and force development in the mesenteric arteries of profilin-I mice via activating Rho/ROCK pathway and MLC20 (Hassona et al., 2010). Activated Rho elevates MLC20 phosphorylation by 1) directly phosphorylating MLC20 and 2) phosphorylation and inhibition of the MBS of MLC20 phosphatase (Higgs & Pollard, 2001; Pollard & Borisy, 2003). This increases myosin contractility and tension contributing to stress fibers. In conclusion, our results indicate that overexpression of profilin-I in smooth muscle cells leads to increased contractility and force development via increasing actin polymerization (Moustafa-Bayoumi et al., 2007) and MLC20 activation(Hassona et al., 2010), which in turn induce mechanical stress that is considered as the main initiator for arterial stiffness and hypertension observed in these mice.

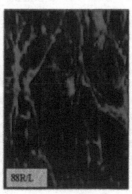

(a)

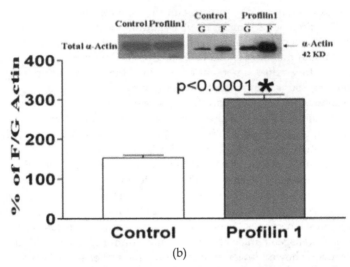

(b)

Fig. 9. Overexpression of profilin-I induced actin polymerization in vascular smooth muscle cells. Rhodamine-phalloidin staining of smooth muscle cell confluent monolayers shows increased stress fibers in vascular smooth muscle cells from profilin-I transgenic mouse as compared with nontransgenic control (a). The 88R/L cells, however, show lower expression of stress fiber formation than the control (a). Analysis of the F-actin/G-actin ratio shows significant increase in F-actin/G-actin ($F/G$) ratio in the aortic smooth muscle cells from profilin-I mice compared with the nontransgenic controls (b) (Moustafa-Bayoumi et al., 2007).

### 3.1.3 Profilin & vascular endothelial cells adhesion

Endothelial cells survival neatly depends on their ability to anchor to extracellular matrix proteins. Suppression of endothelial cell adhesion has been shown to induce apoptosis in these cells (Meredith et al., 1993; Re et al., 1994; Zang et al., 1995). It has been found that transient overexpression of profilin in cultured human aortic endothelial cells using replication-incompetent adenovirus enhances endothelial cells adhesion to the extracellular matrix via promoting the binding of extracellular fibronectin to its receptor on the surface of these cells. Additionally, it was revealed that profilin-mediated enhancement of endothelial cell adhesion has a protective role in situations of focal contacts disruption due to shear, stretch or other focal injuries (Moldovan et al., 1997).

Moreover, the authors, Moldovan et al., (1997) proposed that the profilin-mediated effect seems to be stimulated via recruiting integrins $\alpha_5\beta_1$ to the endothelial cell surface. Numerous mechanisms may explain this later effect. One possibility is that profilin might cause improvement in the access of receptor molecules to the cell surface. Instead, profilin might cause impairment in the internalization of membrane receptors. These effects may be achieved in 1) actin-dependent manner, where profilin might decrease receptor internalization via disrupting actin stress fibers or it might offer a stronger anchor for fibronectin receptor molecules in focal contacts via stabilizing actin filaments that are not stress fibers (Finkel et al., 1994), or 2) actin-independent manner, where profilin interacts with PtdIns 4,5-P2 and inhibits its hydrolysis by phospholipase C (Goldschmidt-Clermont et al., 1990, 1991; Lassing & Lindberg, 1985). Increased concentrations of PtdIns 4,5-P2 could stimulate the stabilization of newly formed focal contacts including the fibronectin receptor via an unknown mechanism or profilin overexpression could overcome other actin-binding proteins for interacting with PtdIns 4,5-P2 and thus enhance their binding to actin filaments (Hartwig et al., 1995).

### 3.1.4 Role of profilin in vascular endothelial cells migration, proliferation & capillary morphogenesis

Vascular endothelial cell (VEC) migration is vital for capillary outgrowth from pre-existing blood vessels during angiogenesis (Bauer, et al., 2005). During cell migration, actin cytoskeleton reorganization is a dynamic process that includes both actin polymerization and depolymerization in an accurate spatiotemporal manner. Regulation of this actin remodeling process is achieved by a large number of actin binding proteins such as those involved in monomer sequestering, nucleating, elongating, severing, depolymerizing, and capping of actin filaments (Pollard & Borisy, 2003). Expression profiles in VEC experiencing capillary morphogenesis identified some of the key actin-binding proteins that have been previously involved in angiogenesis such as thymosin β4, profilin, gelsolin and VASP. Among these proteins, as a minimum thymosin β4 has been established as a proangiogenic molecule *in vivo* (Philp et al., 2004; Salazar et al., 1999). In addition, it has been reported that silencing profilin-I expression in human umbilical vein endothelial cells significantly decreases their capability of forming planar cord-like structures on matrigel (a commonly adopted *in vitro* representation for angiogenesis). These findings proposed for the first time that profilin-I might play a key role in VEC capillary morphogenesis (Ding et al., 2006).

In a more recent report for the same group they adopted a knockdown–knockin experimental system to stably express either fully functional form or mutants of profilin-I that are deficient in binding to actin and proteins containing polyproline domains, in a human dermal microvascular cell line. They showed that silencing endogenous profilin-I expression in this cell line results in slow rate of random migration, decreased membrane protrusion velocity and a significant reduction in matrigel-induced cord formation. These defects were rescued only via re-expression of fully functional but not any of the two ligand-binding deficient mutants of profilin-I. They also showed that loss of profilin-I expression in VEC inhibits three dimensional capillary morphogenesis, MMP2 secretion and ECM invasion. Disruption of actin and polyproline interactions of profilin-I inhibited VEC invasion through ECM, as well. They concluded that profilin-I regulates VEC migration, invasion and capillary morphogenesis through its binding to both actin and proline-rich ligands (Ding et al., 2009). Furthermore, they indicated that these *in vitro* findings pave the way for future *in vivo* studies to investigate the role of profilin-I in angiogenesis.

Interestingly, cutaneous wound healing experiments in our profilin-I and 88R/L transgenic mice showed a significant increase in blood vessel density in profilin-I transgenic mice compared to 88R/L transgenic mice and nontransgenic control at post wound day 7 (Figure 10) (Hassanain HH, unpublished). These data could indicate the importance of profilin-I in angiogenic reponse in VEC.

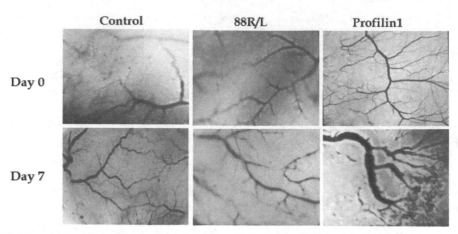

Fig. 10. Stimulation of angiogenesis in the wound area of profilin-I mice. Distribution of capillaries along the margin of the excision wound in Profilin-I, 88R/L and nontransgenic control mice at post wound days 0 and 7. High magnification of capillaries in the skin was obtained with a 2X objective lens light microscope (Hassanain HH, unpublished).

## 3.2 Role of profilin in vascular remodeling & hypertension

Hypertension represents a major risk factor for cardiovascular events such as stroke and myocardial infarction. It is well established that hypertension leads to remodeling of large and small arteries (Folkow, 1982; Simon, 2004). Remodeling of the vasculature is an active process of structural changes that involves alterations in cellular processes, including growth and changes in the extracellular matrix integrin-cytoskeleton axis, resulting in an increase in the media-to-lumen ratio (Gimbrone et al., 1997; Intengan & Schiffrin, 2001). Physiological remodeling is an adaptive process occurring in response to hemodynamic changes and aging. However, when this process becomes maladaptive, it plays a role in hypertension's complications (Ming et al, 2002; Touyz, 2007). Increased mechanical strain/hypertension in the vessel wall triggers the hypertrophic signaling pathway resulting in structural remodeling of vasculature. Increased actin polymerization and stress fiber formation generate mechanical force that represents an important modulator of cellular morphology and function in a variety of tissues and is an important contributor to hypertrophy in the cardiovascular system (Ruwhof &van der Laarse, 2000). Also, it has been shown that actin polymerization within VSMCs in response to increased intravascular pressure is a novel mechanism underlying arterial myogenic behavior. The cytosolic concentration of G-actin is significantly reduced by an elevation in intravascular pressure, demonstrating the dynamic nature of actin within VSMCs and implying a shift in the F:G equilibrium in favor of F-actin. Profilin-I which is a key actin-regulatory protein that plays an essential role in regulating de novo actin polymerization, particularly actin treadmilling (Carlier & Pantaloni, 2007; Suetsugu et al., 1999) could be vital in regulating all of these vascular events. Indeed, our report in the Journal of Biological Chemistry (Moustafa-Bayoumi et al., 2007) established the feasibility of our proposal. We showed that elevated expression of profilin-I gene in VSMCs of profilin-I mice favoring F-actin induces stress fiber formation (Figure 11) and plays an important role in vascular hypertrophy by inducing internal mechanical stress and triggering the hypertrophic signaling pathways, integrins-$\alpha_1\beta_1$/Rho-ROCK/MAPKs e.g. P-ERK and P-JNK, leading to vascular remodeling in both large (e.g. aorta) and small (e.g. mesenteric) arteries (Figure 12A, B) of profilin transgenic mice (Hassona et al., 2010; Moustafa-Bayoumi et al., 2007).

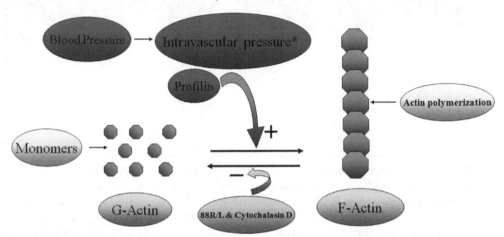

Fig. 11. Hypertension or increased profilin-I expression in VSMCs leads to a shift in the F:G equilibrium in favor of F-actin and an elevation in intravascular pressure. This pathway can be reversed by F-actin inhibitor, cytochalasin D or profilin-I mutant, 88R/L.

Consistent with our finding, very recent studies showed increased profilin-I expression in hypertrophic thoracic aorta and mesenteric arteries of spontaneously hypertensive rats with subsequent elevation in both P-ERK and P-JNK, suggesting that profilin-I may contribute to the vascular remodeling in these rats (Cheng et al., 2011; Zhong et al., 2011). In this context, previous studies suggested that mechanical stretch is closely related to JNK and ERK1/2 activation (Hu et al., 1997; Pyles et al., 1997). These cascades play an important role in remodeling of blood vessels, as well. In addition, this pathway is activated by Ang-II and has been implicated in the pathogenesis of cardiovascular diseases (P.K. Mehta & Griendling, 2007). Interestingly, it has been recently reported that profilin-I is a key component in the Ang-II-induced vascular remodeling (Cheng et al., 2011; Zhong et al., 2011).

As it was mentioned above that hypertension is a major cause of vascular remodeling. The primary aim of anti-hypertensive drugs, particularly Ang-converting enzyme inhibitors and Ang receptor subtype 1 antagonists, is to lower the blood pressure with the hope of reversing this remodeling (Schiffrin, 2001). Importantly, In our profilin-I model we demonstrate that the reverse can be true as well, i.e. alteration in cytoskeleton dynamics favoring increased actin polymerization can contribute to vascular adaptations with aging resulting in increased systolic blood pressure by the time the profilin-I mice were six months old (Figure 12C) (Moustafa-Bayoumi et al., 2007). The blood pressure in the profilin-I mice was elevated 25–30 mm Hg higher than nontransgenic controls. In agreement with our findings, it has been demonstrated that profilin speeds up the actin remodeling and accordingly improves the growth and invasion force of VSMCs resulting in increased vascular resistance and accelerated formation of pulmonary hypertension (Dai et al., 2006).

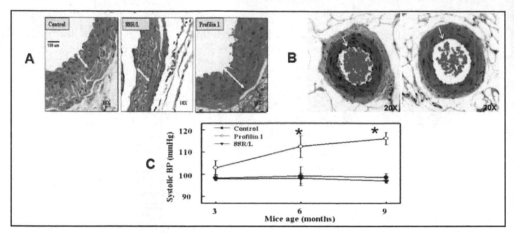

Fig. 12. Profilin overexpression induced vascular hypertrophy and hypertension. Hematoxylin and eosin staining shows clear signs of remodeling and vascular hypertrophy in the aorta of profilin-I transgenic mice (yellow arrows; **A**) and mesenteric arteries (white arrows; **B**). There are no differences, however, between 88R/L and nontransgenic control aortic sections (**A**). Tail cuff measurements of blood pressure show significant increase in the systolic blood pressure (BP) in profilin-I transgenic mice at 6 months and older compared with nontransgenic control mice (C) (Hassona et al., 2010; Moustafa-Bayoumi et al., 2007).

On the other hand, the blood pressure in 88R/L mice was below the control littermates; however, it did not reach statistical significance. The absence of a hypotensive phenotype in the 88R/L mice could be due to the lack of significant vascular remodeling as a result of decreased actin polymerization. Our results showed a decrease in stress fibers formation in 88R/L mice (Figure 9); however, these changes did not translate into significant alterations in the vasculature. This might be due to an activation of a compensatory mechanism to maintain the integrity of vessel structure and thus keep the blood pressure at a survival level. Additionally, our preliminary data showed that inhibition of profilin-I-induced stress fibers by cytochalasin D lowered blood pressure in profilin-I mice. As a pilot study the profilin-I mice were injected with a single dose of cytochalasin D (0.5 µg/gram body weight) which led to lowered blood pressure within 10 minutes in these mice from 140 mmHg to 70 mmHg and the effect was sustained for more than 1.5 hours. Then the mice were recovered without any sign of sickness. To make sure that cytochalasin D had no damaging effect on the endothelium, we assessed the functionality of the endothelium using Ach and wire-myography. Our results showed no damage in the endothelium after cytochalasin D treatment (Hassanain HH, unpublished). We should note that cytochalasin D was used before by other investigators in different studies with much higher doses and no toxicity was observed (Speirs & Kaufman, 1989).

Furthermore, stress fiber formation could affect the relaxation/contraction process of the smooth muscles, making it more constrictive and/or less responsive to vasodilators such as nitric oxide. That could be an important factor contributing to hypertension besides the vascular hypertrophy in the profilin-I transgenic mice. Our recent report in the American Journal of Physiology confirmed this proposal. We showed that vascular hypertrophy-associated hypertension of profilin-I transgenic mice led to functional remodeling of peripheral arteries. Our results showed a significant increase in the contraction response of profilin-I mesenteric arteries toward phenylephrine and significant decreases in the relaxation response toward ACh and sodium nitrite compared with nontransgenic controls (Hassona et al, 2010). Additionally, inhibiting stress fibers formation with cytochalasin D significantly relaxes the phenylephrine-contracted mesenteric arteries, suggesting that the increased constriction of mesenteric arteries to phenylephrine could be because of the increased F- to G actin ratio; however, cytochalasin D treatment reduced this ratio (Hassona et al., 2010).

Moreover, it has been reported that in addition to the role of hypertension in vascular remodeling, there are pressure-independent genes that play a key role in vascular remodeling. This concept is supported by the observation that despite blood pressure control in hypertensive patients, the rate of restenosis (attributable to remodeling) remains high (Gurlek et al., 1995). In harmony with this concept we recently showed that normalization of blood pressure by selected anti-hypertensive agents is not enough to correct the structural and functional remodeling of profilin-I transgenic mice (Hassona et al., 2011). Our results demonstrated that there is only correction in the functional remodeling and signaling cascades of the mesenteric arteries of losartan- and amlodipine-treated, but not those of atenolol-treated profilin-I transgenic mice, where losartan and amlodipine decrease the F-actin and stress fibers formation, proposing that the stress fibers seem to play a major role in the development and progression of the vascular remodeling-associated hypertension. We finally concluded that profilin-I gene, which is the key player controlling stress fiber formation may be a good target to treat not only hypertension but also the vascular remodeling in hypertensive patients (Hassona et al., 2011).

### 3.3 Role of profilin in atherosclerosis & vascular complication in diabetes

Vascular endothelium dysfunction goes before, and may participate in atheroma formation in return to various cardiovascular risk factors such as diabetes (Johnstone et al., 1993; Tesfamariam et al., 1990;), hyperlipidemia (Chikani et al., 2004; Steinberg et al., 1997), and both local and systemic inflammatory mediators (Libby, 2002). Interestingly, Romeo et al., (2004) revealed that profilin-I levels are improved in the endothelium of diabetic aorta of both human and experimental animals. They also demonstrated that profilin overexpression in primary aorta EC was capable of triggering indicators of endothelial dysfunction such as apoptosis, ICAM-1 up-regulation, and decreased VASP phosphorylation. In addition, profilin was found to be required for LDL-mediated ICAM-1 up-regulation and it can be regulated by LDL/cholesterol signaling, but not high glucose (Romeo et al., 2004). Although, Clarkson et al., (2002) reported that exposure to high glucose was able to increase profilin-I mRNA in mesangial cells and in the diabetic rat kidney. Romeo et al., (2004) suggested that the inability of high glucose to enhance profilin-I protein levels in EC is in line with a multifactorial etiology of endothelial dysfunction coupled with the metabolic syndrome and may reveal the inadequate effect of glucose-lowering monotherapy to prevent macrovascular complications in type 2 diabetic patients (U.K. Prospective Diabetes Study (UKPDS) Group, 1998). On the other hand, our preliminary data showed that mouse aortic VSMCs treated with glucose (25 mM/24 hours) increased profilin-I expression (Hassanain HH, unpublished).

Furthermore, Romeo et al., (2004) showed that profilin was clearly increased in EC and macrophages within atherosclerotic lesions of apoE null mice. In a more recent report, the same group specified the significance of profilin-I for atherogenesis *in vivo* as profilin-I heterozygosity resulted in protection from atherosclerosis in LDL receptor-null mice (Romeo et al., 2007). In this report, a variety of atheroprotective indicators were recognized in mice with heterozygous deficiency of profilin-I, as compared to profilin-I wild-type mice. Aortas from these heterozygous mice exhibited preserved activation of endothelial nitric oxide synthase (eNOS) and nitric oxide-dependent signaling, decreased expression of vascular cell adhesion molecule (VCAM)-1 and decreased accumulation of macrophage at the sites of injury. Correspondingly, profilin-I knockdown in cultured aortic ECs was able to protect against endothelial dysfunction induced by oxidized low-density lipoproteins (oxLDL). Additionally, macrophages from bone marrow of profilin-I-deficient heterozygous mice exhibited diminished internalization of oxLDL and oxLDL-induced inflammation. These studies concluded that profilin-I plays a vital role in early atheroma formation and that decreasing profilin-I levels is atheroprotective. Finally, profilin-I atheroprotective effect is mediated via combined mechanisms that depend on both endothelium and macrophages (Romeo et al., 2007).

Moreover, the same group addressed the pathways responsible for profilin-I gene expression in 7-ketocholesterol (oxysterol)-stimulated endothelial cells and in the diabetic aorta. They showed that oxysterol-binding protein-1 (OSBP1) is required for oxysterol-dependent nucleation and activation of the JAK2/STAT3 pathway, which in turn regulates profilin-I gene expression in endothelial cells. Similarly, diabetes increases the activation of STAT3 and its recruitment to the profilin-I promoter in large vessels *in vivo* (Romeo et al., 2008)

Very recently, it has been reported that profilin-I expression is markedly increased in human atherosclerotic plaques compared to the normal vessel wall (Caglayan et al., 2010). A correlation was found between profilin-I serum levels and the degree of atherosclerosis, as well. The atherogenic effects of profilin-I on VSMCs imply an auto-/paracrine role within the plaque. In addition, it was found that profilin-I acts as an extracellular ligand and triggers atherogenic effects in VSMCs including DNA synthesis and migration. Besides, profilin-1 stimulates typical signaling pathways such as the PI3K/AKT and RAS-RAf-MEK-ERK pathways. These findings revealed that profilin-I might play a critical role in atherogenesis and may represent a novel therapeutic target in human patients (Caglayan et al., 2010).

### 3.4 Role of profilin in age-associated vascular problems

Aging is a major risk factor for the development of vascular diseases, such as hypertension and arteriosclerosis, which lead to stroke and heart failure (Spagnoli et al., 1991). Aging is also linked with decreased stress tolerance. Susceptibility to a variety of physiological stresses such as infection, inflammation, and oxidative damage enhances with age and is causally coupled with clinical problems in the elderly (Starr et al., 2011). So far, the mechanism of age-related changes in vasculature has not been completely understood. On the top of that, the role of profilin in these age related changes remains largely unstudied.

Recently, it has been reported that protein nitration levels increased in aged mice compared to young mice. Also, particularly strong nitration was found in the pulmonary vascular endothelium during systemic inflammatory response syndrome (SIRS). Age- and SIRS-dependent increased protein nitration was evident in proteins related to the actin cytoskeleton that are responsible for maintaining pulmonary vascular permeability such as transgelin-2, LASP 1, tropomyosin, myosin and profilin-I. Recognizing the nitrated proteins indicated important modifications to the vascular endothelial cytoskeleton, which potentially participates in the barrier dysfunction, enhanced vascular permeability, and pulmonary edema (Starr et al., 2011).

It has been established that deficiency in plasma fibronectin increases lung vascular permeability (Wheatley et al., 1993); consequently, as adhesion of endothelial cell to fibronectin depends on profilin expression (Moldovan et al., 1997), lack of functional profilin may be to some extent responsible for vascular permeability as a result of inefficient barrier integrity. These data can fairly elucidate the age-associated enhancement in susceptibility to systemic inflammation, acute lung injury, and respiratory failure (Starr et al., 2011).

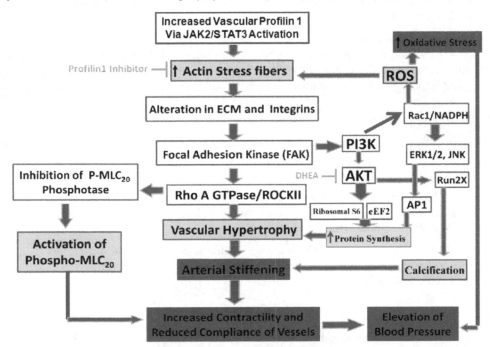

Fig. 13. JAK2/STAT3 pathway activation increases profilin-I (Romeo et al., 2008; Cheng et al., 2011) in the vessel media induced stress fiber formation and increased internal mechanical stress in the vessel walls (Moustafa-Bayoumi et al., 2007) which modulates changes in ECM and integrins (Abouelnaga et al., 2009; Hassona et al., 2010). These changes led to activation of FAK (Abouelnaga et al., 2009) that in turn activate Rho/ROCKII (Hassona et al, 2010; Moustafa-Bayoumi et al., 2007), PI3 kinase and AKT (Caglayan et al., 2010). Activation of Rac1/NADPH pathway (Abouelnaga et al., 2009) results in increased superoxide production and increases oxidative stress (Hassanain HH, unpublished) in vessel walls which could contribute to hypertension. The activation of Rho/ROCKII and AKT result in activation of MLC20 (Hassona et al., 2010), and increases in protein synthesis (Gingras et al., 1998; Kitamura et al., 1998; Ushio-Fukai et al., 1999) and calcification (Byon et al., 2008), respectively. These changes in the media of the vessel walls result in arterial stiffening and hypertension (Moustafa-Bayoumi et al., 2007). Profilin-I inhibitor can block the stress fiber formation in this pathway (Moustafa-Bayoumi et al., 2007 ) and dehydroepiandrosterone (DHEA) can inhibit AKT kinase pathway (Bonnet et al., 2009).

Conversely, other indirect evidence showed that profilin-I increased with age; a recent study using proteomic and genomic analyses of hippocampus from young and old rats showed a significant increase in profilin-I expression in aged rat hippocampus (Weinreb et al., 2007). Another study investigating differential protein expression profiles in chronically stimulated T cell clones found that profilin-I was widely and highly expressed in cytoplasm (Mazzatti et al., 2007). The study concluded that differential expression of profilin-I in aging may contribute directly to immunosenescence via disrupting the intracellular signaling and intercellular communication (Egerton et al., 1992; Witke et al., 1998). Consistent with these findings our preliminary data showed an increase in profilin-I expression in the aortic medial layers of older wild-type mice compared with young mice (Hassanain HH, unpublished).

Taken together, this review shed some light on the important role of profilin-I in vascular diseases. However, more studies need to be done in order to fully understand the profilin-I signaling pathway and its mechanism(s) of regulation. Figure 13 summarize some of the proposed signaling molecules involved in profilin-induced vascular complications.

## 4. Acknowlegements

Figure 1 is reprinted from Krishnan, K. & Moens, P.D.J. (2009). Structure and functions of profilins. *Biophysical reviews*, Vol.1, No.2, pp. 71–81, ISSN 1867-2450, with permission from Springer.

Figures 2, 3 & 4 are reprinted from Witke, W. (2004). The role of profilin complexes in cell motility and other cellular processes. *Trends in cell biology*, Vol.14, No.8, pp. 461-469, ISSN 0962-8924, with permission from Elsevier.

Figures 5 & 6 are reprinted from Schlüter, K.; Jockusch, B.M. & Rothkegel, M. (1997). Profilins as regulators of actin dynamics. *Biochimica et biophysica acta*, Vol.1359, No.2, pp. 97-109, ISSN 0006-3002, with permission from Elsevier.

## 5. References

Abouelnaga, Z.A.; Hassona, M.D.; Awad, M.M; Alhaj, M.A.; Badary, O.A.; Hamada, F.A.; Bergese, S.D. & Hassanain, H.H. (2009). Mechanical Strain in VSM Cells Triggers Vascular Remodeling and Hypertension and Activates Integrins in Profilin1 Transgenic Model. *The FASEB journal*, April 2009 23 (Meeting Abstract Supplement) 704.6, ISSN 0892-6638.

Alberts, A.S. (2002). Diaphanous-related Formin homology proteins. *Current biology*, Vol.12, No. 23, pp. R796, ISSN 0960-9822.

Alvarez-Martinez, M.T.; Mani, J.C.; Porte, F.; Faivre-Sarrailh, C.; Liautard, J.P.; Sri Widada, J. (1996). Characterization of the interaction between annexin I and profilin. *European journal of biochemistry*, Vol.238, No.3, pp. 777–784, ISSN 0014-2956.

Ampe, C. & Vandekerckhove, J. (1994). Actin-actin binding protein interfaces. *Seminars in cell biology*, Vol.5, No.3, pp. 175-182, ISSN 1043-4682.

An, S.S.; Laudadio, R.E.; Lai, J.; Rogers, R.A. & Fredberg, J.J. (2002). Stiffness changes in cultured airway smooth muscle cells. *American journal of physiology. Cell physiology*, Vol.283, No.3, pp. C792–C801, ISSN 0363-6143.

Archer, S.J.; Vinson, V.K.; Pollard, T.D. & Torchia, D.A. (1994). Elucidation of the poly-L-proline binding site in Acanthamoeba profilin I by NMR spectroscopy. *FEBS Letters*, Vol.337, No.2, pp. 145-151, ISSN 0014-5793.

Babcock, G. & Rubenstein, P.A. (1993). Control of profilin and actin expression in muscle and nonmuscle cells. *Cell motility and the cytoskeleton*, Vol.24, No.3, pp. 179-188, ISSN 0886-1544.

Babich, M.; Foti, L.R.; Sykaluk, L.L. & Clark, C.R. (1996). Profilin forms tetramers that bind to G-actin. *Biochemical and biophysical research communications*, Vol.218, No.1, pp. 125-131, ISSN 0006-291X.

Balasubramanian, M.K.; Hirani, B.R.; Burke, J.D. & Gould, K.L. (1994). The Schizosaccharomyces pombe cdc3+ gene encodes a profilin essential for cytokinesis. *The Journal of cell biology*, Vol.125, No.6, pp. 1289-1301, ISSN 0021-9525.

Barany, M.; Barron, J.T.; Gu, L. & Barany, K. (2001). Exchange of the actin-bound nucleotide in intact arterial smooth muscle. *The Journal of biological chemistry*, Vol.276, No.51, pp. 48398–48403, ISSN 0021-9258.

Bauer, S.M; Bauer, R.J. & Velazquez, O.C. (2005). Angiogenesis, vasculogenesis, and induction of healing in chronic wounds. *Vascular and endovascular surgery*, Vol.39, No.4, pp. 293–306, ISSN 1538-5744.

Binette, F.; Bénard, M.; Laroche, A.; Pierron, G.; Lemieux, G. & Pallotta, D. (1990). Cell-specific expression of a profilin gene family. *DNA and cell biology*, Vol.9, No.5, pp. 323-324, ISSN 1044-5498.

Björkegren, C.; Rozycki, M.; Schutt, C.E.; Lindberg, U. & Karlsson, R. Mutagenesis of human profilin locates its poly(L-proline)-binding site to a hydrophobic patch of aromatic amino acids. FEBS Letters, Vol.333, No.(1-2), PP.123-126, ISSN 0014-5793.

Boettner, B.; Govek, E.E.; Cross, J. & Van Aelst, L. (2000). The junctional multidomain protein AF-6 is a binding partner of the Rap1A GTPase and associates with the actin cytoskeletal regulator profilin. *Proceedings of the National Academy of Sciences of the United States of America*, Vol.97, No.16, pp. 9064–9069, ISSN 0027-8424.

Bonnet, S.; Paulin, R.; Sutendra, G.; Dromparis, P.; Roy, M.; Watson, K.O.; Nagendran, J.; Haromy, A.; Dyck, J.R. & Michelakis, E.D. (2009). Dehydroepiandrosterone reverses systemic vascular remodeling through the inhibition of the Akt/GSK3-β/NFAT axis. Circulation. Vol.120, No.13, pp. 1231-1240, ISSN 0009-7322.

Borisy, G.G. & Svitkina, T.M. (2000). Actin machinery: pushing the envelope. *Current opinion in cell biology*, Vol.12, No.1, pp. 104-112, ISSN 0955-0674.

Braun, A.; Aszódi, A.; Hellebrand, H.; Berna, A.; Fässler, R. & Brandau, O. (2002). Genomic organization of profilin-III and evidence for a transcript expressed exclusively in testis. *Gene*, Vol.283, No.1-2, pp. 219–225, ISSN 0378-1119.

Brindle, N.P.; Holt, M.R.; Davies, J.E.; Price, C.J. & Critchley, D.R. (1996). The focal-adhesion vasodilator-stimulated phosphoprotein (VASP) binds to the proline-rich domain in vinculin. *The Biochemical journal*, Vol.318, Pt.3, pp. 753-757, ISSN 0264-6021.

Butt, E.; Abel, K.; Krieger, M.; Palm, D.; Hoppe, V.; Hoppe, J. & Walter, U. (1994). cAMP- and cGMP-dependent protein kinase phosphorylation sites of the focal adhesion vasodilator-stimulated phosphoprotein (VASP) in vitro and in intact human platelets. *The Journal of biological chemistry*, Vol.269, No.20, pp. 14509-14517, ISSN 0021-9258.

Byon, C.H.; Javed, A.; Dai, Q.; Kappes, J.C.; Clemens, T.L.; Darley-Usmar, V.M.; McDonald, J.M. & Chen, Y. (2008). Oxidative stress induces vascular calcification through modulation of the osteogenic transcription factor Runx2 by AKT signaling. *The Journal of biological chemistry*, Vol.283, No.22, pp. 15319-15327, ISSN 0021-9258.

Caglayan, E.; Romeo, G.R.; Kappert, K.; Odenthal, M.; Südkamp, M.; Body, S.C.; Shernan, S.K.; Hackbusch, D.; Vantler, M.; Kazlauskas, A. & Rosenkranz, S. (2010). Profilin-1 is expressed in human atherosclerotic plaques and induces atherogenic effects on vascular smooth muscle cells. *PLoS one*, Vol.5, No.10, pp. e13608, ISSN 1932-6203.

Camera, P.; da Silva, J.S.; Griffiths, G.; Giuffrida, M.G.; Ferrara, L.; Schubert, V.; Imarisio, S. Silengo, L., Dotti, C.G. & Di Cunto, F. (2003). Citron-N is a neuronal Rho-associated protein involved in Golgi organization through actin cytoskeleton regulation. *Nature cell biology*, Vol.5, No.12, pp. 1071-1078, ISSN 1465-7392.

Cao, L.G.; Babcock, G.G.; Rubenstein, P.A. & Wang, Y.L. (1992). Effects of profilin and profilactin on actin structure and function in living cells. *The Journal of cell biology*, Vol.117, No.5, pp. 1023-1029, ISSN 0021-9525.

Carlier, M.F. & Pantaloni, D. (2007). Control of actin assembly dynamics in cell motility. *The Journal of biological chemistry*, Vol.282, No.32, pp. 23005-23009, ISSN 0021-9258.

Castrillon, D.H. & Wasserman, S.A. (1994). Diaphanous is required for cytokinesis in Drosophila and shares domains of similarity with the products of the limb deformity gene. *Development*. Vol.120, No.12, pp. 3367-3377, ISSN 0950-1991.

Chang, F.; Drubin, D. & Nurse, P. (1997). cdc12p, a protein required for cytokinesis in fission yeast, is a component of the cell division ring and interacts with profilin. *The Journal of cell biology*, Vol.137, No.1, pp. 169-182, ISSN 0021-9525.

Cheng, J.F.; Ni, G.H.; Chen, M.F.; Li, Y.J.; Wang, Y.J.; Wang, C.L.; Yuan, Q.; Shi, R.Z.; Hu, C.P. & Yang, T.L. (2011). Involvement of profilin-1 in angiotensin II-induced vascular smooth muscle cell proliferation. *Vascular Pharmacology*, Vol.55, No.1-3, pp. 34-41, ISSN 1537-1891.

Chikani, G.; Zhu, W. & Smart, E.J. (2004). Lipids: potential regulators of nitric oxide generation. *American journal of physiology. Endocrinology and metabolism*, Vol.287, No.3, pp. E386-E389, ISSN 0193-1849.

Cipolla, M.J. & Osol, G. (1998). Vascular smooth muscle actin cytoskeleton in cerebral artery forced dilatation. *Stroke*, Vol.29, No.6, pp. 1223-1228, ISSN 0039-2499.

Cipolla, M.J; Gokina, N.I. & Osol, G. (2002). Pressure-induced actin polymerization in vascular smooth muscle as a mechanism underlying myogenic behavior. *The FASEB Journal*, Vol.16, No.1, pp. 72-76, ISSN 892-6638.

Clarkson, M. R., Murphy, M., Gupta, S., Lambe, T., Mackenzie, H. S., Godson, C., Martin, F., and Brady, H. R. (2002) High glucose-altered gene expression in mesangial cells. Actin-regulatory protein gene expression is triggered by oxidative stress and cytoskeletal disassembly. *The Journal of biological chemistry*, Vol.277, No.12, pp. 9707-9712, ISSN 0021-9258.

Clowes, A.W.; Clowes, M.M.; Fingerle, J. & Reidy, M.A. (1989). Regulation of smooth muscle cell growth in injured artery. *Journal of cardiovascular pharmacology*, Vol.14, Suppl.6, pp. S12–S15, ISSN 0160-2446.

Da Silva, J.S.; Medina, M.; Zuliani, C.; Di Nardo, A.; Witke, W. & Dotti, C.G. (2003). RhoA/ROCK regulation of neuritogenesis via profilin IIa-mediated control of actin stability *The Journal of cell biology*, Vol.162, No.7, pp. 1267-1279, ISSN 0021-9525.

Dai, Y.P.; Bongalon, S.; Tian, H.; Parks, S.D.; Mutafova-Yambolieva, V.N. & Yamboliev, I.A. (2006). Upregulation of profilin, cofilin-2 and LIMK2 in cultured pulmonary artery smooth muscle cells and in pulmonary arteries of monocrotaline-treated rats. *Vascular Pharmacology*, Vol.44, No.5, pp. 275–282, ISSN 1537-1891.

Das, T.; Bae, Y.H.; Wells, A. & Roy, P. (2009). Profilin-1 overexpression upregulates PTEN and suppresses AKT activation in breast cancer cells. *Journal of cellular physiology*, Vol.218, No.2, pp. 436-443, ISSN 0021-9541.

Di Nardo, A.; Gareus, R.; Kwiatkowski, D. & Witke, W. (2000). Alternative splicing of the mouse profilin II gene generates functionally different profilin isoforms. *Journal of cell science*, Vol.113, No.21, pp. 3795–3803, ISSN 0021-9533.

Ding, Z.; Gau, D.; Deasy, B.; Wells, A. & Roy, P. (2009). Both actin and polyproline interactions of profilin-1 are required for migration, invasion and capillary morphogenesis of vascular endothelial cells. *Experimental cell research*, Vol.315, No.17, pp. 2963-2973, ISSN 0014-4827.

Ding, Z.; Lambrechts, A.; Parepally, M. & Roy, P. (2006). Silencing profilin-1 inhibits endothelial cell proliferation, migration and cord morphogenesis. *Journal of cell science*, Vol.119, Pt.19, pp. 4127–4137, ISSN 0021-9533.

Domke, T.; Federau, T.; Schlüter, K.; Giehl, K.; Valenta, R.; Schomburg, D. & Jockusch, B.M. (1997). Birch pollen profilin: structural organization and interaction with poly-(L-proline) peptides as revealed by NMR. *FEBS Letters*, Vol.411, No.2-3, pp. 291-295, ISSN 0014-5793.

Ebner, C.; Hirschwehr, R.; Bauer, L.; Breiteneder, H.; Valenta, R.; Ebner, H.; Kraft, D. & Scheiner, O. (1995). Identification of allergens in fruits and vegetables: IgE cross-reactivities with the important birch pollen allergens Bet v 1 and Bet v 2 (birch profilin). *The Journal of allergy and clinical immunology*, Vol.95, No.5 Pt.1, pp. 962-969, ISSN 0091-6749.

Edwards, K.A.; Montague, R.A.; Shepard, S.; Edgar, B.A.; Erikson, R.L. & Kiehart, D.P. Identification of Drosophila cytoskeletal proteins by induction of abnormal cell shape in fission yeast. *Proceedings of the National Academy of Sciences of the United States of America*, Vol.91, No.10, 4589-4593, ISSN 0027-8424.

Egerton, M.; Ashe, O.R.; Chen, D.; Druker, B.J.; Burgess, W.H. & Samelson, L.E. (1992). VCP, the mammalian homolog of cdc48, is tyrosine phosphorylated in response to T cell. *The EMBO Journal*, Vol.11, No.10, pp. 3533–3540, ISSN 0261-4189.

Evangelista, M.; Blundell, K.; Longtine, M.S.; Chow, C.J.; Adames, N.; Pringle, J.R.; Peter, M. & Boone, C. (1997). Bni1p, a yeast formin linking cdc42p and the actin cytoskeleton during polarized morphogenesis. *Science*, Vol.276, No.5309, pp. 118-122, ISSN 0036-8075.

Ezezika, O.C.; Younger, N.S.; Lu, J.; Kaiser, D.A.; Corbin, Z.A.; Nolen, B.J.; Kovar, D.R. & Pollard, T.D. (2009). Incompatibility with Formin Cdc12p prevents human profilin from substituting for fission yeast profilin: insights from crystal structures of fission yeast profilin. *The Journal of biological chemistry*, Vol.284, No.4, pp. 2088–2097, ISSN 0021-9258.

Fedor-Chaiken, M.; Deschenes, R.J. & Broach, J.R. (1990). SRV2, a gene required for RAS activation of adenylate cyclase in yeast. *Cell*, Vol.61, No.2, pp. 329-340, ISSN 0092-8674.

Fedorov, A.A.; Ball, T.; Mahoney, N.M.; Valenta, R. & Almo, S.C. (1997). The molecular basis for allergen cross-reactivity: crystal structure and IgE-epitope mapping of birch pollen profilin. *Structure*, Vol.5, No.1, pp. 33-45, ISSN 0969-2126.

Fedorov, A.A.; Magnus, K.A.; Graupe, M.H.; Lattman, E.E.; Pollard, T.D. & Almo, S.C. (1994). X-ray structures of isoforms of the actin-binding protein profilin that differ in their affinity for phosphatidylinositol phosphates. *Proceedings of the National Academy of Sciences of the United States of America*, Vol.91, No.18, pp. 8636-8640, ISSN 0027-8424.

Field, J.; Vojtek, A.; Ballester, R.; Bolger, G.; Colicelli, J.; Ferguson, K.; Gerst, J.; Kataoka, T.; Michaeli, T.; Powers, S.; Riggs, M.; Rodgers, L.; Wieland, I.; Wheland, B.; & Wigler, M. (1990). Cloning and characterization of CAP, the S. cerevisiae gene encoding the 70 kd adenylyl cyclase-associated protein. *Cell*, Vol.61, No.2, pp. 319-327, ISSN 0092-8674.

Finkel, T.; Theriot, J.A.; Dise, K.R.; Tomaselli, G.F. & Goldschmidt-Clermont, P.J. (1994). Dynamic actin structures stabilized by profilin. *Proceedings of the National Academy of Sciences of the United States of America*, Vol.91, No.4, pp. 1510-1514, ISSN 0027-8424.

Folkow, B. (1982). Physiological aspects of primary hypertension. Physiological reviews, Vol.62, No.2, pp. 347-504, ISSN 0031-9333.

Freeman, N.L.; Chen, Z.; Horenstein, J.; Weber, A. & Field, J. An actin monomer binding activity localizes to the carboxyl-terminal half of the Saccharomyces cerevisiae cyclase-associated protein. *The Journal of biological chemistry*, Vol.270, No.10, pp. 5680-5685, ISSN 0021-9258.

Gerthoffer, W.T. (2007). Mechanisms of vascular smooth muscle cell migration. *Circulation Research*, Vol.100, No.5, 607-621, ISSN 0009-7330.

Gertler, F.B.; Niebuhr, K.; Reinhard, M.; Wehland, J. & Soriano, P. (1996). Mena, a relative of VASP and Drosophila Enabled, is implicated in the control of microfilament dynamics. *Cell*, Vol.87, No.2, pp. 227-239, ISSN 0092-8674.

Gieselmann, R. & Mann, K. (1992). ASP-56, a new actin sequestering protein from pig platelets with homology to CAP, an adenylate cyclase-associated protein from yeast. *FEBS Letters*, Vol.298, No.2-3, pp. 149-153, ISSN 0014-5793.

Gieselmann, R.; Kwiatkowski, D.J.; Janmey, P.A. & Witke, W. (1995). Distinct biochemical characteristics of the two human profilin isoforms. *European journal of biochemistry*, Vol.229, No.3, pp. 621–628, ISSN 0014-2956.

Giesemann, T.; Rathke-Hartlieb, S.; Rothkegel, M.; Bartsch, J.W.; Buchmeier, S.; Jockusch, B.M. & Jockusch, H. (1999). A role for polyproline motifs in the spinal muscular atrophy protein SMN. Profilins bind to and colocalize with SMN in nuclear gems. *The Journal of biological chemistry*, Vol.274, No.53, pp. 37908–37914, ISSN 0021-9258.

Giesemann, T.; Schwarz, G.; Nawrotzki, R.; Berhörster, K.; Rothkegel, M.; Schlüter, K.; Schrader, N.; Schindelin, H.; Mendel, R.R.; Kirsch, J. & Jockusch, B.M. (2003). Complex formation between the postsynaptic scaffolding protein gephyrin, profilin, and Mena: a possible link to the microfilament system. *The Journal of neuroscience*, Vol.23, No.23, pp. 8330-8339, ISSN 0270-6474.

Gimbrone, M.A.; Nagel, T. & Topper, J.N. (1997). Biomechanical activation: an emerging paradigm in endothelial adhesion biology. *The Journal of clinical investigation*, Vol.99, No.8, pp. 1809–1813, ISSN 0021-9738.

Gingras, A.C.; Kennedy, S.G.; O'Leary, M.A.; Sonenberg, N. & Hay, N. (1998). 4E-BP1, a repressor of mRNA translation, is phosphorylated and inactivated by the Akt(PKB) signaling pathway. *Genes & development*, Vol.12, No.4, pp. 502-513, ISSN 0890-9369.

Goldschmidt-Clermont, P.J.; Kim, J.W.; Machesky, L.M.; Rhee, S.G. & Pollard, T.D. (1991) Regulation of phospholipase C-gamma 1 by profilin and tyrosine phosphorylation. *Science*, Vol.251, No.4998, pp. 1231–1233, ISSN 0036-8075.

Goldschmidt-Clermont, P.J.; Machesky, L.M.; Baldassare, J.J. & Pollard, T.D. (1990). The actin-binding protein profilin binds to PIP2 and inhibits its hydrolysis by phospholipase C. *Science*, Vol.247, No.4950, pp. 1575-1578, ISSN 0036-8075.

Goldschmidt-Clermont, P.J.; Mendelsohn, M.E. & Gibbs, J.B. (1992). Rac and Rho in control. *Current Biology*, Vol.2, No.12, pp. 669-671, ISSN 0960-9822.

Gottwald, U.; Brokamp, R.; Karakesisoglou, I.; Schleicher, M. & Noegel, A.A. (1996). Identification of a cyclase-associated protein (CAP) homologue in Dictyostelium discoideum and characterization of its interaction with actin. *Molecular biology of the cell*, Vol.7, No.2, pp. 261-272, ISSN 1059-1524.

Gürlek, A.; Dağalp, Z.; Oral, D.; Omürlü, K.; Erol, C.; Akyol, T. & Tutar, E. (1995). Restenosis after transluminal coronary angioplasty: a risk factor analysis. *Journal of cardiovascular risk*, Vol.2, No.1, pp. 51-55, ISSN 1350-6277.

Haarer, B.K. & Brown, S.S. (1990). Structure and function of profilin. *Cell motility and the cytoskeleton* Vol.17, No.2, pp. 71-74, ISSN 0886-1544.

Haarer, B.K.; Petzold, A.S. & Brown, S.S. (1993). Mutational analysis of yeast profilin. *Molecular and cellular biology*, Vol.13, No.12, pp. 7864-7873, ISSN 0270-7306.

Haffner, C.; Jarchau, T.; Reinhard, M.; Hoppe, J.; Lohmann, S.M. & Walter, U. (1995). Molecular cloning, structural analysis and functional expression of the proline-rich focal adhesion and microfilament-associated protein VASP. *The EMBO journal*, Vol.14, No.1, pp. 19-27, ISSN 0261-4189.

Halbrügge, M.; Friedrich, C.; Eigenthaler, M.; Schanzenbächer, P. & Walter, U. (1990). Stoichiometric and reversible phosphorylation of a 46-kDa protein in human platelets in response to cGMP- and cAMP-elevating vasodilators. *The Journal of biological chemistry*, Vol.265, No.6, pp. 3088–3093, ISSN 0021-9258.

Hartwig, J.H.; Bokoch, G.M.; Carpenter, C.L.; Janmey, P.A.; Taylor, L.A.; Toker, A. & Stossel, T.P. (1995). Thrombin receptor ligation and activated Rac uncap actin filament barbed ends through phosphoinositide synthesis in permeabilized human platelets. *Cell*, Vol.82, No.4, pp. 643–653, ISSN 0092-8674.

Hassona, M.D.; Abouelnaga, Z.A.; Elnakish, M.T.; Awad, M.M.; Alhaj, M., Goldschmidt-Clermont, P.J. & Hassanain, H.H. (2010). Vascular hypertrophy-associated hypertension of profilin1 transgenic mouse model leads to functional remodeling of peripheral arteries. *American journal of physiology. Heart and circulatory physiology,* Vol.298, No.6, pp. 2112-2120, ISSN 0363-6135

Hassona, M.D.; Elnakish, M.T.; Abouelnaga, Z.A.; Alhaj, M.; Wani, A.A. & Hassanain, H.H. (2011). The Effect of Selective Antihypertensive Drugs on the Vascular Remodeling-associated Hypertension: Insights from a Profilin1 Transgenic Mouse Model. *Journal of Cardiovascular Pharmacology,* Vol.57, No.5, pp. 550-558, ISSN 0160-2446.

Haugwitz, M.; Noegel, A.A.; Karakesisoglou, J. & Schleicher, M. (1994). Dictyostelium amoebae that lack G-actin-sequestering profilins show defects in F-actin content, cytokinesis, and development. *Cell,* Vol.79, No.2, pp. 303–314, ISSN 0092-8674.

Hellstrom, M.; Kalen, M.; Lindahl, P.; Abramsson, A. & Betsholtz, C. (1999). Role of PDGF-B and PDGFR-beta in recruitment of vascular smooth muscle cells and pericytes during embryonic blood vessel formation in the mouse. *Development,* Vol.126, No.14, pp. 3047–3055, ISSN 0950-1991.

Higgs, H.N. & Pollard, T.D. (2001). Regulation of actin filament network formation throughARP2/3 complex: activation by a diverse array of proteins. *Annual review of biochemistry,* Vol.70, pp. 649–662, ISSN 0066-4154.

Hirshman, C.A. & Emala, C.W. (1999). Actin reorganization in airway smooth muscle cells involves Gq and Gi-2 activation of Rho. *American journal of physiology. Lung cellular and molecular physiology,* Vol.277, No.3 Pt.1, pp. L653–L661, ISSN 0002-9513.

Hu, Y.; Cheng, L.; Hochleitner, B.W. & Xu, Q. (1997). Activation of mitogen-activated protein kinases (ERK/JNK) and AP-1 transcription factor in rat carotid arteries after balloon injury. *Arteriosclerosis, thrombosis, and vascular biology,* Vol.17, No.11, pp. 2808-2816, ISSN 1079-5642.

Hubberstey, A.; Yu, G.; Loewith, R.; Lakusta, C. & Young, D. (1996). Mammalian CAP interacts with CAP, CAP2, and actin. *Journal of cellular biochemistry,* Vol.61, No.3, pp. 459-466, ISSN 0730-2312.

Imamura, H.; Tanaka, K.; Hihara, T.; Umikawa, M.; Kamei, T.; Takahashi, K.; Sasaki, T. & Takai, Y. (1997). Bni1p and Bnr1p: downstream targets of the Rho family small G-proteins which interact with profilin and regulate actin cytoskeleton in Saccharomyces cerevisiae. *The EMBO journal,* Vol.16, No.10, pp. 2745-2755, ISSN 0261-4189.

Intengan, H.D. & Schiffrin, E.L. (2001). Vascular remodeling in hypertension: roles of apoptosis, inflammation, and fibrosis. *Hypertension,* Vol.38, No.3 Pt.2, pp. 581–587, ISSN 0194-911X.

Janmey, P.A. (1995). Protein regulation by phosphatidylinositol lipids. *Chemistry & biology,* Vol.2, No. 2, pp. 61–65, ISSN 1074-5521.

Jockusch, B.M.; Bubeck, P.; Giehl, K.; Kroemker, M.; Moschner, J.; Rothkegel, M.; Rüdiger, M.; Schlüter, K.; Stanke, G. & Winkler, J. (1995). The molecular architecture of focal adhesions. *Annual review of cell and developmental biology,* Vol.11, pp. 379-416, ISSN 1081-0706.

Johnstone, M.T.; Creager, S.J.; Scales, K.M.; Cusco, J.A.; Lee, B.K. & Creager, M.A. (1993). Impaired endothelium-dependent vasodilation in patients with insulin-dependent diabetes mellitus. *Circulation,* Vol.88, No.6, pp. 2510-2516, ISSN 0009-7322.

Jones, K.A.; Perkins, W.J.; Lorenz, R.R.; Prakash, Y.S.; Sieck, G.C. & Warner, D.O. (1999). F-actin stabilization increases tension cost during contraction of permeabilized airway smooth muscle in dogs. *The Journal of physiology*, Vol.519, Pt.2, pp. 527–538, ISSN 0022-3751.

Kaiser, D.A.; Goldschmidt-Clermont, P.J.; Levine, B.A. & Pollard, T.D. (1989). Characterization of renatured profilin purified by urea elution from poly-L-proline agarose columns. *Cell motility and the cytoskeleton*, Vol.14, No.2, pp. 251-262, ISSN 0886-1544.

Kelleher, J.F.; Atkinson, S.J. & Pollard TD. (1995). Sequences, structural models, and cellular localization of the actin-related proteins Arp2 and Arp3 from Acanthamoeba. *The Journal of cell biology*, Vol.131, No.2, pp. 385-397, ISSN 0021-9525.

Kitamura, T.; Ogawa, W.; Sakaue, H.; Hino, Y.; Kuroda, S.; Takata, M.; Matsumoto, M.; Maeda, T.; Konishi, H.; Kikkawa, U. & Kasuga, M. (1998). Requirement for activation of the serine-threonine kinase Akt (protein kinase B) in insulin stimulation of protein synthesis but not of glucose transport. *Molecular and cellular biology*, Vol.18, No.7, pp. 3708-3717, ISSN 0270-7306.

Kohno, H.; Tanaka, K.; Mino, A.; Umikawa, M.; Imamura, H.; Fujiwara, T.; Fujita, Y.; Hotta, K.; Qadota, H.; Watanabe, T.; Ohya, Y. & Takai, Y. (1996).Bni1p implicated in cytoskeletal control is a putative target of Rho1p small GTP binding protein in Saccharomyces cerevisiae. *The EMBO journal*, Vol.15, No.22, pp. 6060-6068, ISSN 0261-4189.

Korenbaum, E.; Nordberg, P.; Björkegren-Sjögren, C.; Schutt, C.E.; Lindberg, U. & Karlsson, R. (1998). The role of profilin in actin polymerization and nucleotide exchange. *Biochemistry*, Vol.37, No.26, pp. 9274-9283, ISSN 0006-2960.

Krishnan, K. & Moens, P.D.J. (2009). Structure and functions of profilins. *Biophysical reviews*, Vol.1, No.2, pp. 71–81, ISSN 1867-2450.

Kwiatkowski, D.J. & Bruns, G.A. (1988). Human profilin: Molecular cloning, sequence comparison and chromosomal analysis. *The Journal of biological chemistry*, Vol.263, No.12, pp. 5910–5915, ISSN 0021-9258.

Lambrechts, A.; Braun, A.; Jonckheere, V.; Aszodi, A.; Lanier, L.M.; Robbens, J.; Van Colen, I.; Vandekerckhove, J.; Fässler, R. & Ampe, C. (2000). Profilin II is alternatively spliced, resulting in profilin isoforms that are differentially expressed and have distinct biochemical properties. *Molecular and cellular biology*, Vol.20, No.21, pp. 8209–8219, ISSN 0270-7306.

Lambrechts, A.; Verschelde, J.L.; Jonckheere, V.; Goethals, M.; Vandekerckhove, J. & Ampe, C. (1997). The mammalian profilin isoforms display complementary affinities for PIP2 and proline-rich sequences. *The EMBO journal*, Vol.16, No.3, pp. 484-94, ISSN 0261-4189.

Lassing, I. & Lindberg, U. (1985). Specific interaction between phosphatidylinositol 4,5-bisphosphate and profilactin. Nature, Vol.314, No.6010, pp. 472-474, ISSN 0028-0836.

Lassing, I. & Lindberg, U. (1988). Specificity of the interaction between phosphatidylinositol 4,5-bisphosphate and the profilin:actin complex. *Journal of cellular biochemistry*, Vol. 37, No.3, pp. 255-267, ISSN 0730-2312.

Li, F. & Higgs, H.N. (2003). The mouse Formin mDia1 is a potent actin nucleation factor regulated by autoinhibition. *Current biology*, Vol.13, No.15, 1335-1340, ISSN 0960-9822.

Libby, P. (2002). Inflammation in atherosclerosis. *Nature*, Vol.420, No.6917, pp. 868-874, ISSN 0028-0836.

Lindberg, U.; Schutt, C.E.; Hellsten, E.; Tjäder, A.C. & Hult, T. (1988). The use of poly(L-proline)-Sepharose in the isolation of profilin and profilactin complexes. *Biochimica et biophysica acta*, Vol.967, No.3, pp. 391-400, ISSN 0006-3002.

Lu, P.J.; Shieh, W.R.; Rhee, S.G.; Yin, H.L. & Chen, C.S. (1996). Lipid products of phosphoinositide 3-kinase bind human profilin with high affinity. Biochemistry, Vol.35, No.44, pp. 14027–14034, ISSN 0006-2960.

Luna, E.J. & Hitt, A.L. (1992) Cytoskeleton--plasma membrane interactions. *Science*, Vol.258 No.5084, pp. 955-964, ISSN 0036-8075.

Machesky, L.M. & Pollard, T.D. (1993). Profilin as a potential mediator of membrane-cytoskeleton communication. *Trends in cell biology*, Vol.3, No.11, pp. 381–385, ISSN 0962-8924

Machesky, L.M. (1997). Cell motility: complex dynamics at the leading edge. *Current biology*, Vol.7, No.3, pp. R164-R167, ISSN 0960-9822.

Machesky, L.M.; Atkinson, S.J.; Ampe, C.; Vandekerckhove, J. & Pollard, T.D. (1994). Purification of a cortical complex containing two unconventional actins from *Acanthamoeba* by affinity chromatography on profilin-agarose. *The Journal of cell biology*, Vol.127, No.1, pp. 107–115, ISSN 0021-9525.

Machesky, L.M.; Cole, N.B.; Moss, B. & Pollard, T.D. (1994). Vaccinia virus expresses a novel profilin with a higher affinity for polyphosphoinositides than actin. *Biochemistry*, Vol.33, No.35, pp. 10815–10824, ISSN 0006-2960.

Machesky, L.M.; Goldschmidt-Clermont, P.J. & Pollard, T.D. (1990). The affinities of human platelet and Acanthamoeba profilin isoforms for polyphosphoinositides account for their relative abilities to inhibit phospholipase C. *Cell regulation*, Vol.1, No.12, pp. 937-950, ISSN 1044-2030.

Magdolen, V.; Drubin, D.G.; Mages, G.; Bandlow, W. (1993). High levels of profilin suppress the lethality caused by overproduction of actin in yeast cells. *FEBS Letters*, Vol.316, No.1, pp.41-47, ISSN 0014-5793.

Magdolen, V.; Oechsner, U.; Müller, G. & Bandlow, W. (1988). The introncontaining gene for yeast profilin (PFY) encodes a vital function. *Molecular and cellular biology*, Vol.8, No.12, pp. 5108–5115, ISSN 0270-7306.

Majesky, M.W. & Schwartz, S.M. (1990). Smooth muscle diversity in arterial wound repair. *Toxicologic Pathology*, Vol.18, No.4 Pt. 1, pp. 554 –559, ISSN 0192-6233.

Mammoto, A.; Sasaki, T.; Asakura, T.; Hotta, I.; Imamura, H.; Takahashi, K.; Matsuura, Y.; Shirao, T. & Takai, Y. (1998). Interactions of drebrin and gephyrin with profilin. *Biochemical and biophysical research communications*, Vol.243, No.1, pp. 86–89, ISSN 0006-291X.

Manseau, L.; Calley, J. & Phan, H. (1996). Profilin is required for posterior patterning of the Drosophila oocyte. *Development*, Vol.122, No.7, pp. 2109-2116, ISSN 0950-1991.

Mazzatti, D.J.; Pawelec, G.; Jonath, R. & Forsey, R.J. (2007). SELDI-TOF-MS protein Chip array profiling of T-cell propagated in long-term culture identifies human P potential bio-marker of immunosenescence. *Proteome Science*, 5:7, ISSN 1477-5956.

McLaughlin, P.J. & Weeds, A.G. (1995). Actin-binding protein complexes at atomic resolution. *Annual review of biophysics and biomolecular structure*, Vol.24, pp. 643-675, ISSN 1056-8700.

Mehta, D. & Gunst, S.J. (1999). Actin polymerization stimulated by contractile activation regulates force development in canine tracheal smooth muscle. *The Journal of physiology*, Vol.519, Pt.3, pp. 829–840, ISSN 0022-3751.

Mehta, P.K. & Griendling, K.K. (2007). Angiotensin II cell signaling: physiological and pathological effects in the cardiovascular system. *American journal of physiology. Cell physiology*, Vol.292, No.1, pp. C82-97, ISSN 0363-6143.

Meredith, J.E. Jr.; Fazeli, B. & Schwartz, M.A. (1993). The extracellular matrix as a cell survival factor. *Molecular biology of the cell*, 1993, Vol.4, No.9, pp. 953–961, ISSN 1059-1524.

Metzler, W.J.; Bell, A.J.; Ernst, E.; Lavoie, T.B. & Mueller, L. (1994). Identification of the poly-L-proline-binding site on human profilin. *The Journal of biological chemistry*, Vol.269, No.6, pp. 4620-4625, ISSN 0021-9258.

Metzler, W.J.; Constantine, K.L.; Friedrichs, M.S.; Bell, A.J.; Ernst, E.G.; Lavoie, T.B. & Mueller, L. (1993). Characterization of the three-dimensional structure of human profilin: $^1H$, $^{13}C$ and $^{15}N$ NMR assignments and global folding pattern. *Biochemistry*, Vol.32, No.50, pp. 13818-13829, ISSN 0006-2960.

Miki, H.; Suetsugu, S. & Takenawa, T. (1998). WAVE, a novel WASP-family protein involved in actin reorganization induced by Rac. *The EMBO journal*, Vol.17, No.23, pp. 6932–6941, ISSN 0261-4189.

Ming, X.F.; Viswambharan, H.; Barandier, C.; Ruffieux, J.; Kaibuchi, K.; Rusconi, S. & Yang, Z. (2002). Rho GTPase/Rho kinase negatively regulates endothelial nitric oxide synthase phosphorylation through the inhibition of protein kinase B/Akt in human endothelial cells. *Molecular and cellular biology*, Vol.22, No.24, pp. 8467–8477, ISSN 0270-7306.

Miyagi, Y.; Yamashita, T.; Fukaya, M.; Sonoda, T.; Okuno, T.; Yamada, K.; Watanabe, M.; Nagashima, Y.; Aoki, I.; Okuda, K.; Mishina, M. & Kawamoto S. (2002). Delphilin: a novel PDZ and formin homology domain-containing protein that synaptically colocalizes and interacts with glutamate receptor d2 subunit. *The Journal of neuroscience*, Vol.22, No.3, pp. 803–814, ISSN 0270-6474.

Mogilner, A. & Oster, G. (2003). Polymer motors: pushing out the front and pulling up the back. *Current biology*, Vol.13, No.18, pp. R721–R733, ISSN 0960-9822.

Moldovan, N.I.; Milliken, E.E.; Irani, K.; Chen, J.; Sohn, R.H.; Finkel, T. & Goldschmidt-Clermont, P.J. (1997). Regulation of endothelial cell adhesion by profilin. *Current biology*, Vol.7, No.1, pp. 24-30, ISSN 0960-9822.

Moustafa-Bayoumi, M.; Alhaj, M.A.; El-Sayed, O.; Wisel, S.; Chotani, M.A.; Abouelnaga, Z.A.; Hassona, M.D.H.; Rigatto, K.; Morris, M.; Nuovo, G.; Zweier, J.L.; Goldschmidt-Clermont, P. & Hassanain, H.H. (2007). Vascular hypertrophy and hypertension caused by transgenic overexpression of profilin1. *The Journal of biological chemistry*, Vol.282, No.52, pp. 37632–7639, ISSN 0021-9258.

Murry, C.E.; Gipaya, C.T.; Bartosek, T.; Benditt, E.P. & Schwartz, S.M. (1997). Monoclonality of smooth muscle cells in human atherosclerosis. *The American journal of pathology*, Vol.151, No.3, pp. 697–705, ISSN 0002-9440.

Nobes, C.D. & Hall, A. (1995). Rho, rac, and cdc42 GTPases regulate the assembly of multimolecular focal complexes associated with actin stress fibers, lamellipodia, and filopodia. *Cell*, Vol.81, No.1, pp. 53–62, ISSN 0092-8674.

Nodelman, I.M.; Bowman, G.D.; Lindberg, U. & Schutt, C.E. (1999). X-ray structure determination of human Profilin II: a comparative structural analysis of human profilins. *Journal of molecular biology*, Vol.294, No.5, pp. 1271– 1285, ISSN 0022-2836.

Obermann, H.; Raabe, I.; Balvers, M.; Brunswig, B.; Schulze, W. & Kirchhoff, C. (2005). Novel testis-expressed profilin IV associated with acrosome biogenesis and spermatid elongation. *Molecular human reproduction*, Vol.11, No.1, pp. 53–64, ISSN 1360-9947.

Ostrander, D.B.; Gorman, J.A. & Carman, G.M. (1995). Regulation of profilin localization in Saccharomyces cerevisiae by phosphoinositide metabolism. *The Journal of biological chemistry*, Vol.270, No.45, pp. 27045–27050, ISSN 0021-9258.

Pantaloni, D. & Carlier, M.F. (1993). How profilin promotes actin filament assembly in the presence of thymosin beta 4. *Cell*, Vol.75, No.5, pp. 1007-1014, ISSN 0092-8674.

Parast, M.M. & Otey, C.A. (2000). Characterization of palladin, a novel protein localized to stress fibers and cell adhesions. *The Journal of cell biology*, Vol.150, No.3, pp. 643–656, ISSN 0021-9525.

Perelroizen, I.; Didry, D.; Christensen, H.; Chua, N.H. & Carlier, M.F. (1996). Role of nucleotide exchange and hydrolysis in the function of profilin in action assembly. *The Journal of biological chemistry*, Vol.271, No.21, pp. 12302-12309, ISSN 0021-9258.

Perelroizen, I.; Marchand, J.B.; Blanchoin L.; Didry D. & Carlier, M.F. (1994). Interaction of profilin with G-actin and poly(L-proline). *Biochemistry*, Vol.33, No.28, pp. 8472-8478, ISSN 0006-2960.

Petrella, E.C.; Machesky, L.M.; Kaiser, D.A. & Pollard TD. (1996). Structural requirements and thermodynamics of the interaction of proline peptides with profilin. *Biochemistry*, Vol.35, No.51, pp. 16535-16543, ISSN 0006-2960.

Philp, D.; Goldstein, A.L. & Kleinman, H.K. (2004). Thymosin beta4 promotes angiogenesis, wound healing, and hair follicle development. *Mechanisms of ageing and development*, Vol.125, No.2, pp. 113–115, ISSN 0047-6374.

Pistor, S.; Chakraborty, T.; Walter, U. & Wehland, J. The bacterial actin nucleator protein ActA of Listeria monocytogenes contains multiple binding sites for host microfilament proteins. *Current biology*, Vol.5, No.5, pp. 517-525, ISSN 0960-9822.

Polet, D.; Lambrechts, A.; Ono, K.; Mah, A.; Peelman, F.; Vandekerckhove, J.; Baillie, D.L.; Ampe, C. & Ono, S. (2006). Caenorhabditis elegans expresses three functional profilins in a tissue-specific manner. *Cell motility and the cytoskeleton*, Vol.63, No.1, pp. 14–28, ISSN 0886-1544.

Pollard, T.D & Borisy, G.G. (2003). Cellular motility driven by assembly and disassembly of actin filaments. *Cell*, Vol.112, No.4, pp. 453–465, ISSN 0092-8674.

Pollard, T.D. (1986). Rate constants for the reactions of ATP- and ADP-actin with the ends of actin filaments. *The Journal of cell biology*, Vol.103, No.6 Pt.2, pp. 2747-54, ISSN 0021-9525.

Prass, M.; Jacobson, K.; Mogilner, A. & Radmacher, M. (2006). Direct measurement of the lamellipodial protrusive force in a migrating cell. *The Journal of cell biology*. Vol.174, No.6, pp. 767-772, ISSN 0021-9525.

Pring, M.; Weber, A. & Bubb, M.R. (1992). Profilin-actin complexes directly elongate actin filaments at the barbed end. *Biochemistry*, Vol.31, No.6, pp. 1827-1836, ISSN 0006-2960.

Pyles, J.M.; March, K.L.; Franklin, M.; Mehdi, K.; Wilensky, R.L. & Adam, L.P. (1997). Activation of MAP kinase in vivo follows balloon overstretch injury of porcine coronary and carotid arteries. *Circulation research*, Vol.81, No.6, pp. 904-910, ISSN 0009-7330.

Raghunathan, V.; Mowery, P.; Rozycki, M.; Lindberg, U. & Schutt, C. Structural changes in profilin accompany its binding to phosphatidylinositol, 4,5-bisphosphate. *FEBS Letters*, Vol.297, No.1-2, pp. 46-50, ISSN 0014-5793.

Ramesh, N.; Antón, I.M.; Hartwig, J.H. & Geha, R.S. (1997). WIP, a protein associated with Wiskott–Aldrich syndrome protein, induces actin polymerization and redistribution in lymphoid cells. *Proceedings of the National Academy of Sciences of the United States of America*, Vol.94, No.26, pp. 14671–14676, ISSN 0027-8424.

Re, F.; Zanetti, A.; Sironi, M.; Polentarutti, N.; Lanfrancone, L.; Dejana, E. & Colotta, F. (1994). Inhibition of anchorage-dependent cell spreading triggers apoptosis in cultured human endothelial cells. *The Journal of cell biology*, 1994, Vol.127, No.2, pp. 537–546, ISSN 0021-9525.

Reidy, M.A. (1992). Factors controlling smooth-muscle cell proliferation. *Archives of pathology & laboratory medicine*, Vol.116, No.12, pp. 1276 –1280, ISSN 0003-9985.

Reinhard, M.; Giehl, K.; Abel, K.; Haffner, C.; Jarchau, T.; Hoppe, V.; Jockusch, B.M. & Walter, U. (1995). The proline-rich focal adhesion and microfilament protein VASP is a ligand for profilins. *The EMBO journal*, Vol.14, No.8, pp. 1583–1589, ISSN 0261-4189.

Reinhard, M.; Halbrügge, M.; Scheer, U.; Wiegand, C.; Jockusch, B.M. & Walter, U. (1992). The 46/50 kDa phosphoprotein VASP purified from human platelets is a novel protein associated with actin filaments and focal contacts. *The EMBO Journal*, Vol.11, No.6, pp. 2063-2070, ISSN 0261-4189.

Reinhard, M.; Rüdiger, M.; Jockusch, B.M. & Walter U. (1996). VASP interaction with vinculin: a recurring theme of interactions with proline-rich motifs. *FEBS Letters*, Vol.399, No.1-2, pp. 103-107, ISSN 0014-5793.

Romeo, G.R. & Kazlauskas, A. (2008). Oxysterol and diabetes activate STAT3 and control endothelial expression of profilin-1 via OSBP1. *The Journal of biological chemistry*, Vol.283, No.15, pp.9595-9605, ISSN 0021-9258.

Romeo, G.R.; Frangioni, J.V. & Kazlauskas A. (2004). Profilin acts downstream of LDL to mediate diabetic endothelial cell dysfunction. *The FASEB journal*, Vol.18, No.6, pp. 725–727, ISSN 0892-6638.

Romeo, G.R.; Moulton, K.S. & Kazlauskas, A. (2007). Attenuated Expression of Profilin-1 Confers Protection from Atherosclerosis in the LDL Receptor–Null Mouse. *Circulation research*, Vol.101, No.4, pp. 357-367, ISSN 0009-7330.

Rothkegel, M.; Mayboroda, O.; Rohde, M.; Wucherpfennig, C.; Valenta, R. & Jockusch, B.M. (1996). Plant and animal profilins are functionally equivalent and stabilize microfilaments in living animal cells. *Journal of cell science*, Vol.109, Pt.1, pp. 83-90, ISSN 0021-9533.

Ruwhof, C. & van der Laarse, A. (2000). Mechanical stress-induced cardiac hypertrophy: mechanisms and signal transduction pathways. *Cardiovascular research*, Vol.47, No.1, pp. 23-37, ISSN 0008-6363.

Safer, D.; Elzinga, M. & Nachmias, V.T. (1991). Thymosin beta 4 and Fx, an actin-sequestering peptide, are indistinguishable. *The Journal of biological chemistry*, Vol.266, No.7, pp. 4029-4032, ISSN 0021-9258.

Salazar, R.; Bell, S.E. & Davis, G.E. (1999). Coordinate induction of the actin cytoskeletal regulatory proteins gelsolin, vasodilator-stimulated phosphoprotein, and profilin during capillary morphogenesis in vitro. *Experimental cell research*, Vol.249, No.1, pp. 22–32, ISSN 0014-4827.

Salmon, E.D. (1989). Cytokinesis in animal cells. *Current opinion in cell biology*, Vol.1, No.3, pp. 541-547, ISSN 0955-0674.

Schiffrin, E.L. (2001). Effects of antihypertensive drugs on vascular remodeling: do they predict outcome in response to antihypertensive therapy? *Current opinion in nephrology and hypertension*, Vol.10, No.5, pp. 617-624, ISSN 1062-4821.

Schlüter, K.; Jockusch, B.M. & Rothkegel, M. (1997). Profilins as regulators of actin dynamics. *Biochimica et biophysica acta*, Vol.1359, No.2, pp. 97-109, ISSN 0006-3002.

Schmidt, A. & Hall, M. N. (1998). Signaling to the actin cytoskeleton. *Annual review of cell and developmental biology*, Vol.14, pp. 305-338, ISSN 1081-0706.

Schutt, C.E.; Lindberg, U.; Myslik, J. & Strauss, N. (1989). Molecular packing in profilin: actin crystals and its implications. *Journal of molecular biology*, Vol.209, No.4, pp. 735-746, ISSN 0022-2836.

Schutt, C.E.; Myslik, J.C.; Rozycki, M.D.; Goonesekere, N.C. & Lindberg, U. (1993). The structure of crystalline profilin-beta-actin. *Nature*, Vol.365, No.6449, 810-816, ISSN 0028-0836.

Simon, G. (2004). Pathogenesis of structural vascular changes in hypertension. *Journal of hypertension*, Vol.22, No.1, pp. 3-10, ISSN 0263-6352.

Singh, S.S.; Chauhan, A.; Murakami, N. & Chauhan, V.P. (1996). Profilin and gelsolin stimulate phosphatidylinositol 3-kinase activity. *Biochemistry*, Vol.35, No.51, pp. 16544-16549, ISSN 0006-2960.

Skare, P.; Kreivi, J.P.; Bergström, A. & Karlsson R. (2003) Profilin I colocalizes with speckles and Cajal bodies: a possible role in pre-mRNA splicing. *Experimental cell research*, Vol.286, No.1, pp. 12–21, ISSN 0014-4827.

Smith, G.A.; Theriot, J.A. & Portnoy, D.A. (1996). The tandem repeat domain in the Listeria monocytogenes ActA protein controls the rate of actin-based motility, the percentage of moving bacteria, and the localization of vasodilator-stimulated phosphoprotein and profilin. *The Journal of cell biology*, Vol.135, No.3, pp. 647-660, ISSN 0021-9525.

Sohn, R.H. & Goldschmidt-Clermont, P.J. (1994). Profilin: at the crossroads of signal transduction and the actin cytoskeleton. *BioEssays*, Vol.16, No.7, pp. 465-472, ISSN 0265-9247.

Sohn, R.H.; Chen, J.; Koblan, K.S.; Bray, P.F. & Goldschmidt-Clermont, P.J. (1995). Localization of a binding site for phosphatidylinositol 4,5-bisphosphate on human profilin. *The Journal of biological chemistry*, Vol.270, No.36, pp. 21114-21120, ISSN 0021-9258.

Sonobe, S.; Takahashi, S.; Hatano, S. & Kuroda, K. (1986). Phosphorylation of amoeba G-actin and its effect on actin polymerization. *The Journal of biological chemistry*, Vol.261, No.31, pp. 14837–14843, ISSN 0021-9258.

Spagnoli, L.; Bonanno, L.; Sangiorgi, G. & Mauriello, A. (1991). Role of Inflammation in Atherosclerosis. *Journal of nuclear medicine*, Vol.48, No.11, pp. 1800-1815, ISSN 0161-5505.

Speirs, S. & Kaufman, M.H. (1989). Cytochalasin D-induced triploidy in the mouse. The *Journal of experimental zoology*, Vol.250, No.3, pp. 339-345, ISSN 0022-104X.

Staiger, C.J.; Yuan, M.; Valenta, R.; Shaw, P.J.; Warn, R.M. & Lloyd, C.W. (1994). Microinjected profilin affects cytoplasmic streaming in plant cells by rapidly depolymerizing actin microfilaments. *Current biology*, Vol.4, No.3, pp. 215-219, ISSN 0960-9822.

Starr, M.E.; Ueda, J.; Yamamoto, S.; Evers, B.M. & Saito, H. (2011). The effects of aging on pulmonary oxidative damage, protein nitration, and extracellular superoxide dismutase down-regulation during systemic inflammation. *Free radical biology & medicine*, Vol.50, No.2, 371-380, ISSN 0891-5849.

Steinberg, H.O.; Bayazeed, B.; Hook, G.; Johnson, A.; Cronin, J. & Baron, A.D. (1997). Endothelial dysfunction is associated with cholesterol levels in the high normal range in humans. *Circulation*, Vol.96, No.10, pp. 3287-3293, ISSN 0009-7322.

Stüven, T.; Hartmann, E. & Görlich, D. (2003). Export in 6: a novel nuclear export receptor that is specific for profilin–actin complexes. *The EMBO journal*, Vol.22, No.21, pp. 5928–5940, ISSN 0261-4189.

Suetsugu, S.; Miki, H. & Takenawa, T. (1999). Distinct roles of profilin1 in cell morphological changes: microspikes, membrane ruffles, stress fibers, and cytokinesis. *FEBS Letters*, Vol.457, No.3, pp. 470–474, 1999, ISSN 0014-5793.

Suetsugu, S.; Miki, H.; Takenawa, T. (1998). The essential role of profilin in the assembly of actin for microspike formation. *The EMBO journal*, 17, 6516–6526, ISSN 0261-4189.

Tamura, M.; Yanagihara, N.; Tanaka, H.; Osajima, A.; Hirano, T.; Higashi, K.; Yamada, K.M.; Nakashima, Y. & Hirano, H. (2000). Activation of DNA synthesis and AP-1 by profilin, an actin-binding protein, via binding to a cell surface receptor in cultured rat mesangial cells. *Journal of the American Society of Nephrology*, Vol.11, No.9, pp. 1620-1630, ISSN 1046-6673.

Tanaka, M. & Shibata, H. (1985). Poly(L-proline)-binding proteins from chick embryos are a profilin and a profilactin. *European journal of biochemistry*, Vol.151, No.2, pp. 291-297, ISSN 0014-2956.

Tanaka, M.; Sasaki, H.; Kino, I.; Sugimura, T. & Terada, M. (1992). Genes preferentially expressed in embryo stomach are predominantly expressed in gastric cancer. *Cancer research*, Vol.52, No.12, pp. 3372-3377, ISSN 0008-5472.

Tang, D.D. & Tan, J. (2003). Downregulation of profilin with antisense oligodeoxynucleotides inhibits force development during stimulation of smooth muscle. *American journal of physiology. Heart and circulatory physiology*, Vol.285, No.4, H1528-H1536, ISSN 0363-6135.

Tapon, N. & Hall, A. (1997). Rho, Rac and Cdc42 GTPases regulate the organization of the actin cytoskeleton. *Current opinion in cell biology*, Vol.9, No.1, pp. 86-92, ISSN 0955-0674.

Tellam, R.L.; Morton, D.J. & Clarke, F.M. (1989). A common theme in the amino acid sequences of actin and many actin-binding proteins? *Trends in biochemical sciences*, Vol.14, No.4, pp. 130-133, ISSN 0968-0004.

Tesfamariam, B.; Brown, M.L.; Deykin, D. & Cohen, R.A. (1990). Elevated glucose promotes generation of endothelium-derived vasoconstrictor prostanoids in rabbit aorta. *The Journal of clinical investigation*, Vol.85, No.3, pp. 929-932, ISSN 0021-9738.

Theriot, J.A. & Mitchison, T.J. (1993). The three faces of profilin. *Cell*, Vol.75, No.5, pp. 835-838, ISSN 0092-8674.

Thorn, K.S.; Christensen, H.E.; Shigeta, R., Huddler, D.; Shalaby, L.; Lindberg, U.; Chua, N.H. & Schutt, C.E. (1997). The crystal structure of a major allergen from plants. *Structure*, Vol.5, No.1, pp. 19-32, ISSN 0969-2126.

Tobacman, L.S.; Brenner, S.L. & Korn, E.D. (1983). Effect of Acanthamoeba profilin on the pre-steady state kinetics of actin polymerization and on the concentration of F-actin at steady state. *The Journal of biological chemistry*, Vol.258, No.14, pp. 8806-8812, ISSN 0021-9258.

Toda, T.; Uno, I.; Ishikawa, T.; Powers, S.; Kataoka, T.; Broek, D.; Cameron, S.; Broach, J.; Matsumoto, K. & Wigler, M. (1985). In yeast, RAS proteins are controlling elements of adenylate cyclase. *Cell*, Vol.40, No.1, pp. 27-36, ISSN 0092-8674.

Touyz, RM. (2007). Vascular remodeling, retinal arteries, hypertension. *Hypertension*, Vol.50, No.4, pp. 602–603, ISSN 0194-911X.

Tseng, P.C-H.; Runge, M.S.; Cooper, J.A.; Williams, R.C. Jr. & Pollard, T.D. (1984). Physical, immunochemical, and functional properties of Acanthamoeba proflin. *The Journal of cell biology*, Vol.98, No.1, pp. 214–221, ISSN 0021-9525.

U.K. Prospective Diabetes Study (UKPDS) Group. (1998). Effect of intensive blood-glucose control with metformin on complications in overweight patients with type 2 diabetes (UKPDS 34). *Lancet*, Vol.352, No.9131, pp. 854-865, ISSN 0140-6736.

Ushio-Fukai, M.; Alexander, R.W.; Akers, M.; Yin, Q.; Fujio, Y.; Walsh, K.& Griendling, K.K. (1999). Reactive oxygen species mediate the activation of Akt/protein kinase B by angiotensin II in vascular smooth muscle cells. *The Journal of biological chemistry*, Vol.274, No.32, pp. 22699-22704, ISSN 0021-9258.

Valenta, R.; Breiteneder, H.; Pettenburger, K.; Breitenbach, M.; Rumpold, H.; Kraft, D. & Scheiner, O. (1991a). Homology of the major birchpollen allergen, Bet v I, with the major pollen allergens of alder, hazel, and hornbeam at the nucleic acid level as determined by cross-hybridization. *The Journal of allergy and clinical immunology*, Vol.87, No.3, pp. 677–682, ISSN 0091-6749.

Valenta, R.; Duchene, M.; Breitenbach, M.; Pettenburger, K.; Koller, L.; Rumpold, H.; Scheiner, O. & Kraft, D. (1991b). A low molecular weight allergen of white birch (Betula verrucosa) is highly homologous to human profilin. *International archives of allergy and applied immunology*, Vol.94, No.(1-4), pp. 368–370, ISSN 0020-5915.

Valenta, R.; Duchene, M.; Ebner, C.; Valent, P.; Sillaber, C.; Deviller, P.; Ferreira, F.; Tejkl, M.; Edelmann, H.; Kraft, D. & Scheiner O. (1992). Profilins constitute a novel family of functional plant pan-allergens. *The Journal of experimental medicine*, Vol.175, No.2, pp. 377-385, ISSN 0022-1007.

Valenta, R.; Duchêne, M.; Pettenburger, K.; Sillaber, C.; Valent, P.; Bettelheim, P.; Breitenbach, M.; Rumpold, H.; Kraft, D. & Scheiner O. (1991c). Identification of profilin as a novel pollen allergen; IgE autoreactivity in sensitized individuals. *Science*, Vol.253, No.5019, pp. 557-560, ISSN 0036-8075.

Vandekerckhove, J. (1989). Structural principles of actin-binding proteins. *Current opinion in cell biology*, Vol.1, No.1, pp.15-22, ISSN 0955-0674.

Vinson, V.K.; Archer, S.J.; Lattman, E.E.; Pollard, T.D. & Torchia, D.A. (1993). Three-dimensional solution structure of Acanthamoeba profilin-I. *The Journal of cell biology*, Vol.122, No.6, pp. 1277-1283, ISSN 0021-9525.

Wallar, B.J. & Alberts, A.S. (2003). The formins: active scaffolds that remodel the cytoskeleton. Trends in cell biology, Vol.13, No.8, pp. 435–446, ISSN 0962-8924.

Wang, X.; Kibschull, M.; Laue, M.M.; Lichte, B.; Petrasch-Parwez, E. & Kilimann, M.W. (1999). Aczonin, a 550-kD putative scaffolding protein of presynaptic active zones, shares homology regions with Rim and Bassoon and binds profilin *The Journal of cell biology*, Vol.147, No.1, pp. 151–162, ISSN 0021-9525.

Watanabe, N.; Madaule, P.; Reid, T.; Ishizaki, T.; Watanabe, G.; Kakizuka, A.; Saito, Y.; Nakao, K.; Jockusch, B.M. & Narumiya, S. (1997). p140mDia, a mammalian homolog of Drosophila diaphanous, is a target protein for Rho small GTPase and is a ligand for profilin. *The EMBO journal*, Vol.16, No.11, pp. 3044–3056, ISSN 0261-4189.

Weinreb, O.; Drigues, N.; Yotam Sagi, Y.; Abraham Reznick, A.Z.Z.; Tamar Amit, T. & Youdim, M. (2007).The application of proteomics and genomics to the study of age-related neurodegeneration and neuroprotection. *Antioxidants & Redox Signaling*, Vol.9, No.2, pp. 169–179, ISSN 1523-0864.

Wheatley, E. M.; Vincent, P. A.; McKeown-Longo, P. J. & Saba, T. M. Effect of fibronectin on permeability of normal and TNF-treated lung endothelial cell monolayers. (1993). *The American journal of physiology*, Vol.264, No.1 Pt.2, pp. R90–R96; ISSN 0002-9513.

Widada, J.S.; Ferraz, C. & Liautard, J.P. (1989). Total coding sequence of profilin cDNA from Mus musculus macrophage. *Nucleic acids research*, Vol.17, No.7, pp. 2855, ISSN 0305-1048.

Witke, W. (2004). The role of profilin complexes in cell motility and other cellular processes. *Trends in cell biology*, Vol.14, No.8, pp. 461-469, ISSN 0962-8924.

Witke, W.; Podtelejnikov, A.V.; Di Nardo, A.; Sutherland, J.D.; Gurniak, C.B.; Dotti, C. & Mann, M. (1998). In mouse brain profilin I and profilin II associate with regulators of the endocytic pathway and actin assembly. *The EMBO journal*, Vol.17, No.4, pp. 967–976, ISSN 0261-4189.

Witke, W.; Sutherland, J.D.; Sharpe, A.; Arai, M. & Kwiatkowski, D.J. (2001). Profilin I is essential for cell survival and cell division in early mouse development. *Proceedings of the National Academy of Sciences of the United States of America*, Vol.98, No.7, pp. 3832–3836, ISSN 0027-8424.

Yayoshi-Yamamoto, S.; Taniuchi, I. & Watanabe, T. (2000). FRL, a novel formin-related protein, binds to Rac and regulates cell motility and survival of macrophages. *Molecular and cellular biology*, Vol.20, No.18, pp. 6872–6881, ISSN 0270-7306.

Yeh, E.T.; Zhang, S.; Wu, H.D.; Korbling, M.; Willerson, J.T. & Estrov, Z. (2003). Transdifferentiation of human peripheral blood CD34+-enriched cell population into cardiomyocytes, endothelial cells, and smooth muscle cells in vivo. *Circulation*, Vol.108, No.17, pp. 2070 –2073, ISSN 0009-7322.

Youn, T.; Kim, S.A. & Hai, C.M. (1998). Length-dependent modulation of smooth muscle activation: effects of agonist, cytochalasin, and temperature. *American journal of physiology. Cell physiology*. Vol.274, No.6 Pt.1, pp. C1601–C1607, ISSN 0002-9513.

Zang, Z.; Vuori, K.; Reed, J.C. & Ruoslahti, E. (1995). The alpha 5 beta 1 integrin supports survival of cells on fibronectin and up-regulates Bcl-2 expression. Proceedings of the National Academy of Sciences of the United States of America, Vol.92, No.13, pp. 6161–6165, ISSN 0027-8424.

Zeile, W.L; Purich, D.L. & Southwick, F.S. (1996). Recognition of two classes of oligoproline sequences in profilin-mediated acceleration of actin-based Shigella motility. *The Journal of cell biology*, Vol.133, No.1, pp. 49-59, ISSN 0021-9525.

Zhong, J.C.; Ye, J.Y.; Jin, H.Y.; Yu, X.; Yu, H.M.; Zhu, D.L.; Gao, P.J.; Huang, D.Y.; Shuster, M.; Loibner, H.; Guo, J.M.; Yu, X.Y.; Xiao, B.X.; Gong, Z.H.; Penninger, J.M. & Oudit, G.Y. (2011). Telmisartan attenuates aortic hypertrophy in hypertensive rats by the modulation of ACE2 and profilin-1 expression. *Regulatory Peptides*. Vol.166, No.1-3, pp. 90-97, ISSN 0167-0115

Zou, L.; Jaramillo, M.; Whaley, D.; Wells, A.; Panchapakesa, V.; Das, T. & Roy, P. (2007) Profilin-1 is a negative regulator of mammary carcinoma aggressiveness. *British journal of cancer*, Vol. 97, No.10, pp. 1361-1371, ISSN 0007-0920.

# 3

# Protein Flexibility and Coiled-Coil Propensity: New Insights Into Type III and Other Bacterial Secretion Systems

Spyridoula N. Charova[1,2], Anastasia D. Gazi[1,2], Marianna Kotzabasaki[2],
Panagiotis F. Sarris[1,2], Vassiliki E. Fadouloglou[2,3],
Nickolas J. Panopoulos[1,2] and Michael Kokkinidis[1,2]
*[1]Institute of Molecular Biology & Biotechnology, Foundation of Research & Technology*
*[2]Department of Biology, University of Crete, Vasilika Vouton, Heraklion, Crete*
*[3]Department of Molecular Biology and Biotechnology*
*Democritus University of Thrace, Alexandroupolis*
*Greece*

## 1. Introduction

Secretion in unicellular species is the transport or translocation of molecules, for example proteins, from the interior of the cell to its exterior. In bacteria secretion is a very important mechanism, either modulating their interactions with their environment for adaptation and survival or establishing interactions with their eukaryotic hosts for pathogenesis or symbiosis. To overcome the physical barriers of membranes, Gram-negative bacteria use a variety of molecular machines which have been elaborated to secrete a wide range of proteins and other molecules; their functions include biogenesis of organelles (e.g. pili and flagella), virulence, efflux of toxins etc. As in some cases the secreted proteins are destined to enter host cells (effectors, toxins), some of the secretion systems include extracellular appendices to translocate proteins across the plasma membrane of the host.

With the rapid accumulation of bacterial genome sequences, our knowledge of the complexity of bacterial protein secretion systems has expanded and several secretion systems have been identified. Gene Ontology has been very useful for describing the components and functions of these systems, and for capturing the similarities among the diverse systems (Tseng et al., 2009). These analyses along with numerous biochemical studies have revealed the existence of at least six major mechanisms of protein secretion. These pathways are highly conserved throughout the Gram-negative bacterial species and are functionally independent with respect to outer membrane translocation; commonalities exist in the inner membrane transport steps of some systems, with most of them being terminal branches of the general secretion pathway (Sec). The pathways have been numbered Type I, II, III, IV, V and VI.

In Gram-negative bacteria, some secreted proteins are exported across the inner and outer membranes in a single step via the Type I, III, IV or VI pathways. Other proteins are first exported into the periplasmic space using the universal Sec or two-arginine (Tat) pathways and then translocated across the outer membrane via the Type II, V or less commonly, the Type I or IV machinery. In Gram-positive bacteria, secreted proteins are commonly translocated across the single membrane by the Sec pathway, the two-arginine (Tat) pathway, or the recently identified type VII secretion system (Abdallah et al., 2007). In the following we will briefly survey the six Gram-negative bacterial secretion systems known to modulate interactions with host organisms:

**Type I secretion system:** This system (T1SS) forms a contiguous channel traversing the inner and outer membranes of Gram-negative bacteria. It is a simple system, which consists of only three major components: ATP-binding cassette transporters, Outer Membrane Factors, and Membrane Fusion Proteins (Holland et al., 2005). T1SS transports ions and various molecules including proteins of various sizes (20-900 kDa) and non-proteinaceous substrates like cyclic $\beta$-glucans and polysaccharides.

**Type II secretion system:** This system (T2SS) is encoded by at least 12 genes and supports the transport of a group of seemingly unrelated proteins across the outer membrane. In order for these proteins to enter the type II secretion pathway, they have to first translocate across the cytoplasmic membrane via the Sec-system and then fold into a translocation competent conformation in the periplasm. Proteins secreted by T2SS include proteases, cellulases, pectinases, phospholipases, lipases, and toxins which contribute to cell damage and disease. Although Sec-dependent translocation is universal (Cao & Saier, 2003), the T2SS is found only in Gram-negative proteobacteria phylum (Cianciotto, 2005; Filloux, 2004). A bacterial species may have more than one T2SS (Cianciotto, 2005; Filloux, 2004).

**Type III secretion system:** These systems (T3SS) are essential mediators of the interaction of many Gram-negative pathogenic proteobacteria ($\alpha$, $\beta$, $\gamma$ and $\delta$ subdivisions) with their human, animal, or plant hosts and are evolutionarily related to bacterial flagella. (Dale & Moran, 2006; Tampakaki et al., 2004; Troisfontaines & Cornelis, 2005). The machinery of the T3SS, termed the injectisome, appears to have a common evolutionary origin with the flagellum and translocates a diverse repertoire of effector proteins either to extracellular locations or directly into eukaryotic cells, in a Sec-independent manner (interkingdom protein transfer device). The T3SS effectors (T3EPs) modulate the function of crucial host regulatory molecules and trigger a range of highly dynamic cellular responses which determine pathogen-host recognition, pathogen/symbiont accommodation and elicitation or suppression of defense responses by the eukaryotic hosts. In some cases however, effector proteins are simply secreted out of the cell. T3SS have evolved into seven families (Troisfontaines & Cornelis, 2005). Some bacteria may harbor more than one T3SS, usually from different families. T3SS genes are encoded in pathogenicity islands and/or are located on plasmids, and are commonly subject to horizontal gene transfer.

**Type IV secretion system:** In comparison to other secretion systems, T4SS is unique in its ability to transport nucleic acids in addition to proteins into plant and animal cells, as well as into yeast and other bacteria. Usually T4SS comprises 12 proteins that can be identified as homologs of the VirB1–11 and VirD4 proteins of the *Agrobacterium tumefaciens* Ti plasmid transfer system (Christie & Vogel, 2000). T4SS spans both membranes of Gram-negative bacteria, using a specific transglycosylase, VirB1, to digest the intervening murein (Koraimann, 2003; Baron et al., 1997). While many organisms have homologous type IV secretion systems, not all systems contain the same sets of genes. The only common protein is VirB10 (TrbI) among all T4SS systems (Cao & Saier, 2003).

**Type V secretion system:** T5SS is the simplest protein secretion mechanism. Proteins are secreted via the autotransporter system (type Va or AT-1), the two-partner secretion pathway (type Vb), and the oligomeric autotransporters (type Vc or AT-2 system) (Yu et al., 2008; Desvaux et al., 2004). Proteins secreted via these pathways have similarities in their primary structures as well as striking similarities in their modes of biogenesis.There are three sub-classes of T5SS. The archetypal bacterial proteins secreted via the T5SS (T5aSS subclass) consist of a N-terminal passenger domain of 40-400 kD in size and a conserved C-terminal domain (Henderson et al., 2004). The proteins are synthesized with a N-terminal signal peptide that directs their export into the periplasm via the Sec machinery.

**Type VI secretion system:** In T6SS 13 genes are thought to constitute the minimal number needed to produce a functional apparatus (Boyer et al., 2009). TheT6SS gene clusters (T6SS loci) often occur in multiple, non-orthologous copies per genome and have probably been acquired via horizontal gene transfer (Sarris & Skoulica, 2011; Sarris at al., 2011). Each T6SS probably assumes a different role in the interactions of the harbouring organism with others. Although the T6SS has been studied primarily in the context of pathogenic bacteria-host interactions, it has been suggested that it may also function to promote commensal or mutualistic relationships between bacteria and eukaryotes, as well as to mediate cooperative or competitive interactions between bacterial species. The T6SS machinery constitutes a phage-tail-spike-like injectisome that has the potential to introduce effector proteins directly into the cytoplasm of host cells, analogous to the T3SS and T4SS machineries.

Genetic, structural and biochemical studies of the above bacterial secretion systems along with massive *in silico* analyses of microbial genomes have been used to distinguish pathogens from their non-pathogenic relatives. These studies have established the presence of characteristic conserved features within individual types of secretion systems (e.g. Tampakaki et al., 2004), along with considerable sequence and structural diversities within each system at the level of specific components and effector proteins.

Despite the complexity of these systems however, the problem of identifying conserved features and properties within each secretion system type, or across several types of systems is of particular importance, going beyond a fundamental understanding of how bacterial secretion works. Even for well studied pathogens, not all virulence factors have been identified, making it possible that e.g. effector proteins that are associated with different diseases are still unknown. In less well characterized bacterial species there is certainly a wide spectrum of unknown effectors. This situation may be now changing through new approaches that use advanced machine learning algorithms to identify within individual types of secretion systems common themes for effectors and other system components that go beyond simple amino acid motifs (Arnold et al., 2009; Samudrala et al., 2009), or through the identification of important structural and physicochemical properties as universal signatures of virulence factors (Gazi et al., 2008; 2009).

This review will focus on the well-characterized T3SS proteins where the prevalence of coiled-coil domains along with pronounced structural flexibility/disorder have been proposed to be characteristic properties associated with a protein-protein interaction mode within T3SS and as essential requirements for secretion (Delahay and Frankel, 2002; Pallen et al., 1997; Gazi et al., 2008; 2009). Common themes with other secretion systems (T4SS, T6SS) will be also discussed.

## 2. Overview of the T3SS system: Architecture, conserved features and protein structures

Pathogenic bacterial strains are distinguished from non-pathogenic ones by the presence of specific set of genes that code for toxins, secretion systems, effectors that are meant to act extracellularly or effectors that should be delivered inside the host cell cytoplasm. These genes are usually tightly organized in operones that are located in chromosomal areas with a high distribution of mobile elements or can be found in virulence plasmids. Usually these chromosomal areas are called pathogenicity islands as they possess a different GC content from the rest of the genome, which implies recent acquisition through horizontal gene transfer events. One of the most profound cases was a set of approximately 20-25 genes which together encode one of the best characterized pathogenic mechanisms termed "type III secretion". By this mechanism extracellularly located bacteria that are in a close contact with a eukaryotic cell deliver proteins into the host cell cytosol. While the T3S apparatus is conserved in pathogens across the plant/animal phyllogenetic divide, the secreted proteins differ considerably. The genes coding for what are now recognized as structural T3SS components were first described as a contiguous cluster, designated *"hrp"* (hypersensitive response and pathogenicity) in plant pathogens. Important insights into fundamental questions of bacterial pathobiology came with the recognition, in subsequent years, of the T3SS as a complex multiprotein channel dedicated to translocate the effectors from the pathogen to the host. Although originally linked to pathogenesis, T3SS are also found in members of the phylum proteobacteria that are symbiotic, commensal or otherwise associated with insects, nematodes, fishes, plants, as well as in obligatory bacterial parasites of the phylum *Chlamydiae* (Dale and Moran, 2006; Marie et al., 2001).

T3SS is a multicomponent apparatus with the following characteristics: i) when fully developed it spans both bacterial membranes and the periplasmic space; ii) it possesses a large extracellular appendage (termed 'pilus' in plant pathogenic bacteria or 'needle' in animal pathogenic ones) that reaches the eukaryotic host cell contributing to bacterial adherence; iii) it forms the translocation pore in the host cell membrane to efficiently deliver proteins of bacterial origin inside the host cell; iv) a large number of T3SS cytosolic components form the export gate into the bacterial cytoplasm which sorts and prepares the substrates for secretion (Fig. 1).

The integral bacterial membrane part of the T3S apparatus consists of a series of rings. The protein that oligomerizes and forms the outer membrane and periplasmic rings (yellow parts in Fig. 1) belongs to the secretin family of proteins (which is also common to T2SS) and has a crucial role in T3S biogenesis (Diepold et al., 2010; Korotkov et al., 2011). Secretins consist of various domains with the C-terminal one integrated in the outer membrane. The N-terminal domains are less conserved among secretion systems and are responsible for the formation of the periplasmic rings. An N-terminal signal targets secretins to the periplasmic space through the Sec pathway. From there they are delivered to the outer membrane through a specific small lipidated protein, pilotin (Okon et al., 2008). Pilotins from various secretion systems possess different structures despite their common function, probably due to their interaction with the non-conserved C-terminal tail of various secretins. Thus, for example, the T3SS pilotin of *Shigella flexneri* possess an overall fold which differs from the fold of the T3SS pilotin of *Pseudomonas aeruginosa* or the T2SS pilotins of *Neisseria meningitis* and *P. aeruginosa* (Izore et al., 2011).

The T3SS inner membrane (IM) rings are formed by the proteins SctD and SctJ [orange parts in Fig. 1; the unified nomenclature is followed here as proposed by Hueck (1998)]. SctD is a single-pass inner membrane protein that oligomerizes to form the most external inner membrane ring of the T3SS. Its N-terminal domain is facing the bacterial cytoplasm and its structure is homologous to forkhead-associated (FHA) domains (McDowell et al., 2011). The inner membrane part of the *Salmonella typhimurium* injectisome has been studied by EM (Marlovits et al., 2006; Marlovits et al., 2004). The inner membrane topology of six conserved components (HrcD$^{SctD}$, HrcR$^{SctR}$, HrcS$^{SctS}$, HrcT$^{SctT}$, HrcU$^{SctU}$ and HrcV$^{SctV}$) of the T3SS from *Xanthomonas campestris* was recently studied (Berger et al., 2010) by translational fusions to a dual alkaline phosphatase–β-galactosidase reporter protein. Full IM rings have been modeled for PrgH$^{SctD}$ and PrgK$^{SctJ}$ [the species-specific name is followed by the standard T3SS nomenclature as proposed by Hueck (1998) in superscript] based on docking of atomic structures of individual domains to cryo electron microscopy maps (Schraidt & Marlovits, 2011). The central density observed in the inner membrane rings (socket region) of a T3SS needle complex cryo electron microscopy reconstruction map from *Salmonella enterica* sv. typhimurium (Fig. 1, red parts) is attributed to the SpaP$^{SctR}$, SpaQ$^{SctT}$, SpaR$^{SctS}$, SpaS$^{SctU}$ and InvA$^{SctV}$ proteins (Schraidt & Marlovits, 2011; Wagner et al., 2010).

In the socket region numerous cytosolic components are recruited to orchestrate the secretion of various T3SS substrates, like the ATPase SctN and its various subunits SctO, SctL. As biogenesis of the T3SS must take place before the secretion of the effectors, the first T3SS substrates to be secreted are the proteins that build the needle or pilus (SctF) and the inner rod (SctI), (green part in Fig. 1). The proteins that form the translocator pore in the eukaryotic membrane along with the proteins found in the needle tip are the next substrates to be secreted prior to effector proteins secretion.

An additional cytoplasmic ring is believed to be formed around the T3SS export gate as in the case of the flagellum (Thomas et al., 2006). Although never really observed by electron microscopy, recently Lara-Tejero and colleagues have reported the presence of a large platform in the T3SS of *S. enterica* sv. typhimurium that can sort substrates prior to secretion (Lara-Tejero et al., 2011). This platform consists of SpaO$^{SctQ}$, OrgA$^{SctK}$ and OrgB$^{SctL}$.

Numerous crystal structure determinations of T3SS components have been reported: The structures of the C-terminal domain of HrcQ$_B$$^{SctQ}$ (Fadouloglou et al., 2004; Fadouloglou et al., 2009), the C-terminal domain of FliN (Brown et al., 2005) and the central part of FliM (Park et al., 2006), all members of the SctQ/FliN,Y family and components of the cytoplasmic ring of the T3SS apparatus (C-ring) have been determined. Extended mutational and cross linking studies support a donut-shaped tetramer organization for the FliN protein which is localized at the bottom of the C-ring (Paul and Blair, 2006). A model where the FliN tetramers alterates with the C-terminal domain of FliM (FliM$_C$) seems to be in agreement with the major features observed in electron microscopic reconstructions. The side-wall of C-ring above the FliN$_4$FliM$_C$ array is formed by the middle domain of FliM while the N-terminal domain interacts with the FliG which is localised in proximity with the inner membrane and is the connection unit between the C-ring and the inner membrane, MS-ring (Sarkar et al., 2010; Paul et al., 2011). FliG has no homolog in non-flagellar T3SS and the homolog SctQ proteins are interacting to the T3SS injectisome through the SctD proteins.

The structures of EscU$^{SctU}$ and YscU$^{SctU}$, *EPEC* and *Yersinia* homologs of HrcU$^{SctU}$ respectively (Zarivach et al., 2008; Lountos et al., 2009; Thomassin et al., 2011) provide insights into the properties of conserved core components. The periplasmic domain of PrgH$^{SctC}$ from *Salmonella* (Spreter et al., 2009) and the cytoplasmic domain of MxiD$^{SctC}$ from *Shigella* (McDowell et al., 2011) have been recently determined. Structures of the periplasmic domains of the membrane components EscJ$^{SctJ}$ from the enteropathogenic *Escherichia coli* (EPEC) are also available (Yip et al., 2005b; Spreter et al., 2009 ).

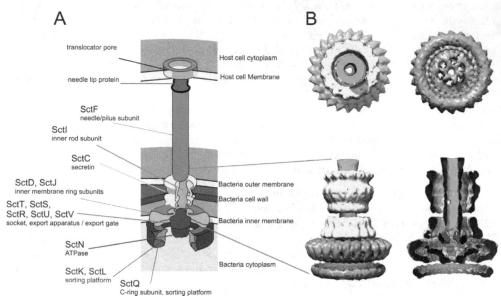

Fig. 1. (A) Overview of the T3S injectisome. (B) Different views of the *S. enterica* T3SS needle complex (Schraidt & Marlovits, 2011). Various parts of the needle complex are depicted in different colors using UCSF Chimera (Goddard et al., 2007). The colouring scheme used in (A) is followed. Top left: The T3SS needle complex viewed from top, Top right: View from the bottom, Lower left: side view, Lower right: a cross section of the side view. The inner membrane rings (orange) possess a 24-fold symmetry axis while the secretin rings (yellow) possess a 15-fold symmetry axis resulting in an overall 3-fold symmetry for the needle complex. The socket region (red parts and part of the orange area under the red parts) has a 6-fold symmetry (top right, icenter of the bottom view), which is also the symmetry of the T3SS ATPase that presumably docks in this area.

The T3SS utilizes an ATPase dedicated to drive secretion substrates through the central channel of the apparatus. Members of the SctN family (HrcN/FliI/YscN homologs) have a demonstrated ATP-hydrolysis activity, and exhibit extensive sequence and structural similarities with the $F_0F_1$-ATPase $\alpha/\beta$ subunits. Biochemical and electron microscopy data suggest that as it is the case with $F_0F_1$-ATPases, also the T3SS ATPases are hexamers anchored at the bacterial inner membrane. The crystal structure of FliI has been determined in the ADP-bound state and extensive structural similarities with to the $\alpha$ and $\beta$ subunits of the $F_0F_1$-ATPsynthase have been found (Imada et al., 2007). The catalytic domain of EscN$^{SctN}$ also shows structural similarity with $F_0F_1$-ATPases (Zarivach et al., 2007). Recently the structure of FliJ, member of the SctO family (HrpO/FliJ/YscO homologs) has been reported as an $\alpha$-helical coiled coil (Ibuki et al., 2011). Its structural similarity to a subunit of the $F_0F_1$-ATPsynthase and its interactions with FliI will be presented in section 6.1.1. Proteins from the SctL family (HrpE/FliH/YscL homologs) interact both with the T3SS ATPase and with structural proteins from the YscQ/FliN family located at the cytoplasm/inner membrane i.e. YscQ$^{SctQ}$, EscQ$^{SctQ}$ and FliN (Blaylock et al., 2006; McMurry et al., 2006; Biemans-Oldhinkel et al., 2011).

The outer supramolecular structure of the needle has been studied (Cordes et al., 2003), while structures of needle subunits from various bacteria have been recently determined (Wang et al., 2007; Deane et al., 2006; Zhang et al., 2006). At the tip of the T3SS needle resides an adaptor structure which mediates the interaction between the needle and the translocation pore at the eukaryotic membrane. The adaptor is formed through polymerization of a single protein. Information is available for the following needle tip proteins from three T3SS families: IpaD (*Shigella flexneri*), SipD (*Salmonella* spp.) and BipD (*Burkholderia pseudomallei*) from the *Inv-Mxi-Spa* T3SS family; LcrV (*Yersinia* spp.), PcrV (*Pseudomonas aeruginosa*, Sato et al., 2011; Sato & Frank, 2011) and AcrV (*Aeromonas salmonicida*) from the *Ysc* T3SS family; EspA (EPEC) from the *Ssa-Esc* family. The structures of IpaD, BipD, LcrV and part of the EspA structure have been elucidated (Espina et al., 2007; Johnson et al., 2007; Yip et al., 2005a; Derewenda et al., 2004), while a 3D-reconstruction of the MxiH filament is available (Deane et al., 2006).

Effectors are a large and structurally diverge group of virulence proteins which usually comprise a domain or a motif with a significant and proven role whithin the host cell during infection (for a review see Dean, 2011). Structures of several T3SS effectors from plant and animal pathogens are known (Desveaux et al., 2006; Stebbins, 2005). In addition, several structures for chaperones and chaperone-substrate complexes have been determined, including class I, class II and class III chaperones (Lilic et al., 2006; Buttner et al., 2008; Quinaud et al., 2007; Sun et al., 2008). Chaperones will be also presented in section 2.1.

## 2.1 The T3SS secretion signal

Type III effector proteins (T3EPs) possess non-cleavable secretion signals in the N-terminal protein regions, but no discernible amino acid or peptide similarities (Buttner and He, 2009) can be found. Three different types of potential secretion signals have been discussed: i) the N-terminus of the effector protein, ii) the ability of a chaperone to bind the effector before secretion, and iii) the 5′-end region of the mRNA; this hypothesis is very controversial (Gauthier et al., 2003; Anderson & Schneewind, 1997; Ramamurthi et al., 2002).

The prevailing view, supported by extensive biocomputing analyses, is that the amino acid composition of the N-terminal region of the effectors serves as secretion signal (Lloyd et al., 2001; Buttner and He, 2009; Arnold et al., 2009; Samudrala et al., 2009). The required N-terminal peptide length for secretion is usually 10–15 residues, whereas the minimum length needed for translocation is 50–60 residues. Additional targeting information is contained within the first 200 residues which provide binding sites for secretion chaperones (Lilic et al., 2006). T3SS chaperones of mammal pathogens interact with their cognate effectors through a chaperone-binding domain (CBD) located within the first 100 amino acids of the effector, after the N-terminal export signal (Cornelis, 2006).

Analyses of effectors from pathogenic bacteria revealed that the 25 N-terminal residues are enriched in Ser and lack Leu (Buttner and He, 2009; Arnold et al., 2009; Samudrala et al.,

2009). The N-terminal regions of T3EPs are probably unfolded, which is an important prerequisite for their transport through the narrow inner T3SS channel of presumably only 2.8 nm in diameter as was previously shown for the T3SS of several animal pathogenic bacteria (Marlovits et al., 2004, 2006; Galan and Wolf-Watz, 2006; Gazi et al., 2009).

For some effectors however, the N-terminal secretion signal is not sufficient for maximal secretion (Buttner & He, 2009) and specific chaperone proteins are needed; these are usually located adjacent to the cognate effector genes, suggesting strong selection for their coexistence in the genome. T3S chaperones are proposed to play a role in targeting secretory cargo to the injectisome, either by providing targeting information (Birtalan et al., 2002), or facilitating the exposure of the N-terminal export signal (Cornelis et al., 2006). Some chaperones are involved in the translocation of many substrate proteins, e.g. the global HpaB chaperone from *Xanthomonas campestris* pv. *vesicatoria* or Spa15 of *S. flexneri* (Hachani et al., 2008; Parsot et al., 2003; 2005; Buttner et al., 2004 ; 2006). Class I chaperones (the chaperones of effectors) are soluble small, usually homodimeric proteins that bind effector proteins. Although diverse in their sequences, they belong to the structural class of $\alpha/\beta$ proteins with a two-layer-sandwich architecture. For the chaperone-effector interaction a $\beta$-strand of the effector is added to extend the $\beta$-sheet layer of the chaperone (Lilic et al., 2006). Class I chaperones have been further subclassified depending on whether they associate with one (class Ia) or several (class Ib) effectors (Page & Parsot, 2002). Class II chaperones are T3SS chaperones of the translocators (Neyt & Cornelis, 1999). Experimental determinations of their structures (Buttner et al., 2008; Lunelli et al., 2009; Job et al., 2010; Priyadarshi & Tang, 2011) have confirmed earlier sequence analyses (Pallen et al., 2003) predicting an all-$\alpha$-helical domain structure, with the bulk of the protein consisting of three tandem tetratricopeptide repeats (TPRs) which are involved in protein-protein interactions. Their substrate is recognised and bound into a concave site of the chaperone. Class III chaperones prevent the premature polymerization of needle components in the bacterial cytoplasm. They are predicted to adopt extended $\alpha$-helical structures; this was confirmed by the crystal structure of the CesA which binds the EspA filament protein (Yip et al., 2005a).

Many functions have been attributed to T3SS chaperones, but the exact role(s) of the entire family of chaperones remain to be determined. However, it has been proposed that one of the main roles of the T3SS chaperones is the stabilization of at least some effector proteins inside bacterial cell, as well as their maintainance in a secretion-competent state, i.e. a partially folded or unfolded conformation.

## 3. The coiled-coil motif in proteins and α-helical bundles

The coiled-coil motif in protein structures consists of amphipathic α-helices that twist around each other to form a supercoiled bundle (Burkhard et al., 2001). It represents one of the efficient geometric solutions to packing helices in a stable way. The motif was one of the earliest protein structures discovered, first described for the hair protein alpha keratin (Crick, 1952). Coiled-coils are associated with all types of protein structure (globular, fibrous, membrane) and frequently provide a structural scaffold linked with molecular recognition interactions and oligomerization. Coiled-coil interactions play a major role in the formation of protein complexes in transcription, cell divisions, host-pathogen interactions etc. (Rackham et al., 2010). Coiled-coil helices may run parallel or antiparallel, and may form homo- or heterocomplexes (Grigoryan and Keating, 2008). The structures range from simple dimers through pentamers to more complex assemblies of many helices or bundles of bundles. Sequences of regular, left-hand twisted coiled-coils are characterized by a seven-residue periodicity (heptad repeat). If the heptad positions are labeled $a$-$g$, then positions $a$ and $d$ are hydrophobic and form the core of the bundle (Fig. 2). Positions $b$, $c$, $e$, $f$ and $g$ are more solvent-exposed and their amino acid preferences reflect constraints which are specific to each type of helical bundle (Paliakasis & Kokkinidis, 1992; Lupas et al., 1991). The hydrophobic residues $a$ and $d$ from one helix form 'knobs' that pack into 'holes' formed by residues $g$, $a$, $d$, $e$ on neighbouring helices. Coiled-coil helices are distinguished from other amphipathic helices by the periodicity of hydrophobic residues (3.5 vs. 3.65 residues per turn), the length (long vs. short) and the packing interactions. Some proteins are induced to form coiled-coils upon association with a binding partner (Lupas, 1996).

Coiled-coil predictions at the genome level have explored the 'coilomes' of individual organisms (Barbara et al., 2007; Newman et al., 2000, Rose et al., 2004). In addition, genomewide analyses of coiled-coils evolution have been performed (Rackham et al., 2010) and shown that coiled-coils do not change their oligomeric state over evolution, and do not evolve from rearrangements of α-helices in protein structures. An analysis of proteomes (Liu and Rost, 2001) showed that twice as many coiled-coils are found in eukaryotes (10%) as in prokaryotes and archaea (4%-5%). The size of coiled-coil proteins ranges from short domains of 6-7 heptad repeats, e.g. Leucine zippers, serving as homo-/heterodimerization motifs in transcription factors (Jakoby et al., 2002; Vinson et al., 2002), to long domains of several hundreds amino acids found in functionally distinct proteins, often involved in attaching protein complexes to larger cellular structures (e.g. the Golgi, centrosomes, centromers, or the nuclear envelope).

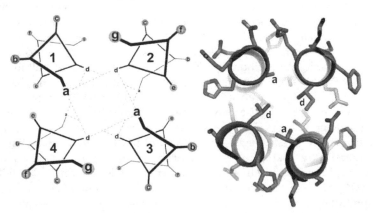

Fig. 2. Left: Antiparallel coiled-coil structure (4-α-helical bundle) and assignment of heptad positions a-g. Hydrophobic and hydrophilic positions are coloured yellow and red respectively. Right: Positions a, d and their packing in the core of the 4-α-helical ColE1 Rop (Banner et al., 1987). Positions a, d form slices perpendicularly to the bundle axis.

## 4. Flexibility and disorder in proteins

Until recently the classical structure–function paradigm which states that protein function is dependent on a defined, if flexible, three-dimensional polypeptide structure was widely accepted in protein science (Anfinsen, 1973). However, even in the early days of structural biology, with only approx. 20 protein crystal structures determined, some protein segments were known which yield weak or non-detectable electron density and yet they may be essential for function (Bloomer et al., 1978; Bode et al., 1978). A common reason (apart from crystal defects) for missing electron density is that the unobserved region fails to scatter X-rays coherently due to variation in position from one atom to the next, i.e. the unobserved atoms are disordered. In addition, during the last decade, many proteins have been described that fail to adopt a stable tertiary structure under physiological conditions and yet display biological activity (Dunker et al., 2008a; Uversky & Dunker, 2010). This state of the proteins, defined as intrinsic disorder, has been found to be rather widespread; disordered regions lacking stable secondary and tertiary structure are often a prerequisite for biological activity, suggesting that structure-function relationships can be frequently only understood in a dynamic context in which function arises from conformational freedom. Fully or partly nonstructured proteins are described as intrinsically disordered (IDPs) or intrinsically unstructured proteins. The term natively unfolded proteins indicates that protein function is associated with a dynamic ensemble of different conformations (Gazi et al., 2008).

Structural plasticity and flexibility is believed to represent a key functional feature of IDPs (Dunker et al., 2008a, 2008b; Dunker & Uversky, 2008; Xie et al., 2007; Cortese et al., 2008), enabling them to interact with numerous binding partners, e.g. proteins, membranes, nucleic acids and small molecules (Durand et al., 2008; Uversky et al., 2009). Because of their functional importance, intrinsically disordered domains are very common in proteomes and play crucial roles in signaling, recognition, regulation and self-assembly (Namba, 2001). The extreme flexibility of IDPs has been suggested to represent a strategy for optimizing the search and interaction with their targets (Sugase et al., 2007). Intrinsically disordered proteins are substantially depleted in W,C,F,Y,V,L,N (order-promoting) and enriched in A,R,G,Q,S,P,E,K (disorder-promoting residues) (Dunker et al., 2002; Uversky, 2010). These biases in the amino acid compositions of IDPs (which result in low overall hydrophobicity and low net charge) are used in various methods for the prediction of the ID propensities (Prilusky et al., 2005). Such analyses suggest that approx. 45% of proteins within a eukaryotic proteome contain a disordered region (Pentony & Jones, 2010). As a result of their frequent node positions in interactoms, many disordered proteins are tightly regulated at the levels of their synthesis, degradation and posttranslational modifications (Gsponer, 2008). It is noteworthy that extreme structural plasticity and ensembles of different conformations has been occasionally observed for coiled-coils and α-helical bundles (Glykos et al., 1999, 2004); as is the case with other proteins, the plasticity of coiled coils may have functional implications, e.g. in the establishment of macromolecular assemblies based on coiled-coil interactions (Gazi et al., 2008).

## 5. Tools for the analysis of coiled-coils and intrinsic disorder

### 5.1 *In silico* prediction and analysis of coiled-coil domains

**Prediction of coiled coils from sequence:** The 'COILS' webserver assesses the probability that a residue in a sequence is part of a coiled-coil structure by comparison of its flaking sequences with sequences of known coiled-coil proteins (Lupas et al., 1991) (http://www.ch.embnet.org/software/COILS_form.html). In the 'Paircoil2' algorithm

(McDonnell et al., 2006), pairwise residue probabilities are used to detect coiled-coil motifs in protein sequences (http://groups.csail.mit.edu/cb/paircoil2/paircoil2.html). 'Matcher' (http://cis.poly.edu/~jps/) determines whether a given sequence contains heptads and assigns heptad positions to residues (Fischetti et al., 1993). To predict the oligomerization states of coiled coils 'Multicoil2' (Trigg et al., 2011) uses pairwise correlations and Hidden Markov Models (HMMs). For distinguishing dimers, trimers and non-coiled-coil oligomerization states the algorithm integrates sequence features through a multinomial logistic regression and devises an optimized scoring function that incorporates pairwise correlations localized in the sequence. A database comprising 2015 sequences with reliable structural annotation from experimental data is used (http://multicoil2.csail.mit.edu). 'SCORER' (Armstrong et al., 2011) also provides predictions of coiled-coil oligomerization (http://coiledcoils.chm.bris.ac.uk/Scorer) .

**Assignment of the coiled coil packing:** COILCHECK (Alva et al., 2008) can be used for analysis and validation of coiled-coil structures through calculation of the strength of interhelical interactions in coiled coils; it can be used to rationalize the behaviour of single residue mutations and to design mutations (http://caps.ncbs.res.in/coilcheck/). SOCKET (Walshaw & Woolfson, 2001) can be used to identify coiled coils through an analysis of the knobs-into-holes side chain packing (http://coiledcoils.chm.bris.ac.uk/socket/).

**Databases:** For genomewide predictions the 'SpiriCoil' algorithm (Rackham et al., 2010) is employed which uses hundreds of HMMs representing coiled-coil-containing domain families. Their results are available through the SpiriCoil Database (http://supfam.org/ SUPERFAMILY/spiricoil). It includes results from all completely sequenced genomes. The CC+ database is a detailed, searchable repository accessible via the SOCKET program (Testa et al., 2009) (http://coiledcoils.chm.bris.ac.uk/ccplus/).

Several of the above tools have been used in sections 6 and 7 of this chapter. In addition, protein sequences were retrieved from the NCBI/GenBank and specialized databases e.g. PPI: *P.syringae* Genome Resources (www.pseudomonas-syringae.org) and the Kyoto Encyclopedia for Genes and Genomes (KEGG) (Kanehisa & Goto, 2000). Secondary structure predictions were performed with 'PSIPRED' (Jones, 1999). Protein structures were retrieved from the Protein Data Bank (PDB).

### 5.2 *In silico* analysis of T3SS effectors and secretion signals

A selection of bioinformatics tools is available for T3SS effector and secretion signal prediction: 'Effective' is an on-line tool for sequence-based prediction of secreted proteins available from the TUM Genome Oriented Bioinformatics, University of Vienna (Arnold et al., 2009; Jehl et al., 2011), which can be used for the effector prediction in bacterial protein-sequences (http://www.effectors.org/). 'Effective' provides pre-calculated predictions on bacterial effectors in all publicly available pathogenic and symbiotic genomes or using sequence data provided by the user. T3SS secretion signal predictions from amino acid sequences, is available from 'moblab' (http://gecco.org.chemie.uni-frankfurt.de/T3SS_prediction/T3SS_prediction.html). The basic concepts of this tool are described by Lower & Schneider (2009). The 'SIEVE' Server (http://www.sysbep.org/sieve/) for the prediction of type III secreted effectors was originally described by Samudrala et al., (2009) and recently reviewed by McDermott et al., (2011). Potential T3SS effectors are scored using a computational model developed via Machine-Learning Methodologies.

### 5.3 Experimental and *in silico* analysis of disordered domains

Disordered regions may be detected in protein structures determined by X-ray crystallography through missing electron density. Heteronuclear multidimensional NMR is a powerful tool for the characterization of protein disorder and provides direct measurement of the mobility of unstructured regions (Eliezer, 2007). Loss of secondary structure may be detected (among other methods) by far-UV CD (Kelly & Price, 1997) and Fourier transform infra-red spectroscopy (FTIR) (Uversky et al., 2000). Hydrodynamic parameters obtained from techniques such as gel filtration, SAXS (Gazi et al., 2008), dynamic and static light scattering provide information on whether a protein is unfolded since the unfolding results in an increase in protein hydrodynamic volume. The degree of globularity, which reflects the presence of a well-packed hydrophobic core may be estimated by a special analysis of small angle X-ray scattering (SAXS) data in form of a Kratky plot. Kratky plots are obtained by plotting $I(s)xs^2$ against $s$ (scattering intensity: $I$; momentum transfer: $s=4\pi\sin(\theta)/\lambda$; $2\theta$: scattering angle; wavelength of X-rays: $\lambda$). They are used to judge the folding of the protein, as the shape of the curve is sensitive to the conformational state of the scattering molecules (Gazi et al., 2008).

Several algorithms have been developed to predict protein disorder on the basis of specific biochemical properties and biased amino acid compositions. These tools include PONDR (Romero et al., 2001; Peng et al., 2005), DisEMBL (Linding et al., 2003), IUPred (Dosztanyi et al., 2005), FoldUnfold (Galzitskaya et al., 2006) and PrDOS (Ishida & Kinoshita, 2007).

The main tool used in sections 6 and 7 for the *in silico* prediction of protein disorder from sequences is FoldIndex© (Prilusky et al., 2005). The propensity of N-termini of proteins for disorder was analyzed on the basis of their biased content of order-/disorder- promoting residues (Dunker et al., 2002).

## 6. The occurrence of coiled-coils and intrinsic disorder in T3SS proteins

Analyses of T3SS protein sequences (Table 1) reveal an unusually frequent occurrence of predicted heptad repeats, which is indicative of a high propensity for coiled-coil formation (Delahay and Frankel, 2002; Pallen et al., 1997; Gazi et al., 2009; Knodler et al., 2011). Structural studies have confirmed the unsual prevalence of coiled-coils among T3SS proteins (Gazi et al., 2009; Ibuki et al., 2009; Lorenzini et al., 2010). In addition, coiled-coil interactions occur frequently in crystal structures of T3SS protein complexes, e.g. in the a macromolecular assembly TyeA-YopN that regulates type III secretion in *Yersinia pestis* (Schubot et al., 2005) or in the complex of the filament protein EspA from the enteropathogenic *E. coli* T3SS with its chaperone CesA (Yip et al., 2005a). In a recent report, the interactions of the *Salmonella typhimurium* needle protein PrgI, an α-helical hairpin, with the tip protein SipD which comprises a long, central coiled coil (Rathinavelan et al., 2011) were studied using NMR paramagnetic relaxation enhancement. A specific region on the SipD coiled-coil was identified as the binding site for the α-helix of PrgI. Crystallographic studies of the PrgI-SipD complex have revealed coiled-coil interactions via the formation of an intermolecular 4-α-helical bundle structure (Lunelli et al., 2011). These studies also showed the importance of the structural flexibility of SipD (introduced by a π-bulge structure) in complex formation. Coiled-coil interactions of HrpO and FliJ with their cognate protein targets have been also reported (Gazi et al., 2008).

Predicted coiled-coil domains have been shown by mutagenesis to enhance membrane association of *Salmonella* T3SS effectors (Knodler et al., 2011). T3SS proteins and coiled-coil domains are frequently predicted to be structurally disordered (Table 1, 2). For many T3SS effectors disorder in their N-terminal region, as well as an increased overall flexibility have been also noted (Table 1, Gazi et al., 2009). In the following, these aspects of T3SS proteins will be elaborated with specific examples from various protein families. Structures of T3SS proteins with increased coiled-coil content are shown in Fig. 3

| Protein | % heptad repeats | % order-promoting aa among the 50 N-terminal residues | % disorder-promoting aa among the 50 N- terminal residues | % overall disorder |
|---|---|---|---|---|
| **EFFECTORS** | | | | |
| AvrPto1 | 27 (~70) | 32 | 52 | 58 |
| HopE1 | 23 | 24 | 44 | 50 |
| HopH1 | 24 | 26 | 49 | 61 |
| HopY1 | 26 | 30 | 62 | 47 |
| **PILUS** | | | | |
| HrpA1 | 43 | 40 | 46 | 26 |
| **OTHER SECRETED/PUTATIVE SECRETED** | | | | |
| HrpJ | 41 | 32 | 54 | 30 |
| HrpF | 60 | 28 | 48 | 57 |
| HrpB | 23 | 26 | 58 | 18 |
| **CYTOPLASMIC** | | | | |
| HrpO | 79 | 22 | 58 | 85 |
| HrpE | 51 | 30 | 54 | 17 |
| HrcN | 21 | 36 | 54 | 21 |
| HrcQ$_B$ | 53 | 14 | 54 | 46 |

Table 1. Heptad repeats prediction and disorder analysis for selected proteins from the T3SS of *P. syringae* pv. tomato DC3000. Only proteins with coiled-coil content above 20% are given. For the AvrPto1 protein the crystalographically determined coiled-coil content is given in parentheses. The overall disorder was calculated using FOLDINDEX. N-terminal protein disorder calculations used Dunker's et al. (2002) definition of order-/ disorder-promoting residues. HrcQ$_B$ does not include the disordered N-terminal domain.

## 6.1 Cytoplasmic proteins

Several cytoplasmic T3SS proteins exhibit a significant coiled coil propensity and intrinsic disorder (Table 1, Fig. 3, 4). Evidence from some cytoplasmic proteins (see section 6.1.1) suggests that these properties might be essential elements in the establishment of key protein-protein interaction networks required for T3SS function (Gazi et al., 2008).

## 6.1.1 The SctO family (HrpO/FliJ/YscO homologs)

The most extensive heptad repeat pattern occurs in the HrpO/FliJ/YscO family of T3SS proteins (Gazi et al., 2008). Despite the absence of significant homologies, the family members share specific characteristics, e.g. increased propensity for coiled coil formation and intrinsic disorder (Gazi et al., 2008). The extreme flexible nature of HrpO$^{SctO}$ from

*Pseudomonas syringae* pv. phaseolicola (Gazi et al., 2008), a property shared with FliJ from *S. typhimurium,* has prevented its crystallization and determination of its 3D-structure by X-ray crystallography. A variant form of the FliJ protein from *S. enterica* sv. typhimurium was crystallized however (Ibuki et al., 2009), and its structure was found to be remarkably similar (Fig. 5) to that of the two-stranded α-helical coiled-coil part of the γ subunit of $F_0F_1$-ATP synthase (Ibuki et al., 2011). A similar coiled coil structure (Fig. 5) consisting of two long α-helices was also reported for the crystal structure of the CT670$^{SctO}$ protein (a YscO homolog) from *Chlamydia trachomatis* (Lorenzini et al., 2010).

| Protein | % total disorder | Protein | % total disorder | Protein | % total disorder |
|---------|------------------|---------|------------------|---------|------------------|
| **NEEDLE** | | **CHAPERONES** | | **TIP** | |
| MxiH | 52 | SycD | 12 | LcrV | 45 |
| PrgI | 21 | PscE | 0 | IpaD | 41 |
| BsaL | 30 | PscG | 7 | BipD | 26 |
| | | YscE | 27 | EspA | 6 |
| | | YscG | 39 | | |
| | | CesA | 66 | | |

Table 2. Disorder analysis for T3SS proteins of known 3D-structures with coiled-coil content exceeding 30%. The overall protein disorder was calculated from sequence data using the FoldIndex program with a window of 21 residues.

Small angle X-ray scattering (SAXS) and circular dichroism (CD) characterization of HrpO$^{SctO}$ from *P. syringae* pv. phaseolicola revealed a high α-helical content with coiled-coil characteristics and molten globule-like properties (Gazi et al., 2008). HrpO$^{SctO}$ like its flagellar counterpart FliJ is essential for export, but its function remains obscure. HrpO$^{SctO}$ interacts, probably via intermolecular coiled-coil formation, with HrpE, a highly α-helical T3SS protein which belongs to the HrpE/FliH/YscL family. FliH, the flagellar counterpart of HrpE is a regulator of the FliI ATPase (Lane et al., 2006). Evidence from HrpO$^{SctO}$ and its analogs in various flagellar or non-flagellar T3S systems suggests that the extreme flexibility (Fig. 4) and propensity for coiled-coil interactions observed in members of the HrpO/FliJ/YscO family might be important factors for increased interactivity and the establishment of functional protein-protein interaction networks in T3SS. This is consistent with the observation that several members of the HrpO/FliJ/YscO family were found to interact with other cytosolic T3SS components or self-associate via coiled-coil interactions: The flagellar FliJ protein, a key player in a chaperone escort mechanism that recruits unloaded chaperones for the minor filament-class subunits of the filament cap and hook-filament junction substructures (Evans et al., 2006) binds to the same chaperone site as the cognate export substrate of the chaperone, albeit with a much lower affinity. Similarly, YscO$^{SctO}$ from *Yersinia enterocolitica* and InvI$^{SctO}$ from *Salmonella typhimurium* do not bind to export substrates but recognize a subset of export chaperones that are specialized to deliver the T3SS translocators to the export apparatus (Evans & Hughes, 2009). In all of these cases the interaction partners of the HrpO/FliJ/YscO family members exhibit a very high α-helical/coiled-coil content (Fig. 5). FliJ was also found to interact with structural cytoplasmic components of the T3SS like the FliM$^{SctQ}$ protein, even in the absence of FliH, suggesting a docking mechanism for export substrates, chaperones and the ATPase to the T3SS machinery (Gonzalez-Pedrajo et al., 2006). The CT670$^{SctO}$ protein exists in monomeric and dimeric forms, with the monomeric form dominating at low protein concentrations. For self-association and dimer formation the involvement of coiled-coil interactions is predicted (Lorenzini et al., 2010). CT670$^{SctO}$ interacts with CT671$^{SctP}$, a T3SS protein, with a predicted coiled-coil domain in its C-terminal region. CT671$^{SctP}$ is a homolog of the YscP protein which has been characterized as a molecular ruler and as a switch for T3SS substrate specificity in *Yersinia* species (Agrain et al., 2005). The two coiled-coil containing proteins CT670$^{SctO}$ and CT671$^{SctP}$ have been suggested to form a chaperone-effector-like pair with CT670$^{SctO}$ acting as chaperone (Lorenzini et al., 2010).

The HrpO/FliJ/YscO family members are encoded by genes located always downstream of the gene coding for T3SS ATPases (the SctN family of T3SS proteins which includes HrcN/FliI/YscN homologs); this implies a close connection between these proteins and the ATPase. In flagellar T3SS the FliI protein is an ATPase that has extensive structural similarity to the α- and β- subunits of the $F_oF_1$-ATP synthase (Imada et al., 2007), while also the structure of FliJ from *S. enterica* sv. typhimurium (Fig. 4, 5) is remarkably similar to that of the two-stranded α-helical coiled-coil part of the γ-subunit of $F_oF_1$-ATP synthase (Ibuki et al., 2001). FliJ promotes the formation of FliI hexamer rings by binding to the center of the ring in a similar way to the γ-subunit penetrating into the central channel of the $α_3β_3$ ring in $F_oF_1$-ATPase. Moreover, the HrpE/FliH/YscL family of proteins (interaction partners of the HrpO/FliJ/YscO family) are distant homologs to both β- and δ- subunits of the $F_oF_1$-ATP synthase (Pallen et al., 2006). In flagellar systems the docking of the ATPase to the T3S machinery is mediated by the FliJ/FliH pair (Minamino et al., 2009). These results strongly suggest that T3SS and F- and V-type ATPases share a similar mechanism and an evolutionary relationship. It is thereby striking that extensive coiled-coil domains (e.g. FliJ, FliH) have been conserved between the two systems.

Overall, the above remarkable findings support our earlier suggestions (Gazi et al., 2008, 2009) that T3SS proteins, and in particular members of the SctO family, with long disordered/flexible coiled coil structures occupy node positions in the T3SS interactome, being capable of interacting with different partners and possess various roles in the secretion mechanism. These roles are to a large extent poorly understood and remain to be elucidated experimentally.

### 6.1.2 The SctL family (HrpE/FliH/YscL homologs)

In terms of predicted heptad repeats content (Gazi et al., 2008) the HrpE/FliH/YscL family of proteins comes second after the HrpO/FliJ/YscO family. These proteins are distant homologs of the second-stalk components of the $F_0$-$F_1$ ATPases (Pallen et al., 2006). FliH is a regulator of the FliI ATPase (Evans et al., 2006) and was found to interact with the 18 N-terminal residues of FliI that are predicted to form an amphipathic α-helix upon interaction with FliH (Lane et al., 2006).

The HrpE/FliH/YscL family members possess glycine-rich repeats of the form AxxxG(xxxG)$_m$xxxA with m representing a non constant value between FliH proteins from different bacteria and x standing for any residue. The amino acid sequence distribution of each of the three x positions was found to differ significantly from the overall amino acid composition of the HrpE/FliH/YscL proteins. The high frequency of Glu, Gln, Lys and Ala residues in the repeat positions suggests the presence of α-helical structure for this motif (Trost et al., 2009). When the Protein Data Bank was searched for GxxxG repeats similar in length to those found in FliH, no helices containing more than three contiguous glycine repeat segments were found implying that long GxxxG repeats are presumably quite rare in nature.

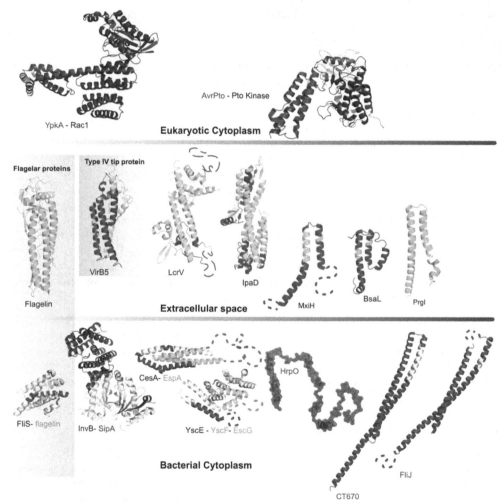

Fig. 3. 3D-structures of T3SS proteins with significant coiled-coil content and their locations. The structure of the intrinsically disordered HrpO is based on SAXS data (Gazi et al., 2008).

## 6.2 T3SS needle and pilus proteins

The major extracellular T3SS component is the needle with a length of 60 nm and an external diameter of 7 nm for animal pathogens; a much longer structure (up to 2μm) named the Hrp pilus is the needle counterpart in phytopathogenic bacteria (Barrett et al., 2008; Cordes et al., 2003; He and Jin, 2003; Alfano & Collmer, 1997; Roine, 1997). The needle appears to play a major role in host sensing and signal transmission from the distal to the basal end of T3SS (Deane et al., 2006). The needle structures are formed through the helical assembly of multiple copies of a small α-helical protein. Along the needle axis runs a narrow (2.5 nm) conduit which is used for the passage of needle components, tip proteins, translocators and effectors, whereby a partially unfolded form of the substrate is required. Structures are available (Wang et al., 2007; Deane et al., 2006; Zhang et al., 2006) for three needle components from animal pathogens: MxiH (*S. flexneri*), BsaL (*B. pseudomallei*) and PrgI (*S. typhimurium*). These structures are highly α-helical, with a central coiled-coil which is essential for needle assembly. Outside this coiled coil, all three proteins have highly mobile N-termini and C-termini, although these regions may retain some degree of helical structure in solution (Blocker et al., 2008). The sequences of the coiled-coil parts of the needle proteins show strong similarities, suggesting that they all share a common fold and pattern of interactions (Wang et al., 2007; Zhang et al., 2006). Analysis of structures and sequences of needle components from various pathogens suggests that the majority has a propensity for structural disorder (Table 2). An analysis of needle/pilus components predicts a mean overall disorder of approx. 30%. The predicted disorder has been confirmed by CD and thermal unfolding studies of MxiH, BsaL and PrgI which reveal that under conditions resembling the physiological ones, all three C-terminally truncated proteins adopt a molten globule-like state; at temperatures above 37° C their tertiary structure collapses while the secondary structure is largely retained (Barrett et al., 2008); this behaviour is strongly reminescent to the one observed for the cytoplasmic HrpO$^{SctO}$ protein (Gazi et al., 2008). A partially unfolded state of the needle components could be functionally important e.g. for transversing the needle channel and for the extracellular assembly. Signal transmission for host cell sensing is suspected to utilize the flexibility of needle subunits (Deane et al., 2006).

The major subunits of the Hrp pilus (HrpA) are generally predicted to be almost entirely α-helical, with the exception of the *Pseudomonas syringae* species, for which the 50 N-terminal amino acids are predicted to contain β-strands (He and Jin, 2003; Koebnik, 2001). The major subunits of the Hrp-dependent pili, like the needle structural proteins, are all small proteins (of 6 to 11 kDa), but their sequences are surprisingly hypervariable, even within *P. syringae* pathovars. This hypervariability may reflect the evolutionary adaptations to evade plant defense systems. The predicted secondary structures of the major pilus subunits however, are remarkably similar, almost entirely α-helical. Insights into the structure of Hrp pilus components have been obtained from the recent investigation of the HrpA protein from the *P. syringae* pv. phaseolicola T3SS pilus (Kotzabasaki & Kokkinidis, unpublished). The C-terminal part of the 11 kDa protein is responsible for the assembly of multiple HrpA copies in the pilus (Roine, 1997). No chaperons for HrpA have been identified. The secondary structure of HrpA is predicted to be highly α-helical, with a propensity for coiled-coil formation in its functionally important C-terminal region. Surprisingly, experimental characterization of the HrpA protein using CD, Raman, FTIR and SAXS provides strong evidence that HrpA does not adopt a helical structure, but rather a highly disordered state with β-strand features. High resolution transmission electron microscopy (TEM) of purified HrpA samples reveals a pronounced propensity for polymerization and formation of two types of fibrils with nano-to micro scale features, one of which has comparable geometrical parameters with the Hrp pilus.

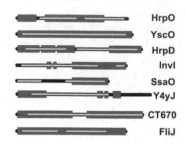

HrpO
YscO
HrpD
InvI
SsaO
Y4yJ
CT670
FliJ

Fig. 4. Predicted coiled-coil regions and structural disorder for the HrpO/FliJ/YscO family. Disordered segments (predicted by FoldIndex) are colored red, coiled-coil domains cyan. All sequences were predicted to be almost entirely α-helical, which was confirmed for HrpO (Gazi et al., 2008), FliJ (Ibuki et al., 2011) and CT670 (Lorenzini et al., 2010). Sequences are drawn to scale, and vary from 166 (HrpD) to 125 residues (SsaO).

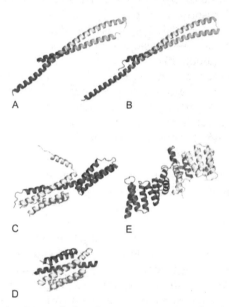

Fig. 5. T3SS proteins involved in coiled-coil interactions: (A) FliJ (Ibuki et al., 2011) from *S. typhimurium*, (B) CT670 SctO (Lorenzini et al., 2010) from *Ch. trachomatis* colored from N-(blue) to C-terminus (red). The dimeric FliT of *S. enterica* sv. typhimurium (C) and *Bordetella bronchiseptica* (D), each monomer differently coloured. In (C) the C-terminal helix adopts a different conformation in each monomer (Imada et al., 2010). (E) The *Y. enterocolitica* SycD (Buttner et al., 2008) which is recognized by YscO SctO (Evans & Hughes, 2009).

## 6.3 T3SS effector flexibility and coiled-coil propensity

Approximately a third of the T3SS effector structures known forms regular coiled-coils with knob-into-holes packing, while several others exhibit short heptad repeat patterns in their sequences which give rise to coiled-coil interactions and short, distorted α-helical bundles: A YpkA subdomain folds in two 3-helical bundles (Prehna et al., 2006). Coiled-coils are also observed in the N-terminal domains of YopH and SptP (Khandelwal et al., 2002; Stebbins and Galan, 2000) and in MxiC and YopN-TyeA (Deane et al., 2008; Schubot et al., 2005). The *S. flexneri* effector IpaH is a E3 ubiquitin ligase with a C-terminal which consists entirely of α-helical bundles and carries the catalytic activity for ubiquitin transfer (Singer et al., 2008). Helical bundle structures are also adopted by AvrPto (Wulf et al., 2004) and AvrPtoB (Dong et al., 2009), although both interact with their target protein kinase Pto via β-strand addition (Xing et al., 2007). AvrPto (PDB ids: 1R5E, 2QKW) displays considerable structural plasticity which is consistent with its predicted increased flexibility (Table 1). In a recent analysis, 49% of the *S. typhimurium* effectors have been predicted to posses at least one coiled-coil domain which enhances membrane association in mammalian cells (Knodler et al., 2011).

Apart from the frequent occurrence of coiled-coils, a further common feature among T3SS effectors are disorder effects established both as localized disorder of their extreme N-terminal peptide where the secretion signal resides, or frequently as an overall structural disorder. Usually, the 15-20 N-terminal residues of effectors are highly disordered. Truncation of N- and C-terminal residues was necessary for the NMR study of the AvrPto effector (Wulf et al., 2004), while in the crystal structure of the AvrPto-Pto kinase complex 28 N-terminal residues are missing (Xing et al., 2007). In the case of ExsE the N- terminal fifteen residues had to be omitted for crystallization and structure determination (Vogelaar et al., 2010). Although, the full length AvrB and AvrPphF ORF2 were crystallized, electron density was not observed for the 27 N-terminal residues due to disorder (Lee et al., 2004; Singer et al., 2004). The N-terminal region of the SipA effector in the complex with the chaperone InvB is highly disordered (Lilic et al., 2006). In the CBD of effectors bound to class IA chaperones, there is a prevalent localized disorder for the part of the effector that crosses the interface of the dimeric chaperone; thus, this part of the effector cannot be modelled, as can be seen not only for SipA but also for ExsE and YopN (Vogelaar et al., 2010). Moreover, a large majority (~75%) of *P.syringae* pv. tomato effectors show a significant propensity for structural disorder in the region of their 50 N-terminal residues based on the ratio of order- vs. disorder- promoting residues (selected effectors are shown in Table 1) which has an average value of 0.45 (M. Kokkinidis, unpublished). Other secreted proteins have an average value of 0.50. The ratio for cytoplasmic T3SS proteins is 0.55; the average ratio in proteomes is 0.58 based on the amino acid frequencies of order- and disorder-promoting residues (Brooks et al., 2002); significantly lower values, as in the case of the N-termini of T3SS effectors, indicate a propensity for disorder. The only structured N-terminal region of a T3SS effector is that of YopH. The 129 residue N-terminal domain has two functions: the first 70 residues contain the CBD domain for chaperone SycH, while the full 129 N-terminal domain binds to phosphotyrosine-containing proteins and adopts an overall globular fold (Khandelwal et al., 2002). The N-terminal region of VirA is partially disordered (Davis et al., 2008). Apart from the N-terminal disorder, a propensity for overall disorder is predicted for T3SS effectors, which may reflect an increased structural flexibility. An average value of 35% of disordered residues is predicted by FOLDINDEX for the T3SS effectors of *P. syringae* pv. tomato DC3000, which is to be contrasted with an average value of 28% for cytoplasmic T3SS proteins, if the extensively disordered members of the HrpO/FliJ/YscO family (some of which could be classified as IDPs) are excluded. For other secreted/putatively secreted T3SS proteins the values are 30-37%.

## 7. Comparison with other secretory systems: T2SS, T4SS and T6SS

To compare with T3SS, earlier analyses for T2SS and T4SS were updated and an analysis of T6SS protein sequences was performed. Coiled-coil predictions and disorder analysis were carried out for *Helicobacter pylori* (T4SS), *Legionella pneumophila* (T2SS and T4SS), the Type-4-pili (T4P) of *L. pneumophila* and *Pseudomonas aeruginosa* and *P. aeruginosa* strain PA14 (T6SS). For 2878 proteins encoded in the *L. pneumophila* genome (GenBank accession number NC_006369), the predicted coiled-coil content is 4%.

Using the Virulence Factor Database (Yang et al., 2008) for the classification of *L. pneumophila* proteins, a coiled-coil content of 14% is predicted for T2SS (11 proteins), 19% for T4SS (50 proteins), and 13% for T4P (3 proteins). For 1573 proteins of *H. pylori* (NC_000915) a coiled-coil content of 3% is predicted, with 26% for T4SS (24 proteins). For 5571 proteins of *P. aeruginosa* (NC_002516) the predicted content is 4%, 8% for T2SS (11 proteins) and 10% for T4P (32 proteins). A high coiled-coil content is predicted for the T4SS of *H. pylori* and the T4SS effectors of *L. pneumophila*. The latter (Table 3) exhibit a particularly high propensity for structural disorder (on the average 46% disordered regions) and coiled-coil content (30%), thus strongly resembling T3SS effectors. The analysis of T6SS protein sequences retrieved from the KEGG database is in Table 4. Hcp and Vgr proteins are proposed effectors (Mougous, et al., 2006), although they may also act as T6SS machine components (Zheng &Leung, 2007). Non-secreted components include ClpV an AAA+ Clp-like ATPase(Cascales, 2008) and various other core components. The analysis suggests a low coiled-coil content for most secreted proteins (on the average 6%) and a higher one (12% on the average) for core proteins. Interestingly, the mean overall disorder of secreted T6SS components is very high (41%) and comparable to T3SS effectors or to T4SS effectors of *L. pneumophila*. Core components of T6SS display a significantly lower degree disorder (28%).

| T4SS effectors | % heptad repeats | % overall disorder | T4SS effectors | % heptad repeats | % overall disorder |
|---|---|---|---|---|---|
| Ceg9 | 0 | 56 | SidJ | 5 | 35 |
| VipA | 29 | 36 | LaiA/SdeA | 28 | 42 |
| AnkX/AnkN/LegA8 | 15 | 35 | YlfA/LegC7 | 60 | 40 |
| LidA | 47 | 68 | DrrA/SidM | 42 | 47 |
| Ceg19 | 33 | 30 | LepB | 49 | 50 |
| LegC3 | 62 | 60 | SidC | 36 | 62 |
| LegC2 | 67 | 61 | SidF | 28 | 56 |
| RalF | 0 | 43 | LepA | 7 | 16 |
| SetA | 26 | 42 | LubX/LegU2 | 10 | 46 |
| AnkB/Cag27/LegAU13 | 20 | 51 | VipD | 42 | 35 |

Table 3. Predictions (using MATCHER ) of the heptad repeats content and disorder analysis (using FOLDINDEX) for effectors from the T4SS of *L. pneumophila* strain Philadelphia 1.

| Locus No (Protein) | % heptad repeats | % overall disorder | Locus No (Protein) | % heptad repeats | % overall disorder |
|---|---|---|---|---|---|
| SECRETED COMPONENTS | | | CORE COMPONENTS | | |
| PA14_01030 (Hcp) | 14 | 37 | PA14_00875 (PpkA) | 7 | 25 |
| PA14_01110 (VgrG) | 6 | 42 | PA14_00890 (PppA) | 0 | 14 |
| PA14_01160 (VgrG) | 6 | 59 | PA14_00910 (ImpL) | 18 | 22 |
| PA14_33960 (VgrG) | 0 | 34 | PA14_00925 (ImpK) | 10 | 29 |
| PA14_34030 (Hcp) | 0 | 20 | PA14_00940 (ImpJ) | 0 | 20 |
| PA14_03220 (VgrG) | 4 | 38 | PA14_00960 (VasD) | 0 | 35 |
| PA14_03240 (Hcp) | 0 | 52 | PA14_00970 unknown | 45 | 29 |
| PA14_18985 (VgrG) | 0 | 45 | PA14_00980 (HFA) | 0 | 30 |
| PA14_21450 (VgrG) | 0 | 41 | PA14_00990 (ImpA) | 12 | 20 |
| PA14_29390 (VgrG) | 8 | 45 | PA14_01060 (ImpF) | 14 | 31 |
| PA14_44900 (VgrG) | 5 | 42 | PA14_01070 (ImpG) | 5 | 28 |
| PA14_67230 (VgrG) | 0 | 46 | PA14_01100 (ClpV) | 18 | 21 |
| PA14_69550 (VgrG) | 0 | 40 | PA14_34050 (ImpC) | 8 | 19 |
| PA14_29190 (tse2) | 44 | 27 | PA14_34070 (ImpB) | 22 | 49 |

Table 4. T6SS proteins (T6S system HSI-I, HSI-III) of the *P. aeruginosa* strain PA14.

## 8. Conclusions

In conclusion, structural studies and *in silico* analyses of bacterial genomes have confirmed the occurrence of coiled-coil domains and protein flexibility in the T3SS and provide a more consolidated insight into the occurrence of such features in other secretory systems, e.g. T4SS and T6SS. In the case of T3SS the occurrence of coiled-coils is considerably higher than the average predicted occurrence in prokaryotic proteomes (Schubot et al., 2005). Coiled-coils occur in all types of T3SS proteins, including in proteins from the T3SSs of plant pathogens, for which in earlier studies no coiled-coils could be predicted (Delahay and Frankel, 2002). Apart from coiled-coils, a further widespread feature in T3SS proteins is a considerable structural flexibility which may range from localized to extensive disorder effects. At the level of experimental observations, disorder manifests itself as missing stretches of electron density in crystallographically determined structures (e.g. in the case of the N-termini of effectors), or occasionally as establishment of a molten-globule-like state at conditions resembling the physiological ones. Examples for the latter include the IDPs HrpO (Gazi et al., 2008) and HrpA from *P. syringae* pv. phaseolicola or the needle subunits MxiH, BsaL and PrgI (Barrett et al., 2008). The flexibility of T3SS proteins is frequently associated with a plasticity of coiled-coil domains; this becomes evident in the case of multiple structural studies of the same protein, e.g. AvrPto (PDB ids: 2QKW, 1R5E) or in differences between subunits of oligomeric proteins, e.g. in the FliT dimer (PDB id: 3A7M).

The combination of coiled-coiled interactions and structural plasticity are frequently essential prerequisites for the establishment of interaction networks within T3SS, as exemplified by the interactions of proteins of the HrpO/FliJ/YscO family with members of the HrpE/FliH/YscL family (Gazi et al., 2008), the SipD/PrgI (Lunelli et al., 2011; Rathinavelan et al., 2011) or the CT670/CT671 interaction (Lorenzini et al., 2010). In addition, the assembly of the T3SS supramolecular structures frequently requires a combination of coiled-coils and conformational flexibility: T3SS needle assembly occurs through the stepwise polymerization of a major subunit (e.g. MxiH, BsaL and PrgI) via a flexible or partially disordered C-terminal helix which exhibits a propensity for coiled-coil interactions. For the IDP HrpA polymerization into pilus-like fibrils has been observed, although no experimental evidence for the involvement of coiled-coil interactions could be obtained, despite the high $\alpha$-helical content predicted by sequence analysis.

The propensity for disorder is frequently reflected the amino acid composition of T3SS protein sequences. The vast majority N-terminal sequences of T3SS effectors and other secreted proteins exhibits specific biases (Table 1) in their composition with respect to order- and disorder-promoting residues (Dunker et al., 2002; Uversky, 2010), from which a disorder propensity can be predicted, usually in agreement with experimental observations. Interestingly, these disorder-associated biases (as reflected in the ratio of order- vs. disorder- promoting residues), result in sequence preferences for the N-termini which are similar to those determined for T3SS effectors from various bacterial species (Greenberg and Vinatzer, 2003). The structural disorder of the N-termini may thus play a role as a secretion signal, a suggestion made earlier by Akeda & Galan (2005) and confirmed by subsequent analyses (Gazi et al., 2009). However, as N-terminal structural disorder does not ensure specificity of substrate recognition (e.g. the cytoplasmic $HrcQ_B$ protein is predicted to possesses a highly flexible N-terminus), it may be assumed that N-terminal flexibility could be one of multiple secretion signals (Marlovits et al., 2006), with other signals, e.g. chaperones, ensuring specificity. Analysis of effectors and other secreted/non-secreted T3SS components strongly suggests that the overall disorder of T3SS proteins is a further parameter strongly correlated with secretion (Table 1, 2). Flexible or disordered T3SS domains could facilitate rapid unfolding which is necessary for secretion. Both N-terminal and overall flexibility might be thus considered in prediction algorithms for the identification of universal T3SS effectors signatures; this would complement recent efforts based on machine learning approaches (Arnold et al., 2009; Samudrala et al., 2009). Interestingly, sequence stretches with coiled-coils propensities are suitable tertiary motifs to provide the necessary flexibility which is proposed to be associated with secretion. In fact, coiled-coil proteins are frequently viewed as a specific set of intrinsically disordered proteins (Gaspari & Nyitray, 2011) and occasionally they have been observed to display molten globule characteristics (Glykos & Kokkinidis, 2004). A further advantage of coiled-coils might be associated with specific features of their disordered state: As shown in the case of the HrpO protein (Gazi et al., 2008), proteins exhibiting coiled-coil propensity are capable of adopting highly non-globular conformations, while maintaining a considerable $\alpha$-helical content. The geometrical dimensions of such non-globular helical conformations permit passage through the narrow needle/pilus channel if the appropriate secretion signal is present. It is intuitive to assume that after passing this conduit, such preformed and folding-competent helices encompassing a few turns may form a nucleation site which promotes fast assembly of a globular coiled-coil domain. Flexible coiled-coil domains are thus particularly suitable as secretion substrates as they can easily unfold into secretion-competent α-helices, which in turn may refold in the host cell into a native structure following a relatively fast pathway, and thus avoid degradation of the unfolded polypeptide by host defences. In addition, coiled-coils of effectors may also be a particularly suitable structural motif for interactions in the host cell, as many key processes in the eukaryotic cell involve coiled-coil domains, a fact already noted by Pallen (1997), and confirmed by recent experiments (Knodler et al., 2011). It might be thus hypothetized that the selective evolutionary pressure for optimization of bacterial effectors favours coiled-coil domains and increased flexibility, and this in turn creates a basis for the overall prevalence of coiled-coil domains in T3SS, as this helps establish interaction networks within the T3SS, which may be exploited by even partially unfolded effectors or other secretion substrates.

The predicted high occurrence of coiled-coil domains and structural disorder in T4SS effectors of *L. pneumophila* (Table 3) indicate that the concepts outlined above for T3SS effectors might also some validity in other Gram-negative secretory mechanisms. In addition, the analysis of T6SS secreted components (Table 4) strongly supports the concept of structural flexibility of proteins being an important prerequisite for bacterial secretion. We still have a long way to go to decipher the full complexity of bacterial secretion, even for extensively studied systems such as T3SS. However, the elegant genetic, biochemical, genetic and computational studies which were reviewed in this contribution may open ways to resolve this issue.

## 9. Acknowledgment

Work was partially funded by the GSRT, Joint Research and Technology Programmes for Greece-France (2010-11). Support for access and use of the SOLEIL synchrotron for SAXS studies is acknowledged. V.E.F is supported by a Marie Curie Reintegration grant. Figure 1 was produced with UCSF Chimera, University of California, San Francisco

## 10. References

Abdallah, A.M., Gey van Pittius, N.C., Champion, P.A., Cox, J., Luirink, J., Vandenbroucke-Grauls, C.M., Appelmelk, B.J., Bitter, W. (2007). Type VII secretion--mycobacteria show the way. *Nat Rev Microbiol.*, Vol. 5, No 11, pp: 883-91

Agrain, C., I. Sorg, I., C. Paroz, C. and Cornelis, G. R. (2005). Secretion of YscP from *Yersinia enterocolitica* is essential to control the length of the injectisome needle but not to change the type III secretion substrate specificity. *Mol. Microbiol.*, Vol. 57, No. 5, pp.1415-27

Akeda, Y. and Galan, J.E. (2005) Chaperone release and unfolding of substrates in type III secretion. *Nature*, Vol. 437, No. 7060, pp. 911-15

Alfano, J.R., and Collmer, A. (1997). The type III (Hrp) secretion pathway of plant pathogenic bacteria: trafficking hairpins, Avr proteins and death. *J. Bacteriol.*, Vol. 179, No. 18, pp. 5655-62

Alva, V., Syamala Devi, D. P., & Sowdhamini, R. (2008). COILCHECK: an interactive server for the analysis of interface regions in coiled coils. *Protein Pept Lett*, Vol. 15, No. 1, pp. 33-38

Anderson, D. M., Schneewind, O. (1997). A mRNA signal for the type III secretion of Yop proteins by *Yersinia enterocolitica. Science*. Vo. 278, No. 5340, pp. 1140-1143

Anfinsen, C.B. (1973). Principles that govern the folding of protein chains. *Science*, Vol. 181, No. 96, pp. 223–30

Armstrong, C. T., Vincent, T. L., Green, P. J., & Woolfson, D. N. (2011). SCORER 2.0: an algorithm for distinguishing parallel dimeric and trimeric coiled-coil sequences. *Bioinformatics*, Vol. 27, No. 14, pp. 1908-14

Arnold, R., Brandmaier, S., Kleine, F., Tischler, P., Heinz, E., Behrens, S., Niinikoski, A., Mewes, H-W., Horn, M. & Rattei, T. (2009). Sequence-based prediction of Type III secreted proteins. *PLoS Pathog., Vol.* 5, No. 4:e1000376

Banner, D.W., Kokkinidis, M. & Tsernoglou, D. (1987). Structure of the ColE1 Rop protein at 1.7 Å resolution. *J. Mol. Biol., Vol.* 196, No. 3, pp. 657–75

Barbara, K. E., Willis, K. A., Haley, T. M., Deminoff, S. J. & Santangelo, G. M. (2007). Coiled coil structures and transcription: an analysis of the *S.cerevisiae* coilome. *Mol. Genet.Genomics*, Vol. 278, No. 2, pp. 135–47

Baron, C., Llosa, M., Zhou, S. & Zambryski, P. C. (1997). VirB1, a component of the T-complex transfer machinery of *Agrobacterium tumefaciens*, is processed to a C-terminal secreted product, VirB1. *J. Bacteriol.*, Vol. 179, No. 4, pp. 1203–1210

Barrett, B.S., Picking, W.L., Picking, W.D., Middaugh, C.R. (2008). The response of type three secretion system needle proteins MxiH(Delta5), BsaL(Delta5), and PrgI(Delta5) to temperature and pH. *Proteins*, Vol. 73, No. 3, pp. 632-43

Berger, B., Wilson, D.B., Wolf, E., Tonchev, Th., Milla, M. & Kim, P.S. (1995) Predicting Coiled Coils by Use of Pairwise Residue Correlations. *Proc. National Academy of Science USA*, Vol 92, No. 18, pp. 8259-63

Berger, C., Robin,G. P., Bonas, U. and Koebnik, R. (2010). Membrane topology of conserved components of the type III secretion system from the plant pathogen Xanthomonas campestris pv. vesicatoria. Microbiology, Vol. 156, No. 7, pp. 1963-74

Biemans-Oldehinkel, E., Sal-Man, N., Deng, W., Foster, L.J. and Finlay, B.B. (2011). Quantitative proteomic analysis reveals formation of an EscL-EscQ-EscN type III complex in enteropathogenic *Escherichia coli. J. Bacteriol.*Vol. 193, No. 19, pp. 5514-19

Birtalan, S.C., Phillips, R.M. & Ghosh, P. (2002). Three-dimensional secretion signals in chaperone-effector complexes of bacterial pathogens. *Mol. Cell*, Vol. 9, No. 5, pp. 971–80

Blaylock, B., Riordan, K.E., Missiakas, D.M. & Schneewind, O. (2006). Characterization of the *Yersinia enterocolitica* type III secretion ATPase YscN and its regulator, YscL. *J. Bacteriol.*, Vol. 188, pp 3525-34

Blocker, A. J., Deane, J. E., Veenendaal, A. K. J., Roversi, P., Hodgkinson, J. L., Johnson, S., and Lea, S. M. (2008). What's the point of the type III secretion system needle? *Proc Natl Acad Sci USA*, Vol. 105, No. 18, pp. 6507-13

Bloomer, A.C., Champness, J.N., Bricogne, G., Staden, R. & Klug, A. (1978). Protein disk of tobacco mosiac virus at 2.8Å resolution showing the interactions within and between subunits. *Nature*, Vol. 276, No. 7363, pp. 362-68

Bode, W., Schwager, P. & Huber, R. (1978). The transition of bovine trypsinogen to a trypsin-like state upon strong ligand binding. The refined crystal structures of the bovine trypsinogen-pancreatic trypsin inhibitor complex and of its ternary complex with Ile-Val at 1.9Å resolution. *J. Mol. Biol.*, Vol. 118, No. 1, pp. 99-112

Boyer, F., Fichant, G., Berthod, J., Vandenbrouck, Y. & Attree, I. (2009). Dissecting the bacterial type VI secretion system by a genome wide *in silico* analysis: what can be learned from available microbial genomic resources? *BMC Genomics*, Vol. 10, No, pp 104

Brooks, D.J., Fresco, J.R., Lesk, A.M. & Singh. M. (2002). Evolution of amino acid frequencies in proteins over deep time: inferred order of introduction of amino acids into the genetic code. *Mol. Biol. Evol.*, Vol. 19, No. 10, pp. 1645-55

Brown, P.N., Mathews, M.A., Joss, L.A., Hill, C.P. & Blair, D.F. (2005). Crystal structure of the flagellar rotor protein FliN from *Thermotoga maritime*. *J. Bacteriol.*, Vol. 187, No 8, pp. 2890-2902

Burkhard, P., Stetefeld, J. & Strelkov, S.V. (2001). Coiled coils: a highly versatile protein folding motif. *Trends Cell Biol.*, Vol. 11, No. 2, pp. 82–88

Buttner, D., Gurlebeck, D., Noel, L. D., Bonas, U. (2004). HpaB from *Xanthomonas campestris* pv. vesicatoria acts as an exit control protein in type III dependent protein secretion. *Mol Microbiol.*, Vol. 54, No. 3, pp. 755–68

Buttner, D., Lorenz, C., Weber, E., Bonas, U. (2006). Targeting of two effector protein classes to the type III secretion system by a HpaC- and HpaBdependent protein complex from Xanthomonas campestris pv. vesicatoria. *Mol Microbiol.*, Vol. 59, No. 2, pp. 513–27

Buttner, C.R., Sorg, I., Cornelis, G.R., Heinz, D.W. & Niemann, H.H. (2008). Structure of the *Yersinia enterocolitica* type III secretion translocator chaperone SycD. *J. Mol. Biol.*, Vol. 375, No. 4, pp. 997-1012

Buttner, D. & He, S-Y. (2009). Type III Protein Secretion in Plant Pathogenic Bacteria. *Plant Physiology*, Vol. 150, No. 4, pp. 1656–64

Cao, T.B. & Saier, M.H. Jr. (2003). The general protein secretory pathway: phylogenetic analyses leading to evolutionary conclusions. *Biochim Biophys Acta*, Vol. 1609, No. 1, pp. 115-25

Cascales, E. (2008). The type VI secretion toolkit. *EMBO Reports*, Vol. 9, No 8, pp 735-41

Christie, P. J. & Vogel, J. P. (2000). Bacterial type IV secretion: conjugation systems adapted to deliver effector molecules to host cells. *Trends Microbiol.*, Vol. 8, No. 8, pp. 354-60

Cianciotto, N.P. (2005). Type II secretion: a protein secretion system for all seasons. *Trends Microbiol.*, Vol. 13, No. 12, pp. 581-88

Cordes, F.S., Komoriya, K., Larquet, E., Yang, S., Egelman, E.H., Blocker, A., Lea, S.M. (2003). Helical structure of the needle of the type III secretion system of *Shigella flexneri*. *J Biol Chem.*, Vol. 278, No. 19, pp. 17103-107

Cornelis, G. R. (2006). The type III secretion injectisome. *Nature Reviews. Microbiology*, Vol. 4, No.11, pp. 811-25

Cortese, M. S., Uversky, V. N. & Dunker, A.K. (2008). Intrinsic disorder in scaffold proteins: getting more from less. *Progress Biophys. Mol. Biol.*, vol. 98, no. 1, pp. 85–106

Crick, F.H. (1952). Is alpha-keratin a coiled coil? *Nature* Vol. 170, No. 4334, pp. 882–83

Dale, C. & Moran, N.A. (2006). Molecular interactions between bacterial symbionts and their hosts. *Cell*, Vol. 126, No. 3, pp. 453-65

Davis, J., Wang, J., Tropea, J. E., Zhang, D., Dauter, Z., Waugh, D. S., Wlodawer, A. (2008). Novel fold of VirA, a type III secretion system effector protein from *Shigella flexneri*. *Protein Sci.*, Vol. 17, No. 12, pp. 2167-73

Dean, P. (2011). Functional domains andmotifs of bacterial type III effector proteins and their roles in infection. *FEMS Microbiol Rev* pp. 1–26

Deane, J.E., Roversi, P., Cordes, F.S., Johnson, S., Kenjale, R., Daniell, S., Booy, F., Picking, W.D., Picking, W.L., Blocker, A.J. & Lea, S.M. (2006). Molecular model of a type III secretion system needle: Implications for host-cell sensing. *Proc Natl Acad Sci USA*, Vol. 103, No. 33, pp. 12529–33

Deane, J.E., Roversi, P., King, C., Johnson, S., Lea, S.M. (2008). Structures of the *Shigella flexneri* type 3 secretion system protein Mxic reveal conformational variability amongst homologues. *J.Mol.Biol.*, Vol. 377, No. 4, pp. 985-92

Delahay, R.M. & Frankel, G. (2002) Coiled-coil proteins associated with type III secretion systems: a versatile domain revisited. *Mol Microbiol.*, Vol. 45, No. 4, pp. 905-16

Derewenda, U., Mateja, A., Devedjiev, Y., Routzahn, K.M., Evdokimov, A.G., Derewenda, Z.S. & Waugh, D.S. (2004) The structure of *Yersinia pestis* V-antigen, an essential virulence factor and mediator of immunity against plague. *Structure*, Vol. 12, No. 2, pp. 301-306

Desvaux, M., Parham, N.J. & Henderson, I.R. (2004). Type V protein secretion: simplicity gone awry? *Curr. Issues Mol. Biol.*, Vol. 6, No. 2, pp. 111–24

Desveaux, D., Singer, A.U. & Dangl, J.L. (2006). Type III effector proteins: doppelgangers of bacterial virulence. *Current Opinion in Plant Biology*, Vol. 9, No. 4, 376–82

Diepold, A., Amstutz, M., Abel, S., Sorg, I., Jenal, U., & Cornelis, G. R. (2010). Deciphering the assembly of the *Yersinia* type III secretion injectisome. *Embo J.*, Vol. 29, No. 11, pp. 1928-40

Dong, J., Xiao, F., Fan, F., Gu, L., Cang, H., Martin, G. B., Chai. (2009). Crystal structure of the complex between *Pseudomonas* effector AvrPtoB and the tomato Pto kinase reveals both a shared and a unique interface compared with AvrPto-Pto. *Plant Cell*, Vol. 21, No. 6, pp. 1846-59

Dosztanyi, Z. , Csizmok, V., Tompa, P. & Simon, I. (2005). IUPred: web server for the prediction of intrinsically unstructured regions of proteins based on estimated energy content. *Bioinformatics*, Vol. 21, No. 16, pp. 3433–34

Dunker, A.K., Brown, C.J., Lawson, J.D., Iakoucheva, L.M. & Obradovic, Z. (2002). Intrinsic disorder and protein function. *Biochemistry*, Vol. 41, No. 21: 6573-82

Dunker, A.K., Oldfield, C.J., Meng, J., Romero, P., Yang, J.Y., Chen, J.W., Vacic, V., Obradovic, Z. & Uversky, V.N. (2008a). The unfoldomics decade: an update on intrinsically disordered proteins. *BMC Genomics*, Vol. 9, suppl. 2, article S1

Dunker, A.K., Silman, I., Uversky, V.N. & Sussman, J. L. (2008b). Function and structure of inherently disordered proteins. *Current Opinion in Structural Biology*, Vol. 18, No. 6, pp. 756–64

Dunker, A.K. & Uversky, V.N. (2008). Signal transduction via unstructured protein conduits. *Nature Chemical Biology*, Vol. 4, No. 4, pp. 229–230

Durand, F., Dagkessamanskaia, A., Martin-Yken, H., Graille, M., Van Tilbeurgh, H., Uversky, V.N. & Francois, J.M. (2008). Structure-function analysis of Knr4/Smi1, a newly member of intrinsically disordered proteins family, indispensable in the absence of a functional PKC1-SLT2 pathway in Saccharomyces cerevisiae. *Yeast*, Vol. 25, No. 8, pp. 563–76

Eliezer, D. (2007). Characterizing residual structure in disordered protein States using nuclear magnetic resonance. *Methods Mol. Biol.*, Vol. 350 pp. 49–67

Espina, M., Ausar, F., Middaugh, C.R., Baxter, M.A., Picking, W.D. & Picking, W.L. (2007) Conformational stability and differential structural analysis of LcrV, PcrV, BipD and SipD from type III secretion systems. *Protein Sci., Vol.* 16, No. 4, pp. 704-714

Evans, L.D., Stafford, G.P., Ahmed, S., Fraser, G.M. & Hughes, C. (2006). An escort mechanism for cycling of export chaperones during flagellum assembly. *Proc. Natl. Acad. Sci. USA*, Vol. 103, No. 46, pp. 17474-79

Evans, L.D. & Hughes, C. (2009). Selective binding of virulence type III export chaperones by FliJ escort orthologues InvI and YscO. *FEMS Microbiol. Lett.*, Vol. 293, No. 2, pp. 292-97

Fadouloglou, V.E., Tampakaki, A.P., Glykos, N.M., Bastaki, M.N., Hadden, J.M, Phillips, S.E., Panopoulos, N.J. & Kokkinidis, M. (2004). Structure of HrcQ$_B$-C, a conserved component of the bacterial type III secretion systems. *Proc. Natl. Acad. Sci. USA, Vol.* 101, No. 1, pp. 70-75

Fadouloglou, V.E., Bastaki, M.N., Ashcroft, A.E., Phillips, S.E., Panopoulos, N.J., Glykos, N.M. & Kokkinidis, M. (2009). On the quaternary association of the type III secretion system HrcQ$_B$-C protein: experimental evidence differentiates among the various oligomerization models. *J. Struct. Biol., Vol.* 166, No. 2, pp. 214-25

Filloux, A. (2004). The underlying mechanisms of type II protein secretion. *Biochimica et Biophysica Acta*, Vol. 1694, No. 1-3, pp. 163-79

Fischetti, V. A., Landau, G. M., Schmidt, J. P., & Sellers, P. (1993). Identifying periodic occurences of a template with applications to protein structure. *Inform Process Let*, Vol. 45, No. 1993, pp. 11-18

Galan, J. E., Wolf-Watz, H. (2006). Protein delivery into eukaryotic cells by type III secretion machines. *Nature*, Vol. 444, No. 7119, pp. 567-73

Galzitskaya, O. V., Garbuzynskiy, S. O. & Lobanov, M. Y. (2006). FoldUnfold: web server for the prediction of disordered regions in protein chain. *Bioinformatics*, Vol. 22, No. 23, pp. 2948-49

Gaspari, Z. & Nyitray, L. (2011). Coiled-coils as possible models of protein structure evolution. *Biomol. Conc.*, Vol. 2, No. 3, pp. 199-210

Gauthier, A., Thomas, N.A. & Finlay, B.B. (2003). Bacterial injection machines. *J. Biol. Chem.*, Vol. 278, No. 28, pp. 25273-76

Gazi, A.D., Bastaki, M., Charova, S.N., Gkougkoulia, E.A., Kapellios, E.A., Panopoulos, N.J. & Kokkinidis, M. (2008). Evidence for a widespread interaction mode of disordered proteins in bacterial type III secretion systems *J. Biol. Chem., Vol.* 283, No. 49, pp. 34062-68

Gazi, A., Charova, S.N., Panopoulos, N.J. & Kokkinidis, M. (2009). Coiled-coils in type III secretion systems: structural flexibility, disorder and biological implications. *Cell. Microbiol., Vol.* 11, No. 5, pp. 719–29

Glykos, N.M., Cesareni, G. & Kokkinidis, M. (1999). Protein plasticity to the extreme: changing the topology of a 4-α-helical bundle with a single amino acid substitution. *Structure,* Vol. 7, No. 6, pp. 597-603

Glykos, N.M. & Kokkinidis, M. (2004). Structural polymorphism of a marginally stable 4-alpha-helical bundle. Images of a trapped molten globule? *Proteins,* Vol. 56, No. 3, pp. 420-25

Glykos, N.M., Papanikolau, Y., Vlassi, M., Kotsifaki, D., Cesareni, G. & Kokkinidis, M. (2006). Loopless Rop: Structure and dynamics of an engineered homotetrameric variant of the repressor of primer protein. *Biochemistry,* Vol. 45, No. 36, pp. 10905-19

Goddard, T. D., Huang, C. C., & Ferrin, T. E. (2007). Visualizing density maps with UCSF Chimera. *J Struct Biol,* Vol. 157, No. 1, pp. 281-287

Gonzalez-Pedrajo, B., Minamino, T., Kihara, M., Namba, K. (2006). Interactions between C ring proteins and export apparatus components: a possible mechanism for facilitating type III protein export. *Mol Microbiol.,* Vol. 60, No. 4, pp. 984-98

Greenberg, J.T. & Vinatzer, B.A. (2003). Identifyng type III efectors of plant pathogens and analysing their interaction with plant cell. *Curr. Opin. Microb.,* Vol. 6, No. 1, pp. 20-28

Grigoryan, G. & Keating, A. E. (2008). Structural Specificity in Coiled-coil Interactions. *Curr. Opin. Struct. Biol.,* Vol. 18, No. 4, pp. 477-483

Gsponer, J., Futschik, M.E., Teichmann, S.A. & Babu, M.M. (2008). Tight regulation of unstructured proteins: from transcript synthesis to protein degradation. *Science,* Vol. 322, No. 5906, pp. 1365–68

Hachani, A., Biskri, L., Rossi, G., Marty, A., Menard, R., Sansonetti, P., Parsot, C., Van Nhieu, G.T., Bernardini, M.L. & Allaoui, A. (2008). IpgB1 and IpgB2, two homologous effectors secreted via the Mxi-Spa type III secretion apparatus, cooperate to mediate polarized cell invasion and inflammatory potential of *Shigella flexenri. Microbes Infect.,* Vol. 10, No. 3, pp. 260-68

He, S.Y. & Jin, Q. (2003). The Hrp pilus: learning from flagella. *Current Opinion in* Microbiology, Vol. 6, No. 1, pp. 15-19

Henderson, I.R., Navarro-Garcia, F., Desvaux, M., Fernandez, R.C. & Ala'Aldeen, D. (2004) Type V protein secretion pathway: the autotransporter story. *Microbiol. Mol. Biol. Rev., Vol.* 68, No. 4, pp. 692-744

Hienonen, E., Romantschuk, M. & Taira, S. (2006). Stabilization of heterologous transcripts with hrpA, mRNA of a type III secretion system component. *Microbial Cell Factories,* Vol. 5 (Suppl 1), P72

Holland, I.B., Schmitt, L. & Young J. (2005). Type 1 protein secretion in bacteria, the ABC-transporter dependent pathway. *Mol. Membr. Biol.,* Vol. 22, No. 1–2, pp. 29-39

Hubber, A., Roy, C.R. (2010). Modulation of host cell function by *Legionella pneumophila* type IV effectors. *Ann. Rev. Cell. Dev. Biol.* Vol. 26, pp. 261-83

Hueck, C.J. (1998). Type III protein secretion systems in bacterial pathogens of animals and plants. *Microbiol. Mol. Biol. Rev.*, Vol. 62, No. 2, pp. 379-433

Ibuki, T., Shimada, M., Minamino, T., Namba, K. & Imada, K. (2009). Crystallization and preliminary X-ray analysis of FliJ, a cytoplasmic component of the flagellar type III protein-export apparatus from Salmonella sp. *Acta. Crystallogr. Sect. F Struct. Biol. Cryst. Commun.*, Vol. 65, No. 1, pp. 47-50

Ibuki, T., Imada, K., Minamino, T., Kato, T., Miyata, T. & Namba, K. (2011). Common architecture of the flagellar type III protein export apparatus and F- and V-type ATPases. *Nat. Struct. Mol. Biol.*, Vol. 18, No. 3, pp. 277-82

Imada, K., Minamino, T., Tahara, A. & Namba, K. (2007). Structural similarity between the flagellar type III ATPase FliI and F1-ATPase subunits. *Proc. Natl. Acad. Sci. USA* Vol 104, No2, pp 485-90

Imada, K., Minamino, T., Kinoshita, M., Furukawa, Y. & Namba, K. (2010). Structural insight into the regulatory mechanisms of interactions of the flagellar type III chaperone FliT with its binding partners. *Proc.Natl.Acad. Sci. USA*, Vol. 107, No. 19, pp. 8812-17

Ishida, T. & Kinoshita, K. (2007). PrDOS: prediction of disordered protein regions from amino acid sequence. *Nucl. Acids Res.*, Vol. 35, Web server issue, pp. W460–W464

Izore, T., Perdu, C., Job, V., Attree, I., Faudry, E., & Dessen, A. (2011). Structural characterization and membrane localization of ExsB from the type III secretion system (T3SS) of *Pseudomonas aeruginosa*. *J. Mol. Biol.*, Vol. 413, No. 1, pp. 236-46

Jakoby, M., Weisshaar, B., Droge-Laser, W., Vincente-Carbajosa, J,. Tiedemann, J., Kroj, T., Parcy, F. (2002). bZIP transcription factors in *Arabidopsis. Trends Plant Sci.*, Vol. 7, No. 3, pp. 106–11

Jehl, M.A., Arnold, R. & Rattei, T. (2011). Effective--a database of predicted secreted bacterial proteins. *Nucl. Acids Res.*, Vol. 39, Database issue, pp. D591-95

Jin, Q., Thilmony, R., Zwiesler-Vollick, J. & He, S.-Y. (2003). Type III protein secretion in *Pseydomonas syringae. Microbes Infect.*, Vol. 5, No. 4, pp. 301-10

Job, V., Matteï, P-J., Lemaire, D., Attree, I. & Dessen, A. (2010). Structural Basis of Chaperone Recognition of Type III Secretion System Minor Translocator Proteins *J. Biol. Chem.*, Vol. 285, No. 30, pp. 23224-23232

Jones, D.T. (1999) Protein secondary structure prediction based on position-specific scoring matrices. *J. Mol. Biol.*, Vol. 292, No. 2, pp. 195-202

Kanehisa, M. & Goto, S. (2000). KEGG: Kyoto encyclopedia of genes and genomes. *Nucl. Acids Res.*, Vol. 28, No. 1, pp. 27-30

Kelly, S.M. & Price, N.C. (1997). The application of circular dichroism to studies of protein folding and unfolding. *Biochim. Biophys. Acta*, Vol. 1338, No. 2, pp. 161–85

Khandelwal, P., Kellikuli, K., Smith, C.L., Saper, M.A., Zuiderweg, E. R.P. (2002). Solution structure and phopsphopeptide binding to the N-terminal domain of *Yersinia* YopH: comparison with a crystal structure. *Biochemistry*, Vol. 41, No. 38, pp. 11425-437

Knodler, L.A., Ibarra, J.A., Perez-Rueda, E., Yip, C.K., & Steele-Mortimer, O. (2011). Coiled-coil domains enhance the membrane association of *Salmonella* type III effectors. *Cell. Microbiol.*, Vol. 13, No. 10, pp. 1497-1517

Koebnik, R. (2001). The role of bacterial pili in protein and DNA translocation. *Trends Microbiol.*, Vol. 9, No. 12, pp. 586-90

Koraimann, G. (2003). Lytic transglycosylases in macromolecular transport systems of Gram-negative bacteria. *Cell Mol. Life Sci.*, Vol. 60, No. 11, pp. 2371–88

Korotkov, K. V., Gonen, T., & Hol, W. G. (2011). Secretins: dynamic channels for protein transport across membranes. *Trends Biochem Sci*, Vol. 36, No. 8, pp. 433-43

Lane, M,C, O'Toole, P.W. & Moore, S.A. (2006). Molecular basis of the interaction between the flagellar export proteins FliI and FliH from Helicobacter pylori. *J. Biol. Chem.*, Vol. 281, No. 1, pp. 508-17

Lara-Tejero, M., Kato, J., Wagner, S., Liu, X., & Galan, J. E. (2011). A sorting platform determines the order of protein secretion in bacterial type III systems. *Science*, Vol. 331, No. 6021, pp. 1188-91

Lee, C.C., Wood, M.D., Ng, K., Luginbuhl, P., Spraggon, G., Katagiri, F. (2004). Crystal Structure of the Type III Effector AvrB from *Pseudomonas syringae*. *Structure*, Vol. 12, No. 3, pp. 487-94

Lilic, M., Vujanac, M., Stebbins, C.E. (2006). A common structural motif in the binding of virulence factors to bacterial secretion chaperones. *Mol.Cell*, Vol. 21, No. 5, pp. 653-64

Linding, R., Jensen, L. J., Diella, F., Bork, P., Gibson, T. J. & Russell, R. B. (2003). Protein disorder prediction: implications for structural proteomics. *Structure*, Vol. 11, No. 11, pp. 1453– 1459

Liu, J. & Rost, B. (2001). Comparing function and structure between entire proteomes. *Protein Sci.*, Vol. 10, No. 10, pp. 1970-79

Lloyd, S. A., Norman, M., Rosqvist, R., and Wolf-Watz, H. (2001). *Yersinia* YopE is targeted for type III secretion by N-terminal, not mRNA, signals. *Mol. Microbiol.*, Vol. 39, No. 2, pp. 520–31

Lorenzini, E., Singer, A., Singh, B., Lam, R., Skarina, T., Chirgadze, N.Y., Savchenko, A. & Gupta, R.S. (2010). Structure and protein-protein interaction studies on *Chlamydia trachomatis* protein CT670 (YscO Homolog). *J. Bacteriol.*, Vol. 192, No. 11, pp. 2746-56

Lountos, G.T., Austin, B.P., Nallamsetty, S., Waugh, D.S. (2009). Atomic resolution structure of the cytoplasmic domain of *Yersinia pestis* YscU, a regulatory switch involved in type III secretion. *Protein Sci.*, Vol. 18, No. 2, pp. 467-74

Löwer, M. & Schneider, G. (2009). Prediction of Type III Secretion Signals in Genomes of Gram-Negative Bacteria. *PLoS ONE, Vol.* 4, No. 6: e5917.doi:10.1371/journal.pone.0005917

Lunelli, M., Lokareddy, R.M., Zychlinsky, A. & Kolbe M. (2009). IpaB-IpgC interaction defines binding motif for type III secretion translocator. *Proc. Natl. Acad. Sci. USA, Vol.* 106, No. 24 pp. 9661-66

Lunelli M, Hurwitz R, Lambers J, Kolbe M. (2011). Crystal Structure of PrgI-SipD: Insight into a Secretion Competent State of the Type Three Secretion System Needle Tip and its Interaction with Host Ligands. *PLoS Pathog.*, Vol. 7, No. 8: e1002163

Lupas, A. (1996). Coiled coils: new structures and new functions. *TRENDS in Biochemical Sciences*, Vol. 21, No. 10, pp. 375-82

Lupas, A., Van Dyke, M. & Stock, J. (1991). Predicting coiled coils from protein sequences. *Science*, Vol. 252, No. 5010, pp. 1162-64

Marie, C., Broughton, W.J. and Deakin, W.J. (2001) *Rhizobium* type III secretion systems: legume charmers or alarmers? *Curr Opin Plant Biol.*, Vol. 4, No. 4, pp. 336-42

Marlovits, T. C., Kubori, T., Sukhan, A., Thomas, D. R., Galan, J. E., Unger, V. M. (2004). Structural insights into the assembly of the type III secretion needle complex. *Science*, Vol. 306, No. 5698, pp. 1040-42

Marlovits, T. C., Kubori, T., Lara-Tejero, M., Thomas, D., Unger, V. M., Galan, J. E. (2006). Assembly of the inner rod determines needle length in the type III secretion injectisome. *Nature*, Vol. 441, No. 7093, pp. 637-40

McDermott, J. E., Corrigan, A., Peterson, E., Oehmen, C., Niemann, G., Cambronne, E. D., Sharp, D., Adkins, J. N., Samudrala, R., Heffron, F. (2011). Computational prediction of type III and IV secreted effectors in gram-negative bacteria. *Infect Immun.*, Vol. 79, No. 1, pp. 23-32

McDonnell, A. V., Jiang, T., Keating, A. E., & Berger, B. (2006). Paircoil2: improved prediction of coiled coils from sequence. *Bioinformatics*, Vol. 22, No. 3, pp. 356-58

McDowell, M.A., Johnson, S., Deane, J.E., Cheung, M. Roehrich, D. A., Blocker, A.J., McDonnell, J. M. & Lea, S. M. (2011). Structural and Functional Studies on the N-terminal Domain of the *Shigella* Type III Secretion Protein MxiG. *J. Biol. Chem.*, Vol. 286, No. 35, pp. 30606-14

McMurry, J.L., Murphy, J.W. & Gonzalez-Pedrajo, B. (2006). The FliN-FliH interaction mediates localization of flagellar export ATPase FliI to the C ring complex. *Biochemistry* Vol 45, pp. 11790-98

Minamino, T., Yoshimura, S. D., Morimoto, Y. V., Gonzalez-Pedrajo, B., Kami-Ike, N., Namba, K. (2009). Roles of the extreme N-terminal region of FliH for efficient localization of the FliH-FliI complex to the bacterial flagellar type III export apparatus. *Mol Microbiol.*, Vol. 74, No. 6, pp. 1471-83

Mougous, J. D., Cuff, M. E., Raunser, S., Shen, A., Zhou, M., Gifford, C. A., Goodman, A. L., Joachimiak, G., Ordonez, C. L., Lory, S., Walz, T., Joachimiak, A. & Mekalanos, J. J. (2006). A virulence locus of *Pseudomonas aeruginosa* encodes a protein secretion apparatus. *Science*, Vol. 312, No 5779, pp 1526-30

Newman, J. R. S., Wolf, E. & Kim, P. S. (2000). A computationally directed screen identifying interacting coiled coils from *Saccharomyces cerevisiae*. *Proc. Natl. Acad. Sci. USA*, Vol. 97, No. 24, pp. 13203-08

Namba, K. (2001). Roles of partly unfolded conformations in macromolecular self-assembly. *Genes to Cells*, Vol. 6, No. 1, pp. 1-12

Neyt, C. & Cornelis, G.R. (1999). Role of SycD, the chaperone of the *Yersinia* Yop translocators YopB and YopD. *Mol. Microbiol.*, Vol. 31, No. 1, pp. 143-156

Okon, M., Moraes, T. F., Lario, P. I., Creagh, A. L., Haynes, C. A., Strynadka, N. C. & MacIntosh, L.P. (2008). Structural characterization of the type-III pilot-secretin complex from *Shigella flexneri*. *Structure*, Vol. 16, No. 10, pp. 1544-54

Page, A. & Parsot, C. (2002). Chaperones of the type III secretion pathway: jacks of all trades. *Mol. Microbiol.*, Vol. 46, No. 1, pp. 1-11

Paliakasis, C.D. & Kokkinidis, M. (1992). Relationships between sequence and structure for the four-alpha-helix bundle tertiary motif in proteins. *Prot Engineering*, Vol. 5, No. 8, pp. 739-49

Pallen, M.J., Dougan, G. & Frankel, G. (1997). Coiled-coil domains in proteins secreted by type III secretion systems. *Mol. Microbiol.*, Vol. 25, No. 2, pp. 423-25

Pallen, M.J., Francis, M.S. & Futterer, K. (2003). Tetratricopeptide-like repeats in type-III-secretion chaperones and regulators. *FEMS Microb.. Lett., Vol.* 223, No. 1, pp. 53–60

Pallen, M.J., Bailey, C.M. & Beatson, S.A. (2006). Evolutionary links between FliH/YscL-like proteins from bacterial type III secretion systems and second-stalk components of the FoF1 and vacuolar ATPases. *Protein Sci.*, Vol. 15, No. 4, pp. 935-41

Park, S. Y., Lowder, B., Bilwes, A. M., Blair, D. F. & Crane, B. R. (2006). Structure of FliM provides insight into assembly of the switch complex in the bacterial flagella motor. *Proc. Natl. Acad. Sci. U S A, Vol.* 103, No. 32, pp. 11886–91

Parsot, C., Hamiaux, C., Page, A. L. (2003). The various and varying roles of specific chaperones in type III secretion systems. *Curr. Opin. Microb.*, Vol. 6, No. 1, pp. 7-14

Parsot, C., Ageron, E., Penno, C., Mavris, M., Jamoussi, K., d'Hauteville, H., Sansonetti, P. & Demers, B. (2005). A secreted anti-activator, OspD1, and its chaperone, Spa15, are involved in the control of transcription by the type III secretion apparatus activity in *Shigella flexneri. Mol Microbiol,* Vol. 56, No. 6, pp. 1627-35

Paul, K. & Blair, D.F. (2006). Organization of FliN subunits in the flagellar motor of Escherichia coli. *J. Bacteriol.*, Vol. 188, No. 7, pp. 2502-11

Paul, K., Gonzalez-Bonet, G., Bilwes, A.M., Crane, B.R. & Blair, D. (2011). Architecture of the flagellar rotor. *EMBO J.*, Vol. 30, No. 14, pp. 2962-71

Peng, K., Vucetic, S., Radivojac, P., Brown, C. J., Dunker, A. K. & Obradovi´c, Z. (2005). Optimizing long intrinsic disorder predictors with protein evolutionary information. *J. Bioinf. Comp. Biol.*Vol. 3, No. 1, pp. 35–60

Pentony, M. M. & Jones, D. T. (2010). Modularity of intrinsic disorder in the human proteome. *Proteins, Vol.* 782010, No. 178, pp. 212–221

Prehna, G., Ivanov, M.I., Bliska, J.B., Stebbins, C.E. (2006). *Yersinia* virulence depends on mimicry of host rho-family nucleotide dissociation inhibitors. *Cell,* Vol. 126, No. 5, pp. 869-80

Prilusky, J., Felder, C.E., Zeev-Ben Mordehai, T., Rydberg, E.H., Man, O., Beckmann, J.S., Silman, I. & Sussman J.L. (2005) FoldIndex©: a simple tool to predict whether a given protein sequence is intrinsically unfolded. *Bioinf.*, Vol. 21, No. 16, pp. 3435-38

Priyadarshi, A. & Tang, L. (2010). Crystallization and preliminary crystallographic analysis of the type III secretion translocator chaperone SicA from *Salmonella enterica. Acta Cryst. F, Vol.* 66, No. 11, pp. 1533-35

Quinaud, M., Ple, S., Job, V., Contreras-Martel, C., Simorre, J.P., Attree, I. & Dessen, A. (2007). Structure of the heterotrimeric complex that regulates type III secretion needle formation. *Proc. Natl. Acad. Sci. U S A*, Vol. 104, No. 19, pp. 7803–08

Rackham, O. J., Madera, M., Armstrong, C. T., Vincent, T. L., Woolfson, D. N., & Gough, J. (2010). The evolution and structure prediction of coiled coils across all genomes. *J Mol Biol*, Vol. 403, No. 3, pp. 480-93

Ramamurthi, K. S., and Schneewind, O. (2002). *Yersinia enterocolitica* type III secretion: mutational analysis of the yopQ secretion signal. *J. Bacteriol., Vol.* 184, No. 12,pp. 3321–28

Rathinavelan, T., Tang, Ch. & De Guzman, R.N. (2011). Characterization of the Interaction between the *Salmonella* Type III Secretion System Tip Protein SipD and the Needle Protein PrgI by Paramagnetic Relaxation Enhancement. *J. Biol. Chem.*, Vol. 286, No. 6, pp. 4922-30

Roine, E., Wei, W., Yuan, J., Nurmiaho-Lassila, E. L., Kalkkinen, N., Romantschuk, M., He, S-Y. (1997). Hrp pilus: an *hrp*-dependent bacterial surface appendage produced by Pseudomonas syringae pv. Tomato DC3000. *Proc Natl Acad Sci USA, Vol.* 94, No. 7, pp. 3459-64

Romero, P., Obradovi´c, Z., Li, X., Garner, E. C., Brown, C. J. & Dunker, A. K. (2001). Sequence complexity of disordered protein. *Proteins*, Vol. 42, No. 1, pp. 38–48

Rose, A., Manikantan, S., Schraegle, S. J., Maloy, M. A., Stahlberg, E. A. & Meier, I. (2004). Genome-wide identification of *Arabidopsis* coiled-coil proteins and establishment of the ARABI-COIL database. *Plant Physiol.*, Vol. 134, No. 3, pp. 927–39.

Samudrala, R., Heffron, F., McDermott, J. E. (2009). Accurate prediction of secreted substrates and identification of a conserved putative secretion signal for type III secretion systems. *PLoS Pathog., Vol.* 5, No. 4:e1000375

Sarris, P. F. & Scoulica, E. V. (2011). *Pseudomonas entomophila* and *Pseudomonas mendocina*: Potential models for studying the bacterial type VI secretion system. *Infection, Genetics and Evolution*, Vol. 11, No 6, pp 1352-60

Sarris, P. F., Zoumadakis, C., Panopoulos, N. J. & Scoulica, E. (2011). Distribution of the putative type VI secretion system core genes in *Klebsiella* spp. *Infection Genetics and Evolution*, Vol. 11, No 1, pp 157–166

Sato, H., Hunt, M.L., Weiner, J.J., Hansen, A.T. & Frank, D.W. (2011). Modified needle-tip PcrV proteins reveal distinct phenotypes relevant to the control of type III secretion and intoxication by *Pseudomonas aeruginosa*. *PLoS One*, Vol 6, No. 3:e18356

Sato, H. & Frank, D.W. (2011). Multi-functional characteristics of the *Pseudomonas aeruginosa* type III needle-tip protein, PcrV; comparison to orthologs in other Gram-negative bacteria. *Front Microbiol.*, Vol. 2:142

Sarkar, M.K., Paul, K. & Blair, D.F. (2010). Subunit organization and reversal-associated movements in the flagellar switch of E. coli. *J Biol Chem.*, Vol. 285, No. 1, pp. 675-84

Schraidt, O. & Marlovits, T. C. (2011). Three-dimensional model of Salmonella's needle complex at subnanometer resolution. *Science*, Vol. 331, No. 6021, pp. 1192-95

Schubot, F.D., Jackson, M.W., Penrose, K.J., Cherry, S., Tropea, J.E., Plano, G.V., Waugh, D.S. (2005). Three-dimensional structure of a macromolecular assembly that regulates type III secretion in *Yersinia pestis*. *J. Mol.Biol.*, Vol 346, No. 4, pp. 1147-61

Singer, A.U., Desveaux, D., Betts, L., Chang, J.H., Nimchuk, Z., Grant, S.R., Dangl, J.L. & Sondek, J. (2004). Crystal structures of the type III effector protein AvrPphF and its chaperone reveal residues required for plant pathogenesis. *Structure*, Vol. 12, No. 9, pp. 1669-81

Singer, A. U., Rohde, J. R., Lam, R., Skarina, T., Kagan, O., Dileo, R., Chirgadze, N. Y., Cuff, M. E., Joachimiak, A., Tyers, M., Sansonetti, P. J., Parsot, C., Savchenko, A. (2008). Structure of the *Shigella* T3SS effector IpaH defines a new class of E3 ubiquitin ligases. *Nat Struct Mol Biol.*, Vol. 15, No. 12, pp. 1293-1301

Spreter, T., Yip, C. K., Sanowar, S., Andre, I., Kimbrough, T. G., Vuckovic, M., Pfuetzner, R. A., Deng, W., Yu, A. C., Finlay, B. B., Baker, D., Miller, S. I., and Strynadka, N. C. (2009). A conserved structural motif mediates formation of the periplasmic rings in the type III secretion system. *Nat. Struct. Mol. Biol., Vol.* 16, pp. 468-76

Stebbins, C.E., Galan, J.E. (2000). Modulation of host signaling by a bacterial mimic: structure of the Salmonella effector SptP bound to Rac1. *Mol.Cell*, Vol. 6, No. 6, pp. 1449-60

Stebbins, C.E. (2005) Structural microbiology at the pathogen–host interface. *Cell. Microbiol.*, Vol. 7, No. 9, pp. 1227–36

Sugase, K., Dyson, H.J. & Wright, P.E. (2007) Mechanism of coupled folding and binding of an intrinsically disordered protein. *Nature*, Vol. 447 , No. 7356, pp. 1021-25

Sun, P., Austin, B.P., Tropea, J.E. & Waugh, D.S. (2008). Structural characterization of the *Yersinia pestis* type III secretion system needle protein YscF in complex with its heterodimeric chaperone YscE/YscG. *J.Mol.Biol., Vol.* 377, No. 3, pp. 819-30

Tampakaki, A. P., Fadouloglou, V. E., Gazi, A. D., Panopoulos, N. J. & Kokkinidis, M. (2004). Conserved features of type III secretion. *Cell. Microbiol.*, Vol. 6, No. 9, pp. 805-16

Testa, O. D., Moutevelis, E., & Woolfson, D. N. (2009). CC+: a relational database of coiled-coil structures. *Nucleic Acids Res*, Vol. 37, Database issue, pp. D315-322

Thomas, D. R., Francis, N. R., Xu, C., & DeRosier, D. J. (2006). The three-dimensional structure of the flagellar rotor from a clockwise-locked mutant of Salmonella enterica serovar Typhimurium. *J. Bacteriol.*, Vol. 188, No. 20, pp. 7039-48

Thomassin, J.L., He, X. & Thomas, N.A. (2011). Role of EscU auto-cleavage in promoting type III effector translocation into host cells by Enteropathogenic Escherichia coli. *BMC Microbiol.*, Vol. 11, No. 205.Troisfontaines, P. & Cornelis, G. R. (2005). Type III secretion: more systems than you think. *Physiol.*, Vol. 20, No. 5, pp. 326-39

Trigg, J., Gutwin, K., Keating, A. E., & Berger, B. (2011). Multicoil2: predicting coiled coils and their oligomerization States from sequence in the twilight zone. *PLoS One*, Vol. 6, No. 8, pp. e23519

Troisfontaines, P. & Cornelis, G. R. (2005). Type III secretion: more systems than you think. *Physiol.*, Vol. 20, No. 5, pp. 326-339Trost, B. & Moore, S.A. (2009). Statistical characterization of the GxxxG glycine repeats in the flagellar biosynthesis protein FliH and its Type III secretion homologue YscL. *BMC Microbiol.*, Vol. 9, No. 1, p. 72

Tseng, T-T.; Tyler, B.M. & Setubal, J.C. (2009). Protein secretion systems in bacterial-host associations, and their description in the Gene Ontology. *BMC Microbiology*, Vol. 9, Suppl 1, article S2

Uversky, V.N., Gillespie, J.R. & Fink, A.L. (2000). Why are "natively unfolded" proteins unstructured under physiologic conditions? *Proteins.* Vol. 41, No. 3, pp. 415–27

Uversky, V.N., Oldfield, C.J., Midic, U., Xie, H., Xue, B., Vucetic, S., Iakoucheva, L.M., Obradovic, Z. & Dunke, A.K. (2009). Unfoldomics of human diseases: linking protein intrinsic disorder with diseases. *BMC Genomics*, Vol. 10, Suppl. 1, article S7

Uversky, V.N. & Dunker, A.K. (2010) Understanding protein non-folding. *Biochim Biophys Acta.*Vol. 1804, No. 6, pp. 1231–64

Uversky, V.N. (2010). The mysterious unfoldome: structureless, underappreciated, yet vital part of any given proteome. *J. Biomed. Biotechnolol.*, Vol. 2010: 568068

Vinson, C., Myakishev, M., Acharya, A., Mir, A.A., Moll, J.R., Bonovich, M. (2002). Classification of human B-ZIP proteins based on dimerization properties. *Mol Cell Biol.*, Vol. 22, No. 18, pp. 6321–35

Vogelaar, N. J., Jing, X., Robinson, H. H., Schubot, F. D. (2010). Analysis of the crystal structure of the ExsC.ExsE complex reveals distinctive binding interactions of the *Pseudomonas aeruginosa* type III secretion chaperone ExsC with ExsE and ExsD. *Biochemistry* Vol. 49, No. 28, pp. 5870-79

Wagner, V.E., Bushnell, D., Passador, L., Brooks, A.I. & Iglewski, B.H. (2003) Microarray analysis of Pseudomonas aeruginosa quorum-sensing regulons: effects of growth phase and environment. *J Bacteriol.*, Vol. 185, No. 7, pp. 2080-95

Wagner, S., Konigsmaier, L., Lara-Tejero, M., Lefebre, M., Marlovits, T. C., & Galan, J. E. (2010). Organization and coordinated assembly of the type III secretion export apparatus. *Proc. Natl. Acad. Sci. U S A*, Vol. 107, No. 41, pp. 17745-50

Walshaw J. & Woolfson D.N. (2001) SOCKET: a program for identifying and analysing coiled-coil motifs within protein structures *J. Mol. Biol.*, *Vol.* 307, No. 5, pp. 1427-50

Wang, Y., Ouellette, A.N., Egan, C.W., Rathinavelan, T., Im, W., De Guzman, R.N. (2007). Differences in the Electrostatic Surfaces of the Type III Secretion Needle Proteins PrgI, BsaL, and MxiH. *J.Mol.Biol.*, Vol. 371, No. 5, pp. 1304-14

Wolf, E., Kim, P.S. & Berger, B. (1997). MultiCoil: A program for predicting two- and three-stranded coiled coils. *Protein Science*, Vol. 6, No. 6, pp. 1179-89

Wulf, J., Pascuzzi, P.E., Martin, G.B., Nicholson, L.K. (2004). The solution structure of type III effector protein AvrPto reveals conformational and dynamic features important for plant pathogenesis. *Structure*, Vol. 12, No. 7, pp. 1257-68

Xie, H., Vucetic, S., Iakoucheva, L.M., Oldfield, C.J., Dunker, A.K., Uversky, V.N., Obradovic, Z. (2007). Functional anthology of intrinsic disorder. 1. Biological processes and functions of proteins with long disordered regions. *Journal of Proteome Research*, Vol. 6, No. 5, pp. 1882–98

Xing, W.M., Zou, Y., Liu, Q., Hao, Q., Zhou, J.M., Chai, J.J. (2007). The structural basis for activation of plant immunity by bacterial effector protein AvrPto. *Nature*, Vol. 449, No. 7159, pp. 243-47

Yang, J., Chen, L.H., Sun, L.L., Yu, J., Jin, Q. (2008). VFDB 2008 release: an enchanced web-based resource for comparative pathogenomics. *Nucleic Acids Res.*, Vol. 36, Database issue, pp. D539-D542

Yip, C.K., Finlay, B.B. & Strynadka, N.C. (2005a). Structural characterization of a type III secretion system filament protein in complex with its chaperone. Nature Struct. Mol. Biol., Vol. 12, No. 1, pp. 75-81

Yip, C.K., Kimbrough, T.G., Felise, H.B., Vuckovic, M., Thomas, N.A., Pfuetzner, R.A., Frey, E.A., Finlay, B.B., Miller, S.I., Strynadka, N.C. (2005b). Structural characterization of the molecular platform for type III secretion system assembly. *Nature*, Vol 435, No. 7042, pp. 702-07

Yu, C., Ruiz, T., Lenox, C., & Mintz, K. P. (2008). Functional mapping of an oligomeric autotransporter adhesin of *Aggregatibacter actinomycetemcomitans*. *J Bacteriol*, Vol. 190, No. 9, pp. 3098-3109

Zarivach, R., Vuckovic, M., Deng, W., Finlay, B.B. & Strynadka, N.C.J. (2007). Structural analysis of a prototypical ATPase from the type III secretion system. *Nat. Struct. Biol.*, Vol 14, No. 2, pp. 131-137

Zarivach, R., Deng, W., Vuckovic, M., Felise, H.B., Nguyen, H.V., Miller, S.I., Finlay, B.B. & Strynadka, N.C. (2008). Structural analysis of the essential self-cleaving type III secretion proteins EscU and SpaS. *Nature*, Vol. 453, No 1, pp. 124-127

Zhang, L., Wang, Y., Picking, W. L., Picking, W. D. & De Guzman R. N. (2006). Solution structure of monomeric BsaL, the type III secretion needle protein of Burkholderia pseudomallei. *J Mol Biol.*, Vol. 359, No. 2, pp. 322–330

Zheng, J. & Leung, K. Y. (2007). Dissection of a type VI secretion system in *Edwardsiella tarda*. *Molecular Microbiology*, Vol. 66, No 5, pp 1192-1206

# 4

# Human ERα and ERβ Splice Variants: Understanding Their Domain Structure in Relation to Their Biological Roles in Breast Cancer Cell Proliferation

Ana M. Sotoca[1,2], Jacques Vervoort[2],
Ivonne M.C.M. Rietjens[1] and Jan-Åke Gustafsson[3,4]
[1]*Toxicology section, Wageningen University, Wageningen*
[2]*Laboratory of Biochemistry, Wageningen University, Wageningen*
[3]*Department of Biology and Biochemistry, Center for Nuclear Receptors and Cell Signaling, University of Houston, Science & Engineering Research Center, Houston*
[4]*Department of Biosciences and Nutrition, Karolinska Institutet, Novum, Huddinge*
[1,2]*The Netherlands*
[3]*USA*
[4]*Sweden*

## 1. Introduction

ERs are members of the nuclear receptor superfamily and have a broad range of biological roles, such as growth, differentiation and physiology of the reproductive system (Pearce & Jordan, 2004). These enzymes also have roles in non-reproductive tissues such as bone, cardiovascular system, brain and liver (Heldring *et al.*, 2007). Until 1996, only one human estrogen receptor (ER) was known. That year Kuiper et al. discovered a novel nuclear estrogen receptor cloned from rat prostate. The known ER was renamed and called ERα to differentiate it from the novel ER, ERβ (Kuiper *et al.*, 1996). The complete human ERβ cDNA sequence was published in 1998 by Ogawa et al (Ogawa *et al.*, 1998a).

### 1.1 Estrogen receptors and signalling function

Estrogen receptors are products of distinct genes localized on different chromosomes; human ERα is encoded on chromosome 6q24-q27 (Gosden *et al.*, 1986), while the gene encoding human ERβ is localized on chromosome 14q22-q24 (Enmark *et al.*, 1997). Despite their distinct localization, the gene organization of the two receptors is well conserved. ESR1 (ERα) and ESR2 (ERβ) genes contain eight exons, separated by seven long intronic sequences. As members of the nuclear receptor superfamily, ERs contain 6 regions in their protein structure common for all nuclear receptors, namely: A, B, C, D, E and F which form functionally different but interacting domains (figure 1). Exon 1 encodes the A/B region in ERα and ERβ, exons 2 and 3 encode part of the C region. Exon 4 encodes the remaining part of region C, the whole of region D and part of region E. Exons 5 to 8 contain the rest of region E and region F is encoded by part of exon 8 [reviewed in (Ascenzi *et al.*, 2006)].

Although ERα and ERβ are encoded separately they share a high degree of homology. The most conserved domain among ERs is the DNA binding domain (DBD) corresponding to the C region, with 96% homology between α and β ER subtypes. The DBD is responsible for binding to specific DNA sequences (Estrogen Responsive Elements or EREs) in target gene promoter regions. High structure similarity in this region suggests similar target promoter sites for both receptors. The A/B region located in the N-terminus of the protein encompasses the AF-1 domain responsible for ligand independent transactivation. The AF-1 domain is the least conserved part among the two ERs with only 30% homology and it is functional only in the ERα subtype (Hall & McDonnell, 1999). The C-terminus of the protein contains the ligand dependent transactivation domain AF-2, the ligand binding domain (LBD) and the homo-/heterodimerization site. Homology between the E/F regions of both proteins is only 53%, explaining differences in ligand binding affinities between the two receptors. The hinge region localized in the D domain contains the nuclear localization signal of the ERs as well as post translational modification sites (Sentis et al., 2005). Information on structure/function relationship of this region is very limited and it appears to be a variable and not well conserved part of the ERs (only 30% homology).

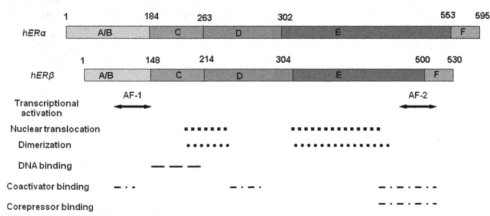

Fig. 1. Proteomic format, domain structure of human ERα (A) and ERβ (B). Based on Matthews and Gustafsson (Matthews & Gustafsson, 2003).

Estrogen (E2) binding to the receptor induces the LBD to undergo a conformational change, upon which the receptor dimerizes, binds to DNA, and stimulates gene expression (Cowley et al., 1997; Katzenellenbogen & Katzenellenbogen, 2000).

## 1.2 Estrogen receptor distribution

The distribution of ERs varies both between and within human tissues (see Table 1). The cardiovascular system, brain, and bones express both receptors. ERβ is predominant in the male reproductive system. Expression of both ERα and ERβ has been found in all major human uterine cell types at every menstrual stage. However, expression varies from cell-type to cell-type with expression of ERα mRNA generally being higher than that of ERβ (Matsuzaki et al., 1999). Changes in expression of estrogen receptors has been found in certain tumour types. Normal mammary tissue in man predominantly expresses ERβ mRNA, whereas most ER-positive breast tumours appear to exhibit increased ratios of ERα/ERβ (Leygue et al., 1998).

Human ERα and ERβ Splice Variants: Understanding Their Domain Structure in Relation to Their Biological
Roles in Breast Cancer Cell Proliferation

109

Likewise, an increased ratio of ERα/ERβ mRNA has been demonstrated in ovarian carcinoma compared with normal tissue or cysts (Bardin *et al.*, 2004). High concentrations of ERβ have also been found within the human gut (Enmark *et al.*, 1997).

Therefore, the ultimate estrogenic effect of a certain compound on cells or tissues will be dependent on the receptor phenotype of these cells or tissues.

| Organ/Tissue | Human ER subtype | | Organ/Tissue | Human ER subtype | |
|---|---|---|---|---|---|
| | ERα | ERβ | | ERα | ERβ |
| Heart | ✓ | ✓ | Adrenal | ✓ | - |
| Lung | - | ✓ | Kidney | ✓ | ✓ |
| Vascular | ✓ | ✓ | Prostate | - | ✓ |
| Bladder | - | ✓ | Testes | - | ✓ |
| Epididymus | - | ✓ | Brain | ✓ | ✓ |
| Pituitary | - | ✓ | Thymus | - | ✓ |
| Liver | ✓ | - | Breast | ✓ | ✓ |
| Muscle | - | - | Uterus | ✓ | ✓ |
| Fat | - | - | Endometrium | ✓ | ✓ |
| Gastrointestinal tract | - | ✓ | Vagina | ✓ | - |
| Colon | - | ✓ | Fallopian tube | - | ✓ |
| Small intestine | - | ✓ | Ovary | ✓ | ✓ |
| Bone | ✓ | ✓ | | | |

Table 1. Tissue distribution of ER subtypes in humans.

## 1.3 Mechanism of estrogen action

Estrogens act on target tissues by binding to ERs. These proteins function as transcription factors when they are activated by a ligand. Biological action of ERs involves complex and broad mechanisms. For the ERs two main mechanisms of action have been described, including a genomic and a non-genomic pathway (Figure 2).

The *genomic action* of ERs occurs in the nucleus of the cell, when the receptor binds specific DNA sequences directly ("direct activation" or classical pathway) or indirectly ("indirect activation" or non-classical pathway). In the absence of ligand, ERs are associated with heat-shock proteins. The Hsp90 and Hsp70 associated chaperone machinery stabilizes the ligand binding domain (LBD) and makes it accessible to the ligand. Liganded ER dissociates from the heat-shock proteins, changes its conformation, dimerizes, and binds to specific DNA sequences called estrogen responsive elements (EREs) in order to regulate transcription (Nilsson *et al.*, 2001). In the presence of the natural ligand E2, ER induces chromatin remodelling and increases transcription of estrogen regulated genes (Berno *et al.*, 2008).

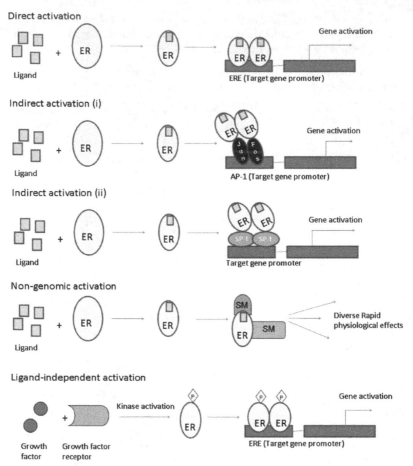

Fig. 2. Mechanisms of estrogen receptor (ER) action. In the *direct activation*, ERs dimerize after ligand binding and attach to the ERE in the promoter of target genes. In the *indirect activation* manner, ligand-bound ER dimers might activate transcription of non-ERE containing genes, by binding to other transcription factors (e.g. AP1 or SP1). In the *non-genomic pathway*, ligand-bound ERs interact directly with and change the function of proteins some of which function as 'second messengers' (SM). ERs can also be activated by phosphorylation in the absence of ER ligands (*ligand-independent activation*). Based on Morani *et al.* 2008 (Morani *et al.*, 2008).

In the non-classical pathway, AP-1 (DeNardo *et al.*, 2005) and SP-1 (Kim *et al.*, 2003) are alternative regulatory DNA sequences used by both isoforms of the receptor, ERα and ERβ, to modulate gene expression. In this case, ER does not interact directly with DNA but interacts with other DNA-bound transcription factors such as c-Jun or c-Fos, or with other proteins (Kushner *et al.*, 2003). Both AF-1 and AF-2 domains of ER are required for the interaction with Fos/Jun complex and both receptors differentially affect AP-1 dependent genes. In the presence of ERα, E2 works as AP-1 agonist by enhancing activity of the proteins at AP-1 sites (Brzozowski *et al.*, 1997), while in the presence of ERβ it antagonizes

AP-1 activity (Nilsson $et$ $al.$, 2001). When both receptors are present, ERβ inhibits the action of ERα on AP-1 promoters (Matthews $et$ $al.$, 2006). Interactions of ERs with other transcription factors might be also selectively modulated by different ligands, such as genistein and quercetin, which are not able to stimulate AP-1 dependent transcription (Figtree $et$ $al.$, 2003; Schreihofer, 2005).

Even though ERs are considered transcription factors they can act through $non$-$genomic$ mechanisms. Rapid ER effects were first observed in 1960s when administration of a physiological dose of E2 was reported to increase uterine cAMP levels in ovariectomized rats within 15 seconds (Szego and Davis, 1967), a time scale that is considered too fast for a genomic action. There is still no agreement if receptors responsible for rapid actions of estrogens are the same proteins as nuclear ERs or distinct G-protein coupled steroid receptors (Funakoshi $et$ $al.$, 2006; Maggiolini $et$ $al.$, 2004; Pedram & Levin, 2006; Warner & Gustafsson, 2006). However, a broad range of other rapid pathways induced by E2 has been identified so far. Some of these pathways include MAPK/ERK pathway, activation of endothelial nitric oxide synthase (eNOS), PLC stimulated IP$_3$ production, calcium influx and PI3K/Akt pathway activation (Stirone $et$ $al.$, 2005; Virgili F, 2004; Ascenzi $et$ $al.$, 2006). Similarly to non-classic mechanisms of activation, phytoestrogens might affect rapid pathways in a different way than E2. Quercetin for example has been shown to fail to phosphorylate ERK-2 kinase (opposite to E2) nor did it stimulate transcription of Cyclin D1, the transcription of which sometimes depends on rapid ER pathways (Virgili F, 2004). The stimulation of eNOS, which plays a role in cardiovascular health effects induced by E2 also seems to be regulated differently by phytoestrogens. Rapid activation of eNOS in the presence of E2 is dependent on ERα (Simoncini $et$ $al.$, 2005), while both receptors are required for prolonged effects. However phytoestrogens do not activate eNOS in a rapid manner but seem to activate it through a prolonged, ERβ dependent transcriptional mechanism (Simoncini $et$ $al.$, 2005).

In addition to ligand dependent mechanisms, ERα has ligand independent activity mediated through AF-1, which has been shown to be associated with stimulation of MAPK through growth factors such as Insulin like Growth Factor – 1 (IGF-1) and Epidermal Growth Factor (EGF). Activity of AF-1 is dependent on phosphorylation of Ser 118. A good example of the cross-talk between ER and growth factor signalling is phosphorylation of Ser 118 by MAPK in response to growth factors, such as IGF-1 and EGF (Kato $et$ $al.$, 1995). The importance of growth factors in ER signalling is well illustrated by the fact that EGF can mimic effects of E2 in the mouse reproductive tract (Nilsson $et$ $al.$, 2001).

## 1.4 Ligand dependent effects and cofactors

The overall biological effects of E2 and other estrogenic compounds are the result of complex interplay between various mechanisms, which largely depend on cellular context, ratio between ER subtypes, expression of coactivators in the cell, sequences of target EREs but also cross-talk with growth factor pathways and activity of kinases and phosphatases. All these factors together enable a precise and targeted response to the natural hormone. However a broad range of pathways involved in ER signaling provides many points of possible signal modulation by estrogens and estrogen-like compounds and small structural changes between different ligands might result in significantly different responses.

Structural differences in the LBD underlie differences in affinity and transcriptional activity of certain ER ligands and provide one of the mechanisms for selective modulation of ER responses. ERβ has an impaired AF-1 domain compared with ERα and the necessary synergy with AF-2 is dramatically reduced (Cowley & Parker, 1999). These differences suggest that it is possible to develop ligands with different affinities, potencies, and agonist vs antagonist behavior for the two ER subtypes.

It has been demonstrated that E2 has higher affinity towards ERα than to ERβ (Bovee et al., 2004; Veld et al., 2006), and certain selective estrogen receptor modulators (SERMs) might exhibit a preference towards one of the receptors (Escande et al., 2006). Plant derived phytoestrogens, which are structurally similar to E2 (Figure 3) provide a good example of ligand selectivity (Kuiper et al., 1998). Genistein is the major isoflavone present in soy and fava beans whereas quercetin is present in red onions, apples, cappers or red grapes among others (Kuiper et al., 1998). In vitro studies with reporter gene assays proved that phytoestrogens are able to stimulate ERE-dependent genes at high concentrations. Therefore they are considered weak ER agonists with the majority of them preferentially binding to ERβ (Chrzan & Bradford, 2007; Harris et al., 2005). The main hypothesis on the positive role of phytoestrogens in modulation of ER signaling is their higher affinity towards the ERβ subtype, which can silence ERα dependent signaling and decrease overall cell sensitivity to E2 (Hall & McDonnell, 1999), which is thought to be significant in cancer prevention.

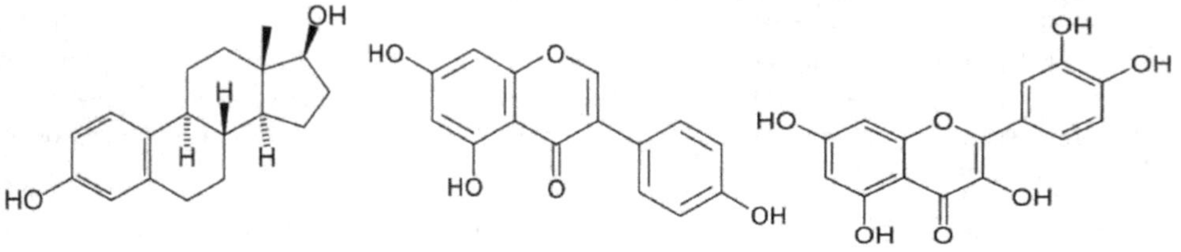

Fig. 3. Chemical structure of estradiol, genistein and quercetin.

ERs can associate with distinct subsets of coactivators and corepressors depending on binding affinities and relative abundance of these factors (Chen & Evans, 1995; Halachmi et al., 1994). Several ER coactivators and corepressors have been described (Nilsson et al., 2001). Differences between ERα and ERβ in coactivator and corepressor recruitment have also been reported (Cowley & Parker, 1999; Suen et al., 1998), and therefore this preferential binding of certain coactivators and corepressors to one of the ERs may have consequences for specific ligand signalling and the ultimate biological effect elicited by ligand binding.

NCoR and SMRT corepressors and the p160 family coactivators are widely expressed (Horlein et al., 1995; Misiti et al., 1998; Oñate et al., 1995). Low levels of SRC-3 have been demonstrated for human proliferating endometrium with increased expression in the late secretory phase (Gregory et al., 2002) while overexpression of SRC-3 is frequently observed in breast, ovarian, and prostate cancers (Anzick et al., 1997; Gnanapragasam et al., 2001; McKenna et al., 1999). Similar expression levels of CBP, p300, AIB1, GRIP1, p300, NCoR, and SMRT have been measured for Ishikawa uterine and MCF-7 breast cancer cells (Shang and Brown, 2002). High levels of SRC-1 expression are found in Ishikawa cells, and this might correlate with the agonist activity of tamoxifen in this cell line (Shang and Brown, 2002). We have seen in our studies (Sotoca et al., 2011), that the T47D breast cancer cells express the ER coactivator PRMT1. Recruitment of this coactivator is accompanied by histone methylation (Huang et al., 2005; Klinge et al., 2004). Recently, PRMT1 gene expression has been used as a marker of unfavourable prognosis for colon cancer patients (Mathioudaki et al., 2008).

Thus, other signalling events within the cell may affect nuclear receptor transcriptional responses via alteration in the expression of certain coregulators, and therefore it is predicted that significant differences in coactivator and corepressor expression found in various cell and tissue types would be important determinants of specific receptor modulator activity.

In addition, distribution of particular splicing variants of both ERs should be taken into account when considering tissue response to estrogens and cofactor recruitment as they have differential and sometimes antagonistic properties and their relative abundance might significantly influence biological responses to hormones. The main physiological role of ER splice variants in breast cancer development is however far from clear and might be a crucial determinant for clinical parameters.

## 2. ER Isoforms: ERα and ERβ

Full length ERα and ERβ proteins are approximately 66 and 59 kDa respectively (Ascenzi et al., 2006; Fuqua et al., 1999), although as a result of alternative splicing both receptors can form different isoforms. ERα has been shown to form over 20 alternative splice variants in breast cancer and other tumors (Poola et al., 2000), three of them with proven functionality, while at least five ERβ variants have been reported in human (Lewandowski et al., 2002).

The function and physiological significance of all isoforms have not been described so far, but some of them are powerful modulators of ER signaling pathways in normal tissues.

### 2.1 ERα splice variants

The two most referenced ERα isoforms that seem to be of particular significance are **ERα46** and **ERα36** as they were reported to oppose genomic actions of full length **ERα66** (figure 4).

The **ERα46** isoform has been identified in the MCF7 breast cancer cell line (Penot et al., 2005) in which it is coexpressed with full length ERα66. The presence of ERα46 has also been confirmed in osteoblasts (Wang et al., 2005) and endothelial cells (Figtree et al., 2003). This isoform is formed by skipping exon 1 encoding the N-terminus (A/B) and it is devoid of AF-1 activity. In contrast with full length ERα66, the truncated isoform ERα46 does not mediate E2 dependent cell proliferation and high levels of this isoform have been shown to be associated with cell cycle arrest in the G0/G1 phase and a state of refraction to E2 stimulated growth, which is normally reached at hyperconfluency of the cells (Penot et al., 2005). Similarly to ERβ, ERα46 is a potent ligand-dependent transcription factor containing AF-2 and a powerful inhibitor of ERα AF-1 dependent transcription (Figtree et al., 2003). By inhibition of ERα66 dependent gene transcription, ERα46 isoform inhibits estrogenic induction of c-Fos and Cyclin D1 promoters, which are involved in cell cycle control. Coexpression of ERα46 with ERα66 in an SaOs osteoblast cell line results in concentration dependent inhibition of E2 stimulated cell proliferation (Ogawa *et al.*, 1998b), an effect similar to the consequence observed with coexpression of ERα with ERβ (Sotoca *et al.*, 2008; Ström *et al.*, 2004).

The second truncated ERα isoform ERα36 was first described recently (Wang *et al.*, 2005), and it has been shown to lack both the AF-1 and AF-2 transactivation functions of full length ERα. However it has functional DBD, partial dimerization and LBD domains. ERα36 contains an exon coding for myristoylation sites, hence predicting an interaction with the plasma membrane. Transcription of this ERα36 isoform is initiated from a previously unidentified promoter in the first intron of the ERα gene and the unique 27 amino acid C-terminal sequence is encoded by a novel ERα exon, localized downstream of exon 8 to replace the last 138 amino acids encoded by exon 7-8 (Wang *et al.*, 2005).

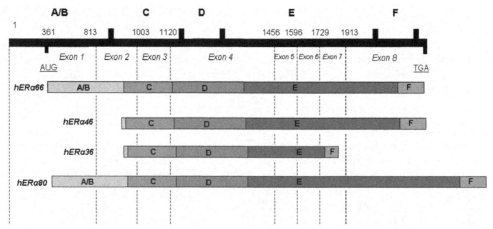

Fig. 4. Schematic comparison between full length ERα and its most referenced truncated isoforms.

This novel isoform has been cloned from a human placenta cDNA library, which indicates that it is a naturally occurring isoform of ERα. With no functional AF-1 and AF-2 ERα36 does not have any direct transcriptional activity. However, it is a robust inhibitor of full length ERα and ERβ dependent transactivation (ZhaoYi Wang *et al.*, 2006). It is mainly localized in the plasma membrane and works in a different way than full length protein. Even though it lacks transcriptional activity it can activate non genomic ER pathways such as MAPK/ERK signaling in response to E2 which is of particular significance in response to antiestrogens such as tamoxifen, 4OH-tamoxifen and ICI-182.780 (ZhaoYi Wang *et al.*, 2006). As a result of MAPK/ERK pathway activation by E2 and these antiestrogens a signal is transduced to the nucleus and consequently Elk1 transcription factor is activated. The effect of MAPK/ERK activation mediated by ERα36 is increased cell proliferation in response to E2 as well as antiestrogens in doses that shut down transcriptional activity of full length ERα and ERβ proteins (ZhaoYi Wang *et al.*, 2006).

The **ERα80** isoform was detected in the MCF7:2A cell line, which is a subclone MCF7 cell line derived from long term growth in the absence of E2. This ERα80 isoform was produced by duplication of exons 6 and 7 (Pink *et al.*, 1996). No evident function has been described so far.

Several other multiple splice variants (ERαΔE2, ERαΔE3, ERαΔE4, ERαΔE5, ERαΔE6, ERαEΔ5,7, ERαEΔ7…) as a result of exon splicing deletions have been confirmed in human (Poola *et al.*, 2000; Zhang *et al.*, 1996) showing a dominant inhibitory effect in normal ER function. A list of selected ERα splice variants and their expression in various breast tissues (normal and tumor) and breast cancer cell lines is given in Table 2.

| Splice variant | Breast | MCF7 | T47D | MDA-MB-231 | MDA-MB-435 | BT-474 | BT20 | ZR-75 | References |
|---|---|---|---|---|---|---|---|---|---|
| ERα36 | + | | | + | | | | | (Shi *et al.*, 2009; Lee *et al.*, 2008; ZhaoYi Wang *et al.*, 2006) |
| ERα46 (or ERαΔ1) | | + | | | | | | | (Penot *et al.*, 2005) |
| ERαΔ2 | + | + | + | | + | | | | (Wang and Miksicek, 1991; Zhang *et al.*, 1996; Bollig and Miksicek, 2000; Poola and Speirs, 2001; Miksicek *et al.*, 1993; Poola *et al.*, 2000) |
| ERαΔ3 | + | + | + | + | | | | | (Wang and Miksicek, 1991; Poola and Speirs, 2001; Bollig and Miksicek, 2000; Zhang *et al.*, 1996; Koduri *et al.*, 2006; Erenburg *et al.*, 1997; Miksicek *et al.*, 1993; Fuqua *et al.*, 1993) |
| ERαΔ4 | | + | | | + | | | + | (Pfeffer *et al.*, 1993; Zhang *et al.*, 1996; Bollig and Miksicek, 2000; Poola *et al.*, 2000; Poola and Speirs, 2001) |
| ERαΔ5 | + | + | + | + | + | + | + | + | (Zhang *et al.*, 1993; Zhang *et al.*, 1996; Bollig and Miksicek, 2000; Poola and Speirs, 2001; Zhang *et al.*, 1996; Fuqua *et al.*, 1991; Daffada *et al.*, 1994) |
| ERαΔ6 | + | | | | | | | | (Poola and Speirs, 2001; Bollig and Miksicek, 2000) |
| ERαΔ7 | + | + | + | + | | | | | (Wang and Miksicek, 1991; Fuqua *et al.*, 1992; Poola and Speirs, 2001; Bollig and Miksicek, 2000; Fuqua *et al.*, 1992; Miksicek *et al.*, 1993) |
| ERαΔ5,7 | + | | | | | | | | (Zhang *et al.*, 1996) |

Table 2. List of selected ERα splice variants and their expression in various breast tissues (normal and tumour) and breast cancer cell lines.

## 2.2 ERβ splice variants

The presence of ERβ isoforms has been confirmed in various human cell lines as well as in a broad range of tissues at different levels (Leung et al., 2006; Moore et al., 1998), which provides another possible mechanism of tissue-dependent modulation of the ER response. Therefore distribution of particular isoforms of both ERs should be taken into account when considering tissue response to estrogens as they have differential and sometimes antagonistic properties and their differential distribution might significantly influence biological response to hormone.

Different isoforms of ERβ have been described (figure 5) with a variable C-terminus, and which were cloned from a testis cDNA library (Moore et al., 1998). At present their functional significance is poorly understood. The ERβ isoform whose function has been described in most detail of all ERβ isoforms studied is **ERβ1**, which is a full length protein with LBD and active AF-2 domain. **ERβ2, 4 and 5** have a shortened Helix 11 and a full length Helix 12 is present only in ERβ1 and β2. In ERβ2, Helix 12 has a different orientation than in ERβ1 due to the shorter Helix 11. It has been reported that the displaced Helix 12 in ERβ2 limits ligand access to the binding pocket. As a consequence of their altered structure, ERβ2, 4 and 5 cannot form homodimers and have no transcriptional activity on their own, although they have been shown to heterodimerize with ERβ1 upon E2 treatment and enhance its AF-2 mediated transcriptional activity (Leung et al., 2006). Studies of interactions between different ERβ isoforms with ERα are very limited. However **ERβ2** (also named **ERβcx**) was shown to limit DNA binding of ERα66 and inhibit its transcriptional activity in similar manner to ERβ1 (Ogawa et al., 1998b).

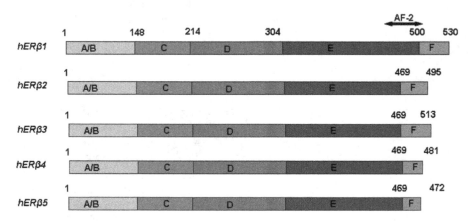

Fig. 5. Comparison between full length ERβ and it is most referenced truncated isoforms.

Two new exon-deleted variants were detected in the cancer cell line MDA-MB-231, **ERβΔ1,2,5** and **ERβΔ1,2,5,6** of approximately 35 and 28 kDa, respectively (Treeck et al., 2008). Both proteins are predicted not to contain AF-1, and to have deletions in the DBD and LBD. Therefore, these two variants are expected to be devoid of or have significantly reduced ligand-dependent and ligand independent activities, and their expression did not affect growth of cancer cell lines tested. A list of selected ERβ splice variants and their expression in various breast tissues (normal and tumor) and breast cancer cell lines is given in Table 3.

Various studies reveal that physiological levels of ERα and ERβ may vary depending on the cell or tissue type (Enmmark *et al.*, 1997; Bonkhoff *et al.*, 1999; Makinen *et al.*, 2001; Pearce *et al.*, 2004) and as a consequence the biological response to endogenous or exogenous ligands can differ significantly.

| Splice variant | Breast | MCF7 | T47D | MDA-MB-231 | MDA-MB-435 | BT20 | References |
|---|---|---|---|---|---|---|---|
| ERβ2 | + | + | + | + | | + | (Davies *et al.*, 2004; Zhao *et al.*, 2007; Girault *et al.*, 2004; Saji *et al.*, 2005; Cappelletti *et al.*, 2006; Leung *et al.*, 2006) |
| ERβ3 | + | | | | | | (Girault *et al.*, 2004) |
| ERβ4 | + | | + | | + | | (Moore *et al.*, 1998; Girault *et al.*, 2004; Poola *et al.*, 2005) |
| ERβ5 | + | + | + | + | + | + | (Davies *et al.*, 2004; Girault *et al.*, 2004; Moore *et al.*, 1998; Fuqua *et al.*, 1999; Leung *et al.*, 2006; Cappelletti *et al.*, 2006) |
| ERβΔ2 | + | | | + | | | (Poola *et al.*, 2002a) |
| ERβΔ3 | + | | | | | | (Poola *et al.*, 2002a; Poola *et al.*, 2002b) |
| ERβΔ4 | + | | | | | | (Poola *et al.*, 2002a; Poola *et al.*, 2002b) |
| ERβΔ5 | + | + | | + | | | (Poola *et al.*, 2002a; Speirs *et al.*, 2000; Leygue *et al.*, 1998) |
| ERβΔ6 | + | | | | | | (Poola *et al.*, 2002a; Leygue *et al.*, 1998) |
| ERβΔ1,2,5 | | + | | + | | | (Treeck *et al.*, 2008) |
| ERβΔ1,2,5,6 | | + | | + | | | (Treeck *et al.*, 2008) |

Table 3. List of selected ERβ splice variants and their expression in various breast tissues (normal and tumour) and breast cancer cell lines.

## 3. Conclusion

Cell proliferation in normal developing breast tissue is stimulated by estrogens and estrogens may prevent osteoporosis by increasing bone mineral density (Douchi et al., 2007). However, as cells can have their own set of ER splice variants that varies in time and abundance the estrogen receptor proteins can be expected to have a role in developmental regulation depending on splice variant and ligand present. ER splice variants are widely expressed in normal, premalignant and cancerous tissues and cell lines [reviewed in (Taylor et al. 2010)]. Co-expression of splice variants remains under investigation to understand its biological implications. Here, we briefly summarize ER expression and its role in positive or negative transcriptional activation in breast cancer.

Several studies have demonstrated that estrogens stimulate the growth of a large proportion of ERα positive breast cancers (Lazennec, 2006; Monroe et al., 2005; Pedram et al., 2006; Weitzmann & Pacifici, 2006). Furthermore, a decreased ERβ expression in cancer tissues as compared to benign tumours or normal tissues has been reported, whereas ERα expression seems to persist (Lazennec et al., 2001, Bardin et al., 2004). Recent progress in cellular experiments confirmed that ERβ opposes ERα actions in breast cancer cell lines (Sotoca et al., 2011; Sotoca et al., 2008; Ström et al., 2004).

The main roles of ER splice variants in breast cancer development are, however, far from clear (Davies et al., 2004; Saji et al., 2005). ERα positivity in breast cancer in vivo is strongly associated with more favourable clinicopathological parameters. ERβ positive patients have been shown to have favourable prognosis and better survival due to better endocrine-treatment response compared with ERβ negative breast tumor patients (Davies et al., 2004; Saji et al., 2005).

When bound to estrogens as homodimers, each receptor activates transcription of certain target genes bearing a classical ERE in their promoter region. However, estrogen binding to ERβ can also inhibit gene transcription via AP-1 sites, while binding to ERα leads to activation. Furthermore, when heterodimers are formed, when the two receptors are co-expressed, ERβ can inhibit ERα function. Given that ER regulates cell proliferation by different mechanisms, we summarize (Table 4 and 5) by which molecular characteristics of ER this proliferation is driven.

Full activation of AF-1 in ERα induces cell proliferation in breast cancer cells (Fujita et al., 2003). AF-1 activity of estrogen-ERβ is weaker compared with that of estrogen-ERα on ERE, whereas their AF-2 activities are similar (Cowley & Parker, 1999). In general ERβ has antiproliferative effects in breast cancer cells. All ERβ variants have negative effect on ERα by heterodimerization and reduce or abrogate both ligand-dependent and ligand-independent activities. Especially the ERβ2 isoform inhibits ERα-mediated estrogen action. In addition, several short ERα isoforms are able to oppose genomics actions of ERβ.

The most important point is that ERα expression induces significant cell proliferation in the absence of ERβ but not the other way around. Cell proliferation is triggered by classical genomic and non-genomic pathways. Only the wild type ERα isoform is able to induce hormone-dependent proliferation. It has been shown that most of the ER variants do not mediate ligand-dependent proliferation.

In conclusion, the overall biological effects of E2 and other estrogenic compounds in breast cancer cells are the result of complex interplay between various mechanisms, which depend on cellular context, balance between ER subtypes, coactivators and corepressors, sequences of target EREs but also cross-talk with growth factor pathways and activity of certain kinases and phosphatases. All these factors taken together enable response to estrogens or antiestrogens.

| Isoform | Feature | AF-1 | DBD | LBD | AF-2 |
|---------|---------|------|-----|-----|------|
| ERα66 | Wild type form<br>Induces cell proliferation | + | + | + | + |
| ERα46 | Does not mediate E2-dependent proliferation<br>Opposes genomic action of ERα66 and ERβ<br>Inhibitor of AF-1 dependent transactivation<br>Potent AF-2 ligand dependent transcription activity | - | + | + | + |
| ERα36 | Opposes genomic actions of ERα66<br>Can activate non-genomic ER pathways via MAPK/ERK<br>No direct transcriptional activity<br>Inhibitor of ERα and β dependent transactivation | - | + | + | - |
| ERα80 | Not described | + | + | +* | +* |
| ERαΔ2 | No transcriptional regulation | - | - | - | - |
| ERαΔ3 | Binds ligand<br>Dominant negative at ERE<br>Interacts with AP-1 sites<br>Suppresses E2-stimulated gene expression | + | - | + | + |
| ERαΔ4 | Dominant negative transcriptional effect | + | - | - | + |
| ERαΔ5 | Dominant positive transcriptional effect<br>Dominant negative at ERE<br>Coexpresses with ERα and enhances ERE-Luc | + | + | - | - |
| ERαΔ7 | Dimerizes with ERα and hERβ<br>Binds to ERE<br>Dominant negative transcriptional effect | + | + | + | - |

Table 4. Summary of ERα mechanism.

| Isoform | Features | AF-1 | DBD | LBD | AF-2 |
|---------|----------|------|-----|-----|------|
| ERβ1 | Wild type form | + | + | + | + |
| ERβ2 | Dimerizes with ERβ1 and ERα<br>Does not bind ligand | + | + | + | - |
| ERβ3 | Dimerizes with ERβ1<br>Does not bind ligand | + | + | + | - |
| ERβ4 | Dimerizes with ERβ1 and ERα<br>Does not bind ligand | + | + | + | - |
| ERβ5 | Dimerizes with ERβ1 and ERα<br>Does not bind ligand | + | + | + | - |
| ERβΔ5 | Negative effect on ERβ1 and ERα | + | + | | |
| ERβΔ1,2,5 | Reduced both ligand-dependent<br>and ligand independent activities | - | + | - | + |
| ERβΔ1,2,5,6 | Reduced both ligand-dependent<br>and ligand independent activities | - | + | - | + |

Table 5. Summary of ERβ mechanism.

## 4. Acknowledgements

This research was partially funded by the Graduate School of Voeding, Levensmiddelentechnologie, Agrobiotechnologie en Gezondheid (VLAG) (project number 61.61.100.040).

## 5. References

Anzick SL, Kononen J, Walker RL, Azorsa DO, Tanner MM, Guan X-Y et al (1997). AIB1, a Steroid Receptor Coactivator Amplified in Breast and Ovarian Cancer. *Science* 277: 965-968.

Ascenzi P, Bocedi A, Marino M (2006). Structure-function relationship of estrogen receptor [alpha] and [beta]: Impact on human health. *Molecular Aspects of Medicine* 27: 299.

Bardin A, Boulle N, Lazennec G, Vignon F, Pujol P (2004). Loss of ER{beta} expression as a common step in estrogen-dependent tumor progression. *Endocrine-Related Cancer* 11: 537-551.

Berno V, Amazit L, Hinojos C, Zhong J, Mancini MG, Dave Sharp Z et al (2008). Activation of Estrogen Receptor-alpha by E2 or EGF Induces Temporally Distinct Patterns of Large-Scale Chromatin Modification and mRNA Transcription. *PLoS ONE* 3: e2286.

Bollig A, Miksicek RJ (2000). An Estrogen Receptor-{alpha} Splicing Variant Mediates Both Positive and Negative Effects on Gene Transcription. *Mol Endocrinol* 14: 634-649.

Bonkhoff H, Fixemer T, Hunsicker I, Remberger K (1999). Estrogen Receptor Expression in Prostate Cancer and Premalignant Prostatic Lesions. *Am J Pathol* 155:641-647.

Bovee TFH, Helsdingen RJR, Rietjens IMCM, Keijer J, Hoogenboom RLAP (2004). Rapid yeast estrogen bioassays stably expressing human estrogen receptors [alpha] and [beta], and green fluorescent protein: a comparison of different compounds with both receptor types. *The Journal of Steroid Biochemistry and Molecular Biology* 91: 99.

Brzozowski AM, Pike ACW, Dauter Z, Hubbard RE, Bonn T, Engstrom O *et al* (1997). Molecular basis of agonism and antagonism in the oestrogen receptor. *Nature* 389: 753-758.

Cappelletti V, Miodini P, Fronzo GD, Daidone MG (2006). Modulation of estrogen receptor-beta isoforms by phytoestrogens in breast cancer cells. *International journal of Oncology* 28: 1185-91.

Chen JD, Evans RM (1995). A transcriptional co-repressor that interacts with nuclear hormone receptors. *Nature* 377: 454-457.

Chrzan BG, Bradford PG (2007). Phytoestrogens activate estrogen receptor beta1 and estrogenic responses in human breast and bone cancer cell lines. *Molecular Nutrition & Food Research* 51: 171-177.

Cowley SM, Hoare S, Mosselman S, Parker MG (1997). Estrogen Receptors alpha and beta Form Heterodimers on DNA. *Journal of Biological Chemistry* 272: 19858-19862.

Cowley SM, Parker MG (1999). A comparison of transcriptional activation by ER alpha and ER beta. *Journal of Steroid Biochemistry and Molecular Biology* 69: 165-175.

Daffada AAI, Johnston SRD, Nicholls J, Dowsett M (1994). Detection of wild type and exon 5-deleted splice variant oestrogen receptor (ER) mRNA in ER-positive and -negative breast cancer cell lines by reverse transcription/polymerase chain reaction. *Molecular Endocrinology* 13: 265-273.

Davies MPA, O'Neill PA, Innes H, Sibson DR, Prime W, Holcombe C *et al* (2004). Correlation of mRNA for oestrogen receptor beta splice variants ER{beta}1, ER{beta}2/ER{beta}cx and ER{beta}5 with outcome in endocrine-treated breast cancer. *J Mol Endocrinol* 33: 773-782.

DeNardo DG, Kim H-T, Hilsenbeck S, Cuba V, Tsimelzon A, Brown aPH (2005). Global Gene Expression Analysis of Estrogen Receptor Transcription Factor Cross Talk in Breast Cancer: Identification of Estrogen-Induced/Activator Protein-1-Dependent Genes. *Molecular Endocrinology* 19: 362-378.

Douchi T, Yonehara Y, Kosha S, Iwamoto I, Rai Y, Sagara Y *et al* (2007). Bone mineral density in breast cancer patients with positive estrogen receptor tumor status. *Maturitas* 57: 221-225.

Enmark E, Pelto-Huikko M, Grandien K, Lagercrantz S, Lagercrantz J, Fried G *et al* (1997). Human Estrogen Receptor {beta}-Gene Structure, Chromosomal Localization, and Expression Pattern. *J Clin Endocrinol Metab* 82: 4258-4265.

Erenburg I, Schachter B, Lopez RMy, Ossowski L (1997). Loss of an Estrogen Receptor Isoform (ER{alpha}{Delta}3) in Breast Cancer and the Consequences of Its Reexpression: Interference with Estrogen-Stimulated Properties of Malignant Transformation. *Mol Endocrinol* 11: 2004-2015.

Escande A, Pillon A, Servant N, Cravedi J-P, Larrea F, Muhnd P *et al* (2006). Evaluation of ligand selectivity using reporter cell lines stably expressing estrogen receptor alpha or beta. *Biochemical Pharmacology* 71: 1459-1469.

Figtree GA, McDonald D, Watkins H, Channon KM (2003). Truncated Estrogen Receptor {alpha} 46-kDa Isoform in Human Endothelial Cells: Relationship to Acute Activation of Nitric Oxide Synthase. *Circulation* 107: 120-126.

Fujita T, Kobayashi Y, Wada O, Tateishi Y, Kitada L, Yamamoto Y *et al* (2003). Full activation of estrogen receptor alpha (ER alpha) activation function-1 (AF-1) induces proliferation of breast cancer cells. *J. Biol. Chem.*: M301031200.

Funakoshi T, Yanai A, Shinoda K, Kawano MM, Mizukami Y (2006). G protein-coupled receptor 30 is an estrogen receptor in the plasma membrane. *Biochemical and Biophysical Research Communications* 346: 904.

Fuqua SAW, Allred DC, Elledge RM, Krieg SL, Benedix MG, Nawaz Z *et al* (1993). The ER-positive / PgR-negative breast cancer phenotype is not associated with mutations within the DNA binding domain. *Breast Cancer Research and Treatment* 26: 191-202.

Fuqua SAW, Fitzgerald SD, Allred DC, Elledge RM, Nawaz Z, McDonnell DP *et al* (1992). Inhibition of Estrogen Receptor Action by a Naturally Occurring Variant in Human Breast Tumors. *Cancer Research* 52: 483-486.

Fuqua SAW, Fitzgerald SD, Chamness GC, Tandon AK, McDonnell DP, Nawaz Z *et al* (1991). Variant Human Breast Tumor Estrogen Receptor with Constitutive Transcriptional Activity. *Cancer Research* 51: 105-109.

Fuqua SAW, Schiff R, Parra I, Friedrichs WE, Su J-L, McKee DD *et al* (1999). Expression of Wild-Type Estrogen Receptor {{beta}} and Variant Isoforms in Human Breast Cancer. *Cancer Res* 59: 5425-5428.

Girault I, Andrieu C, Tozlu S, Spyratos F, Bièche I, Lidereau R (2004). Altered expression pattern of alternatively spliced estrogen receptor [beta] transcripts in breast carcinoma. *Cancer Letters* 215: 101-112.

Gnanapragasam VJ, Leung HY, Pulimood AS, Neal DE, Robson CN (2001). Expression of RAC 3, a steroid hormone receptor co-activator in prostate cancer. *Br J Cancer* 85: 1928-1936.

Gosden J, Middleton P, Rout D (1986). Localization of the human oestrogen receptor gene to chromosome 6q24----q27 by in situ hybridization. *Cytogenetics and cell genetics* 43: 218-20.

Gregory CW, Wilson EM, Apparao KBC, Lininger RA, Meyer WR, Kowalik A *et al* (2002). Steroid Receptor Coactivator Expression throughout the Menstrual Cycle in Normal and Abnormal Endometrium. *J Clin Endocrinol Metab* 87: 2960-2966.

Halachmi S, Marden E, Martin G, MacKay H, Abbondanza C, Brown M (1994). Estrogen receptor-associated proteins: possible mediators of hormone-induced transcription. *Science* 264: 1455-1458.

Hall JM, McDonnell DP (1999). The Estrogen Receptor {beta}-Isoform (ER{beta}) of the Human Estrogen Receptor Modulates ER{alpha} Transcriptional Activity and Is a Key Regulator of the Cellular Response to Estrogens and Antiestrogens. *Endocrinology* 140: 5566-5578.

Harris DM, Besselink E, Henning SM, Go VLW, Heber aD (2005). Phytoestrogens induce differential estrogen receptor alpha- or Beta-mediated responses in transfected breast cancer cells. *Experimental Biology and Medicine* 230: 558-568.

Heldring N, Pike A, Andersson S, Matthews J, Cheng G, Hartman J *et al* (2007). Estrogen Receptors: How Do They Signal and What Are Their Targets. *Physiological Reviews* 87: 905-931.

Human ERα and ERβ Splice Variants: Understanding Their Domain Structure in Relation to Their Biological
Roles in Breast Cancer Cell Proliferation

123

Horlein AJ, Naar AM, Heinzel T, Torchia J, Gloss B, Kurokawa R et al (1995). Ligand-independent repression by the thyroid hormone receptor mediated by a nuclear receptor co-repressor. *Nature* 377: 397-404.

Huang S, Litt M, Felsenfeld G (2005). Methylation of histone H4 by arginine methyltransferase PRMT1 is essential in vivo for many subsequent histone modifications. *Genes & Development* 19: 1885-1893.

Kato S, Endoh H, Masuhiro Y, Kitamoto T, Uchiyama S, Sasaki H et al (1995). Activation of the Estrogen Receptor Through Phosphorylation by Mitogen-Activated Protein Kinase. *Science* 270: 1491-1494.

Katzenellenbogen BS, Katzenellenbogen JA (2000). Estrogen receptor transcription and transactivation: Estrogen receptor alpha and estrogen receptor beta - regulation by selective estrogen receptor modulators and importance in breast cancer. *Breast Cancer Res* 2: 335 - 344.

Kim K, Thu N, Saville B, Safe aS (2003). Domains of Estrogen Receptor alpha (ERalpha) Required for ERalpha/Sp1-Mediated Activation of GC-Rich Promoters by Estrogens and Antiestrogens in Breast Cancer Cells *Molecular endocrinology* 17: 804-817.

Klinge CM, Jernigan SC, Mattingly KA, Risinger KE, Zhang aJ (2004). Estrogen response element-dependent regulation of transcriptional activation of estrogen receptors alpha and beta by coactivators and corepressors. *Journal of Molecular Endocrinology* 33: 387-410.

Koduri S, Goldhar A, Vonderhaar B (2006). Activation of vascular endothelial growth factor (VEGF) by the ER-α variant, ERΔ3. *Breast Cancer Research and Treatment* 95: 37-43.

Kuiper GGJM, Enmark E, Pelto-Huikko M, Nilsson S, Gustafsson J-A (1996). Cloning of a novel estrogen receptor expressed in rat prostate and ovary. *PNAS* 93: 5925-5930.

Kuiper GGJM, Lemmen JG, Carlsson B, Corton JC, Safe SH, van der Saag PT et al (1998). Interaction of Estrogenic Chemicals and Phytoestrogens with Estrogen Receptor {beta}. *Endocrinology* 139: 4252-4263.

Kushner PJ, Webb P, Uht RM, Liu M-M, and Richard H. Price J (2003). Estrogen receptor action through target genes with classical and alternative response elements. *Pure Applied Chemistry* 75, Nos: 1757-1769.

Lazennec G (2006). Estrogen receptor beta, a possible tumor suppressor involved in ovarian carcinogenesis. *Cancer Letters* 231: 151-7.

Lazennec G, Bresson D, Lucas A, Chauveau C, Vignon F (2001). ER beta inhibits proliferation and invasion of breast cancer cells. *Endocrinology* 142: 4120-30.

Lee LMJ, Cao J, Deng H, Chen P, Gatalica Z, Wang Z-Y (2008). ER-α36, a Novel Variant of ER-α, is Expressed in ER-positive and -negative Human Breast Carcinomas. *Anticancer research* 28: 479-483.

Leung Y-K, Mak P, Hassan S, Ho S-M (2006). Estrogen receptor (ER)-beta isoforms: A key to understanding ER-beta signaling. *PNAS* 103: 13162-13167.

Lewandowski S, Kalita K, Kaczmarek L (2002). Estrogen receptor [beta]: Potential functional significance of a variety of mRNA isoforms. *FEBS Letters* 524: 1-5.

Leygue E, Dotzlaw H, Watson PH, Murphy LC (1998). Altered Estrogen Receptor {alpha} and {beta} Messenger RNA Expression during Human Breast Tumorigenesis. *Cancer Res* 58: 3197-3201.

Maggiolini M, Vivacqua A, Fasanella G, Recchia AG, Sisci D, Pezzi V et al (2004). The G Protein-coupled Receptor GPR30 Mediates c-fos Up-regulation by 17{beta}-Estradiol and Phytoestrogens in Breast Cancer Cells. *J. Biol. Chem.* 279: 27008-27016.

Makinen S, Makela S, Weihua Z, Warner M, Rosenlund B, Salmi S, Hovatta O, Gustafsson J-A (2001). Localization of oestrogen receptors alpha and beta in human testis. *Mol Hum Reprod* 7:497-503

Mathioudaki K, Papadokostopoulou A, Scorilas A, Xynopoulos D, Agnanti N, Talieri M (2008). The PRMT1 gene expression pattern in colon cancer. *Br J Cancer* 99: 2094-2099.

Matsuzaki S, Fukaya T, Suzuki T, Murakami T, Sasano H, Yajima A (1999). Oestrogen receptor {alpha} and ß mRNA expression in human endometrium throughout the menstrual cycle. *Mol. Hum. Reprod.* 5: 559-564.

Matthews J, Gustafsson J-A (2003). Estrogen Signaling: A Subtle Balance Between ER{alpha} and ER{beta}. *Mol. Interv.* 3: 281-292.

Matthews J, Wihlen B, Tujague M, Wan J, Strom A, Gustafsson J-A (2006). Estrogen Receptor (ER) {beta} Modulates ER{alpha}-Mediated Transcriptional Activation by Altering the Recruitment of c-Fos and c-Jun to Estrogen-Responsive Promoters. *Mol Endocrinol* 20: 534-543.

McKenna NJ, Lanz RB, O'Malley BW (1999). Nuclear Receptor Coregulators: Cellular and Molecular Biology. *Endocr Rev* 20: 321-344.

Miksicek RJ, Lei Y, Wang Y (1993). Exon skipping gives rise to alternatively spliced forms of the estrogen receptor in breast tumor cells. *Breast Cancer Research and Treatment* 26: 163-174.

Misiti S, Schomburg L, M. Yen P, Chin WW (1998). Expression and Hormonal Regulation of Coactivator and Corepressor Genes. *Endocrinology* 139: 2493-2500.

Monroe DG, Secreto FJ, Subramaniam M, Getz BJ, Khosla S, Spelsberg TC (2005). Estrogen Receptor {alpha} and {beta} Heterodimers Exert Unique Effects on Estrogen- and Tamoxifen-Dependent Gene Expression in Human U2OS Osteosarcoma Cells. *Mol Endocrinol* 19: 1555-1568.

Moore JT, McKee DD, Slentz-Kesler K, Moore LB, Jones SA, Horne EL *et al* (1998). Cloning and Characterization of Human Estrogen Receptor [beta] Isoforms. *Biochemical and Biophysical Research Communications* 247: 75-78.

Morani A, Warner M, Gustafsson JÅ (2008). Biological functions and clinical implications of oestrogen receptors alfa and beta in epithelial tissues. *Journal of Internal Medicine* 264: 128-142.

Nilsson S, Makela S, Treuter E, Tujague M, Thomsen J, Andersson G *et al* (2001). Mechanisms of Estrogen Action. *Physiol. Rev.* 81: 1535-1565.

Ogawa S, Inoue S, Watanabe T, Hiroi H, Orimo A, Hosoi T *et al* (1998a). The Complete Primary Structure of Human Estrogen Receptor [beta] (hER[beta]) and Its Heterodimerization with ER [alpha]in Vivoandin Vitro. *Biochemical and Biophysical Research Communications* 243: 122-126.

Ogawa S, Inoue S, Watanabe T, Orimo A, Hosoi T, Ouchi Y *et al* (1998b). Molecular cloning and characterization of human estrogen receptor betacx: a potential inhibitor ofestrogen action in human. *Nucl. Acids Res.* 26: 3505-3512.

Oñate SA, Tsai SY, Tsai M-J, O'Malley BW (1995). Sequence and Characterization of a Coactivator for the Steroid Hormone Receptor Superfamily. *Science* 270: 1354-1357.

Pearce ST, Jordan VC (2004). The biological role of estrogen receptors alpha and beta in cancer. *Critical Reviews in Oncology/Hematology* 50: 3-22.

Pedram A, Levin MRaER (2006). Nature of Functional Estrogen Receptors at the Plasma Membrane. *Molecular Endocrinology* 20: 1996-2009.

Pedram A, Razandi M, Wallace DC, Levin ER (2006). Functional Estrogen Receptors in the Mitochondria of Breast Cancer Cells. *Mol. Biol. Cell* 17: 2125-2137.

Penot G, Le Peron C, Merot Y, Grimaud-Fanouillere E, Ferriere F, Boujrad N et al (2005). The Human Estrogen Receptor-{alpha} Isoform hER{alpha}46 Antagonizes the Proliferative Influence of hER{alpha}66 in MCF7 Breast Cancer Cells. Endocrinology 146: 5474-5484.

Pfeffer U, Fecarotta E, Castagnetta L, Vidali G (1993). Estrogen Receptor Variant Messenger RNA Lacking Exon 4 in Estrogen-responsive Human Breast Cancer Cell Lines. Cancer Research 53: 741-743.

Pink JJ, Wu SQ, Wolf DM, Bilimoria MM, Jordan VC (1996). A novel 80 kDa human estrogen receptor containing a duplication of exons 6 and 7. Nucl. Acids Res. 24: 962-969.

Poola I, Abraham J, Baldwin K (2002a). Identification of ten exon deleted ER[beta] mRNAs in human ovary, breast, uterus and bone tissues: alternate splicing pattern of estrogen receptor [beta] mRNA is distinct from that of estrogen receptor [alpha]. FEBS Letters 516: 133-138.

Poola I, Abraham J, Baldwin K, Saunders A, Bhatnagar R (2005). Estrogen receptors beta4 and beta5 are full length functionally distinct ERβ isoforms. Endocrine 27: 227-238.

Poola I, Abraham J, Liu A (2002b). Estrogen receptor beta splice variant mRNAs are differentially altered during breast carcinogenesis. The Journal of Steroid Biochemistry and Molecular Biology 82: 169-179.

Poola I, Koduri S, Chatra S, Clarke R (2000). Identification of twenty alternatively spliced estrogen receptor alpha mRNAs in breast cancer cell lines and tumors using splice targeted primer approach. The Journal of Steroid Biochemistry and Molecular Biology 72: 249-258.

Poola I, Speirs V (2001). Expression of alternatively spliced estrogen receptor alpha mRNAs is increased in breast cancer tissues. The Journal of Steroid Biochemistry and Molecular Biology 78: 459-469.

Saji S, Hirose M, Toi M (2005). Clinical significance of estrogen receptor β in breast cancer. Cancer Chemotherapy and Pharmacology 56: 21-26.

Schreihofer DA (2005). Transcriptional regulation by phytoestrogens in neuronal cell lines. Molecular and Cellular Endocrinology 231: 13-22.

Sentis S, Le Romancer M, Bianchin C, Rostan M-C, Corbo L (2005). Sumoylation of the Estrogen Receptor {alpha} Hinge Region Regulates Its Transcriptional Activity. Mol Endocrinol 19: 2671-2684.

Shang Y, Brown M (2002). Molecular Determinants for the Tissue Specificity of SERMs. Science 295: 2465-2468.

Shi L, Dong B, Li Z, Lu Y, Ouyang T, Li J et al (2009). Expression of ER-{alpha}36, a Novel Variant of Estrogen Receptor {alpha}, and Resistance to Tamoxifen Treatment in Breast Cancer. J Clin Oncol 27: 3423-3429.

Simoncini T, Fornari L, Mannella P, Caruso A, Garibaldi S, Baldacci C et al (2005). Activation of nitric oxide synthesis in human endothelial cells by red clover extracts. Menopause:The Journal of The North American Menopause Society 12: 69-77.

Sotoca AM, Sollewijn Gelpke MD, Boeren S, Stöm A, Gustafsson J-Å, Murk AJ et al (2011). Quantitative proteomics and transcriptomics addressing the estrogen receptor subtype-mediated effects in T47D breast cancer cells exposed to the phytoestrogen genistein. Molecular and cellular proteomics 10: M110.002170.

Sotoca AM, van den Berg H, Vervoort J, van der Saag P, Ström A, Gustafsson J-Å et al (2008). Influence of Cellular ER{alpha}/ER{beta} Ratio on the ER{alpha}-Agonist Induced Proliferation of Human T47D Breast Cancer Cells. Toxicological sciences 105: 303-311.

Speirs V, Adams IP, Walton DS, Atkin SL (2000). Identification of Wild-Type and Exon 5 Deletion Variants of Estrogen Receptor {beta} in Normal Human Mammary Gland. *J Clin Endocrinol Metab* 85: 1601-1605.

Stirone C, Duckles SP, Krause DN, Procaccio V (2005). Estrogen Increases Mitochondrial Efficiency and Reduces Oxidative Stress in Cerebral Blood Vessels. *Mol Pharmacol* 68: 959-965.

Ström A, Hartman J, Foster JS, Kietz S, Wimalasena J, Gustafsson aJ-Å (2004). Estrogen receptor beta inhibits 17beta-estradiol-stimulated proliferation of the breast cancer cell line T47D. *Proc Natl Acad Sci U S A* 101: 1566-71.

Suen C-S, Berrodin TJ, Mastroeni R, Cheskis BJ, Lyttle CR, Frail DE (1998). A Transcriptional Coactivator, Steroid Receptor Coactivator-3, Selectively Augments Steroid Receptor Transcriptional Activity. *Journal of Biological Chemistry* 273: 27645-27653.

Szego CM, Davis JS (1967). Adenosine 3',5'-monophosphate in rat uterus: acute elevation by estrogen. *Proc Natl Acad Sci USA* 58: 1711–1718.

Taylor SE, Martin-Hirsch PL, Martin FL (2010). Oestrogen receptor splice variants in the pathogenesis of disease. *Cancer Letters* 288: 133-148

Treeck O, Juhasz-Boess I, Lattrich C, Horn F, Ortmann RGaO (2008). Effects of exon-deleted estrogen receptor b transcript variants on growth, apoptosis and gene expression of human breast cancer cell lines. *Breast Cancer Reserach Treatment* 110: 507–520.

Veld MGRt, Schouten B, Louisse J, Es DSv, Saag PTvd, Rietjens IMCM *et al* (2006). Estrogenic Potency of Food-Packaging-Associated Plasticizers and Antioxidants As Detected in ER and ER Reporter Gene Cell Lines. *Journal or Agriculture and Food Chemistry* 54 4407 -4416.

Virgili F AF, Ambra R, Rinna A, Totta P, Marino M. (2004). Nutritional flavonoids modulate estrogen receptor alpha signaling. *IUBMB Life* 56: 145-151.

Wang Y, Miksicek RJ (1991). Identification of a Dominant Negative Form of the Human Estrogen Receptor. *Molecular Endocrinology* 5: 1707-1715.

Wang Z, Zhang X, Shen P, Loggie BW, Chang Y, Deuel TF (2005). Identification, cloning, and expression of human estrogen receptor-[alpha]36, a novel variant of human estrogen receptor-[alpha]66. *Biochemical and Biophysical Research Communications* 336: 1023-1027.

Warner M, Gustafsson J-A (2006). Nongenomic effects of estrogen: Why all the uncertainty? *Steroids* 71: 91-95.

Weitzmann MN, Pacifici R (2006). Estrogen deficiency and bone loss: an inflammatory tale. *J. Clin. Invest.* 116: 1186-1194.

Zhang Q-X, Borg A, Fuqua SAW (1993). An Exon 5 Deletion Variant of the Estrogen Receptor Frequently Coexpressed with Wild-Type Estrogen Receptor in Human Breast Cancer. *Cancer Research* 53: 5882-5884.

Zhang Q-X, Hilsenbeck SG, Fuqua SAW, Borg Å (1996). Multiple splicing variants of the estrogen receptor are present in individual human breast tumors. *The Journal of Steroid Biochemistry and Molecular Biology* 59: 251-260.

Zhao C, Matthews J, Tujague M, Wan J, Strom A, Toresson G *et al* (2007). Estrogen Receptor {beta}2 Negatively Regulates the Transactivation of Estrogen Receptor {alpha} in Human Breast Cancer Cells. *Cancer Res* 67: 3955-3962.

ZhaoYi Wang, XinTian Zhang, Peng Shen, Brian W Loggie, YunChao Chang, Deuel. TF (2006). A variant of estrogen receptor-{alpha}, hER-{alpha}36: Transduction of estrogen- and antiestrogen-dependent membrane-initiated mitogenic signaling. *PNAS* 103: 9063-9068.

# 5

# *E. coli* Alpha Hemolysin and Properties

Bakás Laura, Maté Sabina, Vazquez Romina and Herlax Vanesa
*Instituto de Investigaciones Bioquímicas La Plata (INIBIOLP), CCT- La Plata, CONICET,
Facultad de Ciencias Médicas. Universidad Nacional de La Plata. La Plata, Buenos Aires
Argentina*

## 1. Introduction

Protein toxins are prominent virulence factors in many pathogenic bacteria. While toxins of Gram-positive bacteria do not generally require activation, many toxins of the Gram-negatives are translated into an inactive form and require a processing step.

The most common such step involves a proteolytic cleavage to generate the active form, especially in those toxins with enzymatic activity. Toxins are activated by proteolysis in a variety of ways: As examples, the anthrax toxin is proteolyzed after its interaction with the receptor on the target cell to promote the formation of a prepore (van der Goot & Young, 2009); the toxic subunit of the *Vibrio cholerae* toxin (CT) is posttranslationally modified through the action of a *V. cholerae* protease that generates two fragments, one containing the toxic activity and the other serving to interact with the binding domain (Sanchez & Holmgren, 2011); finally, the toxins that are synthesized as a single polypeptides must be separated by proteolytic cleavage to generate a catalytic, a transmembrane, and a receptor-binding domain—a salient example here being the diphtheria toxin (Murphy, 1996).

Another processing step involves the acylation of proteins, which substitution is achieved by various mechanisms that differ according to the particular fatty acid transferred, the modified amino acid, and the fatty-acyl donor. Myristate and palmitate are the most common fatty acids cross-linked to proteins. Proteins sorted to the bacterial outer membrane or to the eukaryotic plasma membrane undergo processing in which an acyl group is attached to the N-terminal amino acid. In prokaryotes, acyltransferase, lipases, or esterases use catalytic mechanisms involving ester-linked acyl groups attached to serine and cysteine residues; while eukaryotic proteins utilize ester-linked palmitoylation and ether-linked prenylation of cysteine residues for membrane sorting and protein-protein interaction (Stanley *et al.*, 1998).

The pore-forming α-hemolysin (HlyA) of *Escherichia coli*, a member of the RTX toxins, represents a unique class of bacterial toxins that require for activation a posttranslational modification involving a covalent amide linkage of fatty acids to two internal lysine residues (Stanley *et al.*, 1998). In general, protein acylation is divided into labile modifications of internal regions and stable modifications at the N and C termini. By contrast, the mechanism of stable internal acylation of HlyA represents a unique example among prokaryotic proteins, thus generating interest in its study and discussion. After introducing HlyA, its synthesis, posttranslational modification, secretion, and activity; this chapter will focus on the role that covalently bound fatty acids play in the toxin's mechanism of action.

In recent decades, scientific advances have permitted the manipulation of toxins by using different strategies for directing toxic moieties to diseased cells and tissues. The end of the chapter will involve a discussion of this so-called *toxin-based therapy* and the potential use of HlyA in that modality.

## 2. The alpha-hemolysin (HlyA) of *E. coli*

Extraintestinal pathogenic *Escherichia coli* (ExPEC) is the causative agent of at least 80% of all uncomplicated urinary-tract infections (UTIs), which pathologies currently rank among the most common of infectious diseases worldwide (Marrs et al. 2005), (Foxman & Brown, 2003). ExPEC strains that cause a UTI are called uropathogenic *E. coli* (UPEC). This unique group of *E. coli* strains can reside in the lower gastrointestinal tract of healthy adults (Foxman *et al.*, 2002), (Yamamoto *et al.*, 1997), but upon entry into the urinary tract can ascend to and colonize the bladder, causing cystitis. The infection may be confined to the bladder, or bacteria may ascend into the ureters to infect the kidneys and cause pyelonephritis. In severe cases, bacteria can further disseminate across the proximal-tubular and capillary endothelia to the bloodstream, causing bacteremia (Mobley *et al.*, 2009.). A significant proportion of UTIs occur in patients with no known abnormalities of the urinary tract—the so-called *uncomplicated UTIs*. Certain host characteristics, however, such as a congenital defect in urinary-tract anatomy, are considered complicating factors for UTI and accordingly increase susceptibility to this infection as well as affect its diagnosis and management (Foxman, 2002.). Finally, colonization of the bladder in high numbers may occur without eliciting symptoms in the host, a condition known as asymptomatic bacteriuria (Hooton *et al.*, 2000.). In recent years, an enormous amount of information has accrued through sequencing the genomes of several ExPEC patients. These data, together with epidemiological analyses, have confirmed that different ExPEC pathotypes share many known as well as putative virulence factors. These latter include a number of secreted toxins, iron-acquisition systems, adhesins, and capsular antigens (Wiles *et al.*, 2008). Secreted toxins—which proteins include **HlyA**, the cytotoxic necrotizing factor-1 (CNF-1), and the secreted autotransporter—can alter host signaling cascades, disrupt inflammatory responses, and induce host-cell death; whereas bacterial siderophores like aerobactin, bacteriocin, and enterobactin allow the ExPEC to sequester iron away from the host (Guyer *et al.*, 2002), (Wiles *et al.*, 2008). Adhesive organelles can mediate ExPEC interaction with, and entry into, host cells and tissues; while the expression of encapsulation may enable ExPEC to more effectively avoid professional phagocytes (Wiles *et al.*, 2008), (Dhakal *et al.* 2008).

Experiments in murine and cell-culture model systems have demonstrated that high levels of HlyA can cause the osmotic lysis of host cells, while sublytic concentrations of this pore-forming toxin can modulate host-survival pathways by interfering with phagocyte chemotaxis (Wiles *et al*, 2008),(Jonas *et al.*, 1993), (Cavalieri & Snyder, 1982), (Chen *et al.*, 2006). Both HlyA and CNF-1 may in addition stimulate the breakdown of tissue barriers and the release of nutrients (Smith *et al.* 2008), (Bauer & Welch, 1996), but through the use of the zebrafish infection model phagocytes were found that appeared to be the primary targets of these toxins (Wiles $et\ al.$,2009).

HlyA represents the prototype of the first RTX family of proteins characterized by Rodney Welch (Welch 1991). Produced by a variety of Gram-negative bacteria, these proteins exhibit two common features: The first is the presence of arrays of glycine- and aspartate-rich nonapeptide repeats, which sequences are located at the C-terminal portion. The second is the unique mode of secretion via the type-I system (an ABC-binding–cassette transporter). This first group of RTX toxins consists of toxins — mostly exhibiting cytotoxic pore-forming activity — that often are first detected as a hemolytic halo surrounding bacterial colonies grown on blood-agar plates (Muller $et\ al.$, 1983), (Welch, 1991), $(Felmlee\ et\ al.$, 1985). Recently, a subgroup of very large RTX toxins (>3200 residues) were discovered with multiple activities, such as protease and lipase. These pathogens were named the multifunctional autoprocessing RTX toxins, with the $Vibrio\ cholerae$ toxin being the prototype of this group. In summary, the RTX proteins form a large and diverse family with a broad spectrum of biological and biochemical activities (Linhartova, $et\ al.$, 2010).

## 2.1 Synthesis and structure of HlyA

The synthesis, maturation, and secretion of $E.\ coli$ HlyA are determined by the $hlyCABD$ operon ((Felmlee $et\ al$, 1985), (Issartel $et\ al.$, 1991), (Koronakis $et\ al.$, 1997), (Nieto $et\ al.$ ,1996)). The membrane-associated export proteins are synthesized at a lower level than the cytosolic HlyC and pro-HlyA, in part because of transcription termination within the $hlyCABD$ operon (Felmlee $et\ al$, 1985). This termination is suppressed by the elongation protein RfaH and a short 59-bp, $ops$ (operon polarity suppressor) (Bailey $et\ al.$, 1992, 1996), (Cross $et\ al.$, 1990), (Nieto $et\ al.$, 1996) that act in concert to allow the transcription of long operons such as $hly$, $rfa$, and $tra$ encoding the synthesis and export of extracellular components key in the virulence and fertility of Gram-negative bacteria (Bailey $et\ al.$, 1992, 1997).

The structural gene $hlyA$ produces a single 110-kDa polypeptide. The estimated pI of the toxin is 4.5, with this characteristic being common among the RTX toxins. The N-terminal hydrophobic domain is predicted to contain nine amphipathic α-helices (Soloaga $et\ al.$, 1999). Using photoactivable liposomes, Hyland $et\ al$ (2001) demonstrated that the region comprised between residues 177-411 is the one that becomes inserted into membranes. The C-terminal calcium-binding domain contains 11-17 of the glycine- and aspartate-rich nonapeptide β-strand repeats. Although the membrane interaction of HlyA is assumed to occur mainly through the amphipathic α-helical domain, that both major domains of HlyA are directly involved in the membrane interaction of HlyA has recently been proposed, with the calcium-binding domain in particular being responsible for the early stages of the HlyA's docking to the target membrane (Sanchez-Magraner $et\ al.$, 2007).

The topic of the existence of a receptor for the toxin in erythrocytes remains quite controversial. Nevertheless, Cortajarena $et\ al$ (2003) observed that a short sequence from the C-terminal domain (between residues 914–936) was the main HlyA segment that bound to the glycophorin A on erythrocytes.

The last 60 C-terminal amino acids consist of 2 α-helices separated by 8-10 charged residues. This domain is implicated in the transport of the toxin to the extracellular medium (Hui $et\ al.$, 2000). Fig. 1 shows a scheme of the HlyA structure.

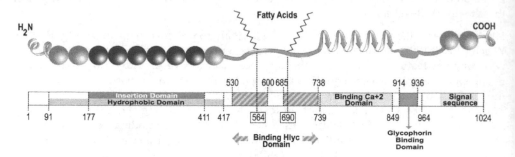

Fig. 1. A scheme of the HlyA structure.

The more relevant domains of HlyA are indicated.

## 2.2 The posttranslational activation of HlyA

The proHlyA protoxin is matured in the cytosol to the active form by HlyC-directed fatty acylation before export from the toxin-producing bacteria. This process consists in a posttranslational modification of the ε-amino groups of internal lysine residues by covalent attachment of amide-linked fatty-acyl residues. This reaction is catalyzed by the HlyC acyltransferases expressed together with the protoxins (Goebel & Hedgpeth, 1982). The mechanism of this novel type of protein acylation was extensively analyzed for HlyA (Issartel et al., 1991), (Stanley et al., 1994). HlyC uses the fatty-acyl residues carried by acyl-carrier protein (ACP) to form a covalent acyl-HlyC intermediate, which species then transfers the fatty-acyl residues to the ε-amino groups of the Lys 564 and Lys 690 residues of proHlyA (Worsham et al., 2001, 2005). ACPs carrying various fatty-acyl residues— including palmitate (16:0) and palmitoleate (16:1), the most common in E. coli—could be efficiently used in vitro as acyl donors for the modification of HlyA ((Issartel et al., 1991), (Trent et al., 1998)). In vivo, however, HlyC exhibits a high selectivity for myristic acid (14:0), which species was found to constitute about 68% of the acyl chains covalently linked to Lys 564 and Lys 690 of the native HlyA (Lim et al., 2000). Contrary to expectations, the extremely rare odd-carbon saturated fatty-acyl residues 15:0 and 17:0 were found to constitute the rest of the in-vivo acylation of HlyA in two different clinical E. coli isolates (Lim et al., 2000). Both acylation sites in the HlyA genome function independently of one another with respect to the kinetics of their interaction with acyl-HlyC (Langston et al., 2004). By using deleted protoxin variants and protoxin peptides as substrates in an in-vitro maturation reaction dependent on only HlyC and acyl-ACP, two independent HlyC-recognition domains were identified on the HlyA protoxin, each of which spanned one of the target lysine residues (Stanley et al., 1996). Each domain required 15 to 30 amino acids for basal recognition and 50 to 80 for full wild-type acylation, but HlyC recognized a large topology rather than a linear sequence. The loss of the Lys 564 acylation site either by mutation or structural deletion affected the thermodynamics of the acylation reaction at Lys 690, implying an undefined connectivity between the two acylation sites (Worsham et al., 2005). Nevertheless, the intact acylation at Lys 690 is essential for HlyA activity.

No other HlyA sequences are required for toxin maturation, including the immediately C-terminal $Ca^{+2}$ binding repeats. Indeed, *in vitro*, $Ca^{+2}$ ions prevent acylation at both sites (Stanley *et al.*, 1996). The extreme sensitivity of the proHlyA activation reaction to free $Ca^{+2}$ supports the view that intracellular $Ca^{+2}$ levels in *E. coli* are too low to affect toxin activity and that $Ca^{+2}$ binding does not occur until the toxin is outside the cell.

This posttranslational modification is remarkable because the behavior of the protein is changed by lipid modification from a benign protein to a frank toxin—part of this transformation being an exclusive mechanism in prokaryotes since in only a few eukaryotic proteins is this type of acylation found (for example, in the nicotinic acetylcholine receptor; the insulin receptor; and cytokines such as TNF-$\alpha$, IL-1$\alpha$ and IL-1$\beta$) (Stanley *et al.*, 1998). In the following section we discuss the role that these covalently bound fatty acids play in the toxin's mechanism of action.

### 2.3 The secretion of HlyA into the extracellular medium

Maturation increases the hydrophobicity of the protein, but that property is not required for export (Ludwig et al. 1987). *E. coli* HlyA-related toxins are all secreted across both membranes by the type-I export process employing an uncleaved C-terminal recognition signal (Nicaud *et al.*, 1986), (Stanley *et al.*, 1991), but no N-terminal leader peptide (Felmlee *et al.*, 1985) or periplasmic intermediate (Felmlee & Welch, 1988), (Koronakis *et al.*, 1989). The HlyA secretory apparatus comprises HlyB (an inner-membrane traffic ATPase, the ATP-binding cassette), HlyD (a membrane-fusion protein), and TolC (an outer-membrane protein) (Schulein *et al.*,1992), (Wandersman & Delepelaire, 1990), (Wang *et al.*, 1991). In *E. coli* and most other pathogens, TolC is encoded by a separate gene from *hlyCABD*. As mentioned before (*cf*. **Section 2.0**) the type-I-secretion–signal sequences have been located within the last 60 C-terminal amino acids, consisting of 2 $\alpha$-helices separated by 8-10 charged residues (Hui *et al.*, 2000).

The mechanism of exportation of HlyA is as follows: The trimeric accessory protein HlyD has been proposed to form a substrate-specific complex with the inner-membrane protein HlyB, which latter species subsequently recognizes the C-terminal signal peptide of HlyA. Upon the binding of HlyA, the HlyD trimer interacts with the trimeric TolC protein of the outer membrane, inducing a conformational change and the consequent export of HlyA. This assembly between the complex HlyB-HlyD with TolC very likely occurs because, as has been demonstrated by X-ray crystallography, the trimeric complex of TolC is very similar in size to the trimeric structure of HlyD, thus facilitating the formation of a continuous transperiplasmic export channel through which HlyA can pass (Koronakis *et al.*, 2000). This complex appears to be transient, with it disengaging and reverting to a resting state once the substrate has been transported (Thanabalu *et al.*, 1998). The energy necessary for the secretion process depends not only on ATP hydrolysis mediated by HlyB but also on the proton motive force exerted on the inner membrane (Koronakis *et al.*, 1991, 1995). Type-I secretion is generally assumed to involve the translocation of unfolded proteins (Young & Holland, 1999), although Pimenta *et al* (2005) have suggested that contact with HlyD directly or indirectly affects the folding of HlyA either during the latter's transit through the translocator or afterwards.

In the last decade many researchers have been interested in this type of secretion machinery because of its potential use in the export of chimeric proteins and in vaccine production (Gentschev *et al.*,1996, 2002).

Although HlyA has its own machinery for export from the bacteria, the presence of a physiologically active HlyA in the outer-membrane vesicles (OMVs) of clinical-hemolytic (Balsalobre *et al.*, 2006) as well as laboratory-recombinant strains of *E. coli* (Herlax *et al.*, 2010) has recently been demonstrated.

OMVs are constantly being discharged from the surface of Gram-negative bacteria during bacterial growth. All Gram-negative bacteria studied to date, including *E. coli*, produce OMVs; and their release is increased when the bacteria are exposed to stressful conditions such as antibiotics or serum. Even though the release of OMVs could not be demonstrated *in vivo*, the presence of particles resembling those vesicles has, in fact, been detected in plasma from patients with different infectious processes (Beveridge, 1999). OMVs serve as secretory vehicles for the proteins and lipids of Gram-negative bacteria and in this manner play roles in establishing a colonization niche for carrying or transmitting virulence factors into host cells or otherwise modulating the host defense and response, thus acting as well as long-range virulence factors that can protect luminal cargo from extracellular host proteases and so penetrate into tissues more readily than the larger bacteria (Kuehn & Kesty, 2005). In addition to toxin-protein delivery, other roles have been characterized for OMVs—namely, interspecies interaction and communication during multispecies infections plus DNA uptake and transfer (Mashburn-Warren & Whiteley, 2006). In the particular example of HlyA, we have demonstrated that the toxin secreted in this way is transferred to the target cell in a concentrated manner and as such is more hemolytically efficient than the free HlyA (Herlax *et al.*, 2010). Moreover, Balsalobre *et al.* (2006) demonstrated that the HlyA associated with OMVs is protected from the attack of proteases, thus facilitating the survival of the toxin within the adverse medium of a patient's plasma.

### 2.4 The mechanism of action of HlyA

HlyA belongs to one class of a wide range of host-cell-specific toxins. HlyA acts on a variety of cell types from several species—*e. g.*, red blood cells, embryo and adult fibroblasts, granulocytes, lymphocytes, and macrophages (Cavalieri *et al.*, 1984)—and also binds to and disrupts protein-free liposomes (Ostolaza *et al.*, 1993).

The host environments encountered by the ExPEC are extremely nutrient-poor; and the function of HlyA has generally been thought to be primarily the destruction of host cells, thereby facilitating the release of nutrients and other factors, such as iron, that are critical for bacterial growth. The lytic mechanism of HlyA is a complex process. Three stages seem to be involved that ultimately lead to cell lysis: binding, insertion, and oligomerization of the toxin within the membrane.

Studies that have explored the binding of HlyA to membranes and the characterization of a putative toxin-specific receptor have produced contradictory results. First, the lymphocyte function-associated antigen (LFA-1) (CD11a/CD18; $\alpha_1\beta_2$ integrin), was reported to serve as the receptor for HlyA on polymorphonuclear neutrophils (Lally *et al.*, 1997) and HlyA was found to recognize and bind the N-linked oligosaccharides to their $\beta$2-integrin receptors (Morova *et al.*, 2008). This finding raises the possibility that the initial binding of the toxin to various cells might occur through the recognition of glycosylated membrane components, such as glycoproteins and gangliosides. Recently, Cortajarena *et al.* (2001) found that HlyA binds to the glycophorin of horse erythrocytes and that this binding was abolished by a trypsinization of the membranes. In addition, these authors found that the glycophorin purified from erythrocyte ghosts and reconstituted in liposomes significantly increased liposomal sensitivity to HlyA. Amino acids 914-936 of HlyA were subsequently hypothesized to be responsible for binding to the ghost receptor (Cortajarena et al., 2003).

Other studies, however, indicated that the binding of HlyA to cells occurred in a nonsaturable manner and that the toxin did not interact with a specific protein receptor either on granulocytes or erythrocytes (Valeva et al., 2005). Nevertheless, HlyA produces protein-free liposome disruption. Ostolaza et al. have reported that HlyA causes the release of fluorescent solutes following a so-called all-or-none mechanism. Using large unilamellar vesicles of different lipid compositions, the authors found that the vesicles composed of phosphatidylcholine, phosphatidylethanolamine, and cholesterol at a molar ratio of 2:1:1 were the most sensitive (Ostolaza et al., 1993). These results demonstrated that the presence of a receptor was not necessary for hemolysis to occur. These contradictory findings regarding the presence or absence of a toxin-specific receptor might be related to the different amounts of toxin and/or the different types and animal species of target cells used in the various studies. At all events, the interaction of HlyA with a target-cell membrane devoid of any specific proteinaceous receptor appears to occur in two steps: an initial reversible adsorption of the toxin that is sensitive to electrostatic forces followed by an irreversible membrane insertion (Bakás et al., 1996), (Ostolaza et al., 1997). Studies with the isolated calcium-binding domain of HlyA revealed that that part of the protein may be adsorbed onto the membrane during the early stages of HlyA-membrane interaction (Sanchez-Magraner et al., 2007).

The next step in the hemolytic process is the insertion of the toxin into the membrane. Hyland et al. (2001) demonstrated that the major region of HlyA that inserts into the membrane is located between residues 177 and 411. The insertion is furthermore independent of membrane lysis since HlyA-protein mutants that are completely nonlytic can insert into lipid monolayers (Sanchez-Magraner et al., 2006). In addition, a binding of calcium to the toxin was shown to induce a protein conformational change that made the insertion process irreversible (Sanchez-Magraner et al., 2006), (Bakás et al. 1998). Once the toxin is inserted, an oligomerization process occurs. We previously found that the fatty acids covalently bound to the toxin induce conformational changes that expose intrinsically disordered regions so as to promote protein-protein interactions. Thus, the oligomerization process of the toxin is facilitated by microdomains within the membrane (Herlax & Bakas, 2007), (Herlax et al., 2009).

The HlyA pore that is formed is highly dynamic because the size depends on both the interaction time and the concentration of the toxin (Welch, 2001). We recently demonstrated that the pore is of a proteolipidic nature since the conductance and membrane lifetime are dependent on membrane composition (Bakas et al., 2006).

Nevertheless, what is not clear is how often HlyA reaches levels that are high enough to lyse host target cells during the course of an infection. In fact, sublytic concentrations of HlyA may even be more physiologically relevant. Indeed, recent studies have demonstrated that sublytic concentrations of a number of pore-forming toxins can modulate a variety of host signaling pathways, including the transient stimulation of calcium oscillations, the activation of MAP-kinase signaling, and the alteration of histone-phosphorylation and -

acetylation patterns (Hamon *et al.*, 2007), (Ratner *et al.*, 2006). In addition, sublytic concentrations of HlyA have been recently found to potently stimulate the inactivation of the serine/threonine protein kinase B (PKB), which enzyme plays a central role in host cell-cycle progression, metabolism, vesicular trafficking, survival, and inflammatory-signaling pathways (Wiles *et al.*, 2008). These findings may help to explain previously published results implicating sublytic concentrations of HlyA in the inhibition of chemotaxis and in bacterial killing by phagocytes in addition to the HlyA-mediated stimulation of host apoptotic and inflammatory pathways (Cavalieri & Snyder, 1982), (Koschinski *et al.*, 2006), (Mansson *et al.*, 2007), (Tran Van Nhieu *et al.*,2004), (Uhlen *et al.*, 2000).

## 3. Role of the fatty acids covalently bound to HlyA

In general, lipid moieties play central roles in protein function—*e. g.*, the targeting into membranes, an increase in the affinity for biological membranes, and an enhancement of protein-protein interactions (Stanley *et al.*, 1998), (Chow *et al.*, 1992).

After a brief introduction to the general aspects of HlyA in the following section, we will describe the role that covalently bound fatty acids play in the mechanism of action of the toxin, from its initial activation to its final functioning in the target cell. This posttranslational modification must be critical since the presence of fatty acids transforms the innocuous proHlyA into the virulent toxin HlyA.

### 3.1 Exposure of intrinsically disordered regions

After the initial activation of HlyA by acylation, the toxin is exported into the extracellular medium by the type-I secretion system and by OMVs. None of the secretion routes are acylation-dependent, although the extracellular transport yield was found to be lower for proHlyA compared to that for HlyA. In addition, a high concentration of ProHlyA was found in inclusion bodies (Sanchez-Magraner *et al.*, 2006). For comparative studies where acylated and nonacylated proteins were used, proHlyA was obtained from *E. coli* DH1—it having been transformed by a recombinant plasmid, pSF4000ΔBamHI, in whose DNA a portion of the *hlyC* gene had been deleted. This strain secreted a full-length, but inactive hemolysin. Fatty acids were not necessary for the secretion of the toxin by OMVs, or by the bacteria's own export machinery; but they were essential for the toxin's hemolytic activity (Boehm *et al.*, 1990).

Several steps are involved in the lytic mechanism of the toxin: a binding of calcium previous to the toxin's interaction with membranes, the binding to and insertion into membranes, and the oligomerization of the toxin to form the final lytic pore. We will discuss below to what extent covalently bound fatty acids influence the different steps.

In the extracellular medium, HlyA must associate with calcium in order to bind to membranes in the lytically active form (Ostolaza & Goñi 1995), (Bakás *et al.*, 1998). This second activation step is acylation-dependent because the calcium-binding capacity is lower in the unacylated protein (Soloaga *et al.*, 1996). Once HlyA is calcium-activated, the toxin appears to have a two-stage interaction with membranes: first, a reversible adsorption that is sensitive to electrostatic forces; and second, an irreversible insertion (Bakás *et al.*, 1996). The inserted HlyA behaves as an integral protein because this form of the toxin cannot be extracted without the use of detergents (Soloaga *et al.*, 1999).

Nevertheless, proHlyA, though nonacylated, also interacts with membranes. This observation is not surprising because the amino-acid sequence of the polypeptide shows amphipathic helices in the 250–400 amino-acid region. Despite the amphipathic stretches known to be essential for lytic activity, however, proHlyA is unable to alter the bilayer permeability (Soloaga et al., 1999). Experiments on protein adsorption at an air-water interface suggested that the fatty acids present in HlyA, unlike those in proHlyA, did not modify the surface-active properties of the protein and that the main difference between the precursor and the mature protein was that the proHlyA was virtually unable to insert itself into lipid monolayers (Sanchez-Magraner et al., 2006). Furthermore, we found that the presence of two acyl chains in HlyA confers on this protein the property of irreversible binding to membranes, which feature is essential for the lytic process to take place (Herlax & Bakas, 2003). In summary, although fatty acids covalently bound to HlyA help the toxin to bind calcium in order to adopt a competitive conformation for interaction with membranes, the absence of these fatty acids does not modify that interaction of the toxin, so that these fatty acids must play some other relevant role. The answer is that the fatty acids expose intrinsically disordered regions of the toxin that are involved in a different step within the mechanism of action.

HlyA has a molten-globule conformation promoted by the presence of acyl chains, as demonstrated by a lower denaturing concentration of guanidinium-chloride. Other characteristics demonstrating this conformation were the binding of a higher number of 8-anilinonaphtalene-1-sulfonate (ANS) molecules to HlyA with a weaker affinity, a higher efficiency of energy transfer from tryptophan to the bound ANS, and a faster digestion of HlyA with trypsin compared to the same reactions with proHlyA (Herlax & Bakas, 2007).

The acylated protein was more stable in the absence of denaturant than the unacylated form, as demonstrated by the higher $\Delta G°H_2O$ value for HlyA compared to proHlyA. Acyl chains covalently bound to the protein, however, promote a steric hindrance that contributes to a more relaxed structure, which acylated form can thus be denatured at a lower guanidinium-chloride concentration.

ANS binding to ordered regions can be distinguished from the binding to molten-globule-like regions by differences in the apparent binding constant. The exceptionally high value of ANS bound to HlyA and proHlyA might result from amphipathic regions in both forms, but the presence of fatty acids has been observed to double this value because of the molten structure those lipids impart. The binding of a large number of ANS molecules in a weak manner is characteristic of the loose structure of the molten conformation. ANS binding to pockets in ordered or molten-globule proteins operationally gives apparent $Kd$ values that differ by more than a factor of 5; thus, despite the uncertainties involved, these apparent $Kd$ values serve as a diagnostic probe to distinguish ordered from molten proteins (Bailey et al., 2001). This structural difference was also observed between the HlyA and proHlyA $Kd$ values, demonstrating by an independent means that the fatty acids on the former induce a molten structure. Moreover, the higher fluorescence-transfer efficiency for HlyA compared to that for proHlyA indicated that the quenching of tryptophan fluorescence was more effective when the binding of ANS to the molten-globule conformation took place, where the accessibility of both the surface and inner tryptophan residues was increased. Thus, the capability of ANS to quench tryptophan fluorescence was seen to be correlated with the ANS-binding behavior.

Proteins with molten-globule-like regions are included in the category of intrinsically disordered proteins, as recently reviewed elsewhere (Dunker et al., 2001). Most of the disordered regions of proHlyA that were predicted through the use of the predictor of naturally disordered regions (PONDR) were located in the C-terminal half of the protein (Fig. 2). These domains could be related to the different steps in this toxin's mechanism of action from its export from the bacterium to pore formation in the target cell.

HlyA carries a carboxy-terminal–secretion signal located within the last 50–60 amino acids (Jarchau et al., 1994). This region is predicted to be disordered; and although export of the toxin has been observed to be acylation-independent (Ludwig et al., 1987), as mentioned above, the yield from extracellular transport for proHlyA was lower than that for HlyA. Consequently, covalently bound acyl chains can expose these signal regions and thus facilitate transport.

Intrinsically unstructured proteins can bind in several different patterns through a process termed *binding promiscuity*. The intrinsic lack of structure can confer functional advantages, including the ability to bind — perhaps in various conformations — to several different target cells. This binding promiscuity would furthermore explain the previously mentioned ambiguity in experimental determinations of the presence of a specific receptor for HlyA published to date (Lally et al., 1997), (Cortajarena et al., 2001), (Valeva et al., 2005).

Many studies have searched for the presence of a receptor for HlyA in different target cells. For example, CD11a and CD18, the two subunits of $\beta$2-integrin, were identified as cell-surface receptors that mediate HlyA toxicity in the human target cells HL60 (Lally et al., 1997). This receptor was found in most circulating leukocytes (lymphocytes, neutrophils, monocytes, and macrophages). Despite the absence of studies identifying the protein region responsible for the interaction with this receptor, studies on the adenylate-cyclase-containing hemolysin of *Bordetella pertusis* (CyaA) — another RTX toxin — revealed that the main integrin-interacting domain of CyaA is located in its glycine/aspartate-rich repeat region; which stretch is characteristic of all protein members of this family. These results allowed the identification of region 1166–1287 as a major CD11b-binding motif (Azami-El-Idrissi et al., 2003). Because this domain is involved in calcium binding, the authors proposed that CyaA shifts from a disordered structure to an R-helical conformation upon calcium binding to the RTX motifs (Rose et al., 1995); therefore, the speculation that the calcium-binding domain composed of glycine-rich tandem repeats corresponding to amino acids 550–850 of HlyA might be involved in the binding to $\beta$2-integrin is tempting. That these regions also match the disordered regions predicted and that acyl chains might be implicated in the exposure since the calcium-binding capacity of proHlyA is lower than that of HlyA, should also be borne in mind (Soloaga et al., 1996).

As cited above in **Section 2.4**, another protein identified as a receptor of HlyA in horse erythrocytes is the glycoprotein glycophorin (Cortajarena et al., 2001). A glycophorin-binding region between residues 914 and 936 accordingly has been identified (Cortajarena et al., 2003). Previous sequence analyses of several RTX toxins had revealed that this stretch was a conserved region. If this region was deleted, the specific binding of HlyA to the cell-surface receptors on erythrocytes was lost without affecting its nonspecific binding (adsorption) to lipid bilayers. This region was also predicted to be intrinsically disordered.

The role of fatty acids in the exposure of disordered regions is supported by results published for the D12-monoclonal-antibody–epitope reactivity. The D12 epitope maps to amino acids 673–726. Since the D12 monoclonal antibody reacts with HlyA, but not with proHlyA; the acylation of the former is directly responsible for the exposure of the epitope within this region (Pellett *et al.*, 1990), (Rowe *et al.*, 1994).

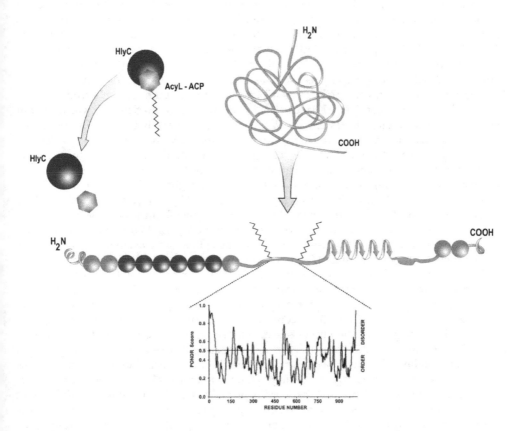

Fig. 2. ProHlyA is activated by acylation.

This process is catalyzed by HlyC, which species transfers a fatty acid from ACP to two internal lysines of ProHlyA (Lys 564 and Lys 690). Once covalently bound, these fatty acids induce a molten-globule conformation in HlyA that exposes intrinsically disordered regions, the existence of which zones was predicted by the predictor of naturally disordered regions PONDR. The amino-acid sequence is represented on the $x$-axis, and the prediction of disorder on the $y$-axis. Peaks >0.5 are strongly predicted to be disordered (Dunker *et al.*, 2005).

## 3.2 Promotion of protein oligomerization

Lipid binding to proteins can also be a determinant of specific protein-protein interactions such as the assembly of proteins into oligomeric complexes. This circumstance obtains for HlyA, where an oligomer was found at lytic concentrations in sheep-erythrocyte ghosts. In contrast, no oligomeric structure was found for proHlyA (Herlax et al., 2009).

Fluorescence-Resonance-Energy Transfer (FRET) is a photochemical process whereby one fluorescent molecule or fluorophore, the "donor", upon excitation by an initial photon of light, spontaneously transfers its energy to another molecule, the "acceptor", by a nonradioactive dipole-dipole interaction (Forster, 1959). The distance over which energy can be transferred depends on the spectral characteristics of the fluorophores, but is generally within the 10–100-Å range. Hence, FRET can be used for measuring structure (Lakowicz et al., 1990), conformational changes (Heyduk, 2002), and interactions between molecules (Parsons et al., 2004). Since HlyA does not contain cysteine residues in its sequences, lysine 344 was replaced by a cysteine (HlyA K344C) and the same point mutation introduced into the unacylated protein (proHlyA K344C). The aim of this point mutagenesis was to permit the binding of only one fluorescent probe per protein, where that mutation—hopefully located in the insertion region of the toxin into membranes (Hyland et al., 2001)—would not affect the hemolytic activity of the toxin. To carry out this study, two populations of HlyA K344C mutant proteins, one labelled with donor (Alexa-488) and the other with acceptor fluorophores (Alexa-546), were bound to sheep-erythrocyte ghosts. Our report showed that an oligomer was involved in the hemolytic mechanism of HlyA (Herlax et al., 2009). FRET can be used to study the distribution of molecules in membranes because the average spacing between molecules of interest will depend primarily on their lateral arrangement. Molecules may be within FRET distance either because they are clustered or because they are randomly distributed at such high surface densities that a fraction of them is within FRET proximity. The latter possibility was avoided in our experiments by using a high lipid/protein molar ratio ($10^9$) to insure that the observed FRET corresponded to oligomerization of the toxin on the erythrocyte surface. In comparison, the absence of FRET in the mutant protein, proHlyA K344C confirmed the participation of the covalently bound fatty acids in the oligomerization process. Fig. 3 shows the fluorescence spectra obtained in the FRET experiments for both proteins. Prima facie, this absence of FRET could be attributed to a reduced binding of the mutant protein to the erythrocyte ghosts, but this possibility was discarded because the percentage of binding to the membranes of both proteins was similar. We need to underscore here that fatty acids are essential for hemolytic activity; and considering that they are needed for oligomerization, we can state that oligomerization is necessary for hemolysis. We thus feel tempted to propose that the presence of fatty acids covalently bound to the protein leads to the exposure of regions that are implicated in protein-protein interactions.

In addition, a critical role of acylation in the oligomerization process to form hemolytic pores has been proposed for the adenylate-cyclase toxin from Bordetella pertussis (cf. **Section 3.1**) (Hackett et al., 1995).

Finally, if we consider that pores formed by HlyA are sensitive to proteases on the cis side of the planar lipid membranes (Menestrina et al., 1987), we could propose the possibility that the part of the toxin remaining external to the membrane is involved in the protein-protein interaction responsible for oligomerization and thus participates in pore formation.

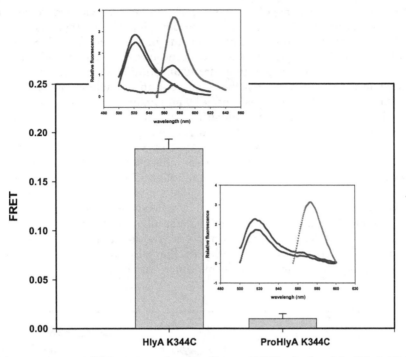

Fig. 3. Oligomerization of HlyA in erythrocyte ghosts. FRET calculated for HlyA K344C and proHlyA K344C bound to erythrocyte ghosts. The lipid/protein ratio was $10^9$. *Left inset*: Example of spectra measured for HlyA K344C. Fluorescence-emission spectrum of erythrocyte ghosts containing donor/acceptor, $F_{D/A}^{(480,\lambda em)}$ (excited at 480 nm; *blue triangle*), emission spectrum of erythrocyte ghosts labelled only with donor, $F_D^{(480,\lambda em)}$ (*violet circle*), emission spectrum of erythrocyte ghosts containing D/A, $F_{D/A}^{(530,\lambda em)}$ (excited at 530 nm where only the acceptor absorbs; *light blue square*), and emission spectrum of erythrocyte ghosts labelled only with acceptor, $F_A^{(480,\lambda em)}$ (*purple square*). *Right inset*: The same emission spectrum as in the left inset but measured for ProHlyA K344C.

### 3.3 Contrary to expectations, fatty acids do not facilitate the interaction of HlyA with membrane microdomains

A variety of pathogens and toxins have been recognized as interacting with microdomains in the plasma membrane. These microdomains are enriched in cholesterol and sphingolipids and probably exist in a liquid-ordered phase, in which lipid acyl chains are extended and ordered (Brown & London, 1998). Many proteins are targeted to these membrane microdomains by their favorable association with ordered lipids. Interestingly, such proteins are linked to saturated acyl chains, which species partition well into those domains (Pike, 2003). Although covalently bound fatty acids had not been implicated in the targeting of HlyA to membranes, their involvement in the targeting to membrane microdomains was studied (Herlax et al., 2009). For this purpose—and taking into account that these microdomains are enriched in cholesterol and sphingolipids—the hemolytic activity of the toxin on sheep erythrocytes was compared with the activity on cholesterol-depleted erythrocytes. The hemolysis rate of the cholesterol-poor erythrocytes was lower than that of the control erythrocytes at each HlyA concentration tested, thus pointing to the participation of cholesterol-enriched microdomains in the oligomerization process. For cholesterol-depleted erythrocytes, at low toxin concentrations, the kinetics of hemolysis seemed to be more complex, suggesting that toxin diffusion in membranes is the rate-limiting step. In order to determine if the decrease in the hemolytic rate observed in the cholesterol-depleted erythrocytes was caused by an impairment of toxin oligomerization, we repeated the FRET experiments comparing control and cholesterol-depleted sheep-erythrocyte ghosts. We demonstrated that cholesterol depletion led to a decrease in FRET of 75% compared to the control sheep ghosts. This result indicated that cholesterol-enriched microdomains played a significant role in the oligomerization process. To obtain more information about the effect of cholesterol-enriched microdomains within the oligomerization process, we performed FRET-kinetics experiments. The role of cholesterol was confirmed by the results of FRET kinetics, where the biphasic behavior of FRET suggested the initial formation of small oligomers, followed by their assembly to form multimeric structures (Fig. 4). The concentration of the small oligomers was favored by the cholesterol-enriched microdomains, where the diffusion time in the membrane became diminished. The number of HlyA molecules that became associated to form the

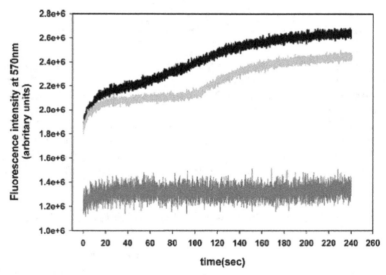

Fig. 4. *FRET kinetics.* Measurement of acceptor fluorescence at 570 nm as a function of time in a mixture composed of HlyA K344C labelled with fluorescent donor and acceptor plus either control erythrocytes (black line) or cholesterol-depleted erythrocytes (light gray line). Measurement of a mixture of unlabelled and labelled with acceptor HlyA K344C with control erythrocytes (*dark gray line*) was done as FRET-negative control. Assays were performed at a ratio of 5 µg of total toxin per 100 µg of phospholipids (as erythrocyte membranes). The excitation and emission monochromators were set at 480 nm and 570 nm, respectively. Alexa-546 emission was measured at a rate of 25 samples/s for 240 s, at 37°C. The *curves* represent the average value of three independent experiments containing five replicates each.

pore was uncertain; nevertheless, the assumption that several molecules could oligomerize to form a pore was not unreasonable. An extension of this reasoning suggested that at high doses a progressive oligomerization of HlyA leads to the fusion of the pore and rapid destruction of the cell membrane with little time for activation of the central apoptotic pathway. By contrast, at lower concentrations, the pores would be smaller and fewer in number so that the cells, though injured, would survive long enough for apoptosis to be observed (Lally et al., 1997). These results can explain why toxin association with erythrocytes at 0–2°C is characterized as a prelytic state, whereas following a shift to 23°C – and after a lag period – lysis begins (Moayeri & Welch, 1997). In conclusion, the fusion of oligomers may be the rate-limiting step in pore formation, and the integrity of the cholesterol-enriched microdomains is necessary for the concentration of HlyA to induce hemolysis. This notion agrees with the findings of Moayeri and Welch (Moayeri & Welch, 1994), who observed that the degree of osmotic protection of erythrocytes afforded by protectants of varying sizes depended on the amount of toxin applied and the duration of the assay. These authors suggested that HlyA creates a lesion with a very small initial size that then increases in apparent diameter over time. Consequently, the larger the oligomer is, the bigger the pore size becomes.

That the terms "membrane microdomains" and "detergent-resistant microdomains" (DRMs) are not synonymous is essential to remember because the two have different origins and conceptual meanings (Lichtenberg et al., 2005). The DRM technique, though, is widely used in the current literature to investigate the interaction between a protein and membrane microdomains. This technique takes advantage of the selective solubilization of different lipids that occurs when a biomembrane is submitted to the action of a nonionic detergent such as Triton X-100. When erythrocyte ghosts were incubated with HlyA and the DRMs were separated by sucrose-gradient ultracentrifugation, the immunoblot analysis revealed that most of the ghost-associated HlyA was localized in the DRMs, indicating that the binding of HlyA to the erythrocyte membranes was mediated by membrane microdomains that served as concentration platforms for the toxin's oligomerization. That proHlyA colocalizes with HlyA and flotillin (a microdomain protein marker) in DRMs emphasizes our hypothesis that the main role of the saturated acyl chain covalently bound to HlyA is a participation in the oligomerization process, and not the targeting to cholesterol-enriched membranes (Fig 5).

A key feature of cholesterol-enriched microdomains is the tight packing of lipid acyl chains in the liquid-ordered phase, where the lipid acyl chains are extended and ordered (Brown & London, 1998). Because of the difficulty in packing membrane-spanning helices into the ordered lipid environment, some proteins are linked to saturated acyl chains and partition well into those microdomains (de Planque & Killian, 2003). Shogomori et al. (2005) found, however, that acylation did not measurably enhance microdomain association, and they concluded that the acylated linker for the activation of T-cell transmembrane domains had a low inherent affinity for cholesterol-enriched microdomains. The possible inferrence is that acylation is not sufficient for the targeting of any transmembrane protein and that therefore a second mechanism – such as protein-protein interactions for microdomain associations – is required (Fragoso et al., 2003), (Cherukuri et al., 2004).

To conclude, we propose that fatty acids covalently bound to HlyA and membrane microdomains are implicated in the hemolysis process. Fatty acids are essential because they induce the exposure of intrinsic disordered regions in the toxin so as to enhance protein-protein interactions in order to form the oligomer, while the membrane microdomains act as platforms for the concentration of the toxin during the oligomerization process.

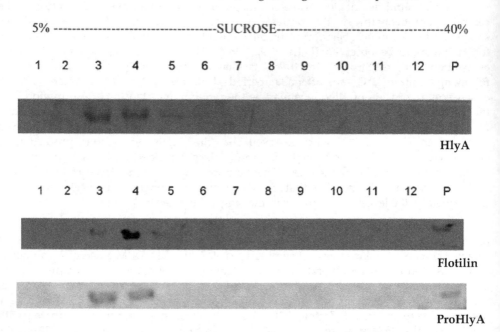

Fig. 5. *Interaction of HlyA with DRMs.*Thirty μg of HlyA were incubated with 100 μl of erythrocyte ghosts for 30 min at 37 °C. Cells were lysed with 1% (v/v) Triton X-100 and insoluble cell components separated by sucrose-density-gradient centrifugation. The gradient fractions were analyzed by immunoblotting with anti-HlyA antibodies. HlyA was present in fractions 3 and 4. Gradient fractions were also analyzed by immunoblotting with anti-Flotillin-1 antibodies. Flotillin-1 appears mainly in fractions 3 and 4. ProHlyA was incubated under the same conditions as HlyA. The gradient fractions were analyzed by immunoblotting with anti-HlyA antibodies. ProHlyA colocalizes with HlyA.

In summary, fatty acids covalently bound to HlyA induce a molten-globule structure in the toxin, exposing intrinsically disordered regions involved in the different steps in the toxin's mechanism of action. Fatty acids expose specific regions that induce protein-protein interaction in the oligomerization process that takes place within the membrane microdomains of erythrocytes. The irreversibility of the toxin's membrane binding promoted by fatty acids might result from the formation of the HlyA oligomeric structure (Herlax & Bakas, 2003,2007), (Herlax *et al.*, 2009).

## 4. Toxin-based theraphy

Bacterial toxins have been defined as "soluble substances that alter the normal metabolism of host cells with deleterious effects on the host" (Schlessinger & Schaechter, 1993). Nonetheless, during the last decade, taking advantage of advances in toxin research, investigators have sought ways of obtaining benefits from toxins. In the present section we will discuss these toxin-based therapies and the possible relevant use of HlyA.

## 4.1 Immunotoxins

Conventional cancer treatments such as surgery, chemotherapy, and radiotherapy often fail to achieve complete cancer remission. Moreover, radiotherapy and/or chemotherapy are almost always cause significant—and sometimes long-lasting—side effects. These considerations have prompted the development of many new approaches for the treatment of cancer. One such example involves the use of immunotoxins (Bernardes et al., 2010).

The term "immunotoxin" classically refers to chimeric proteins with a cell-selective ligand chemically linked or genetically fused to a toxin moiety that can target cancer cells overexpressing tumor-associated antigens, membrane receptors, or carbohydrate antigens. In the 1970s the first therapeutic agents composed of toxins conjugated to antibodies against cell-surface antigens started to emerge as tumor-cell killers (Moolten & Cooperband, 1970), (Moolten et al., 1976). Since then, many hybrid molecules consisting of a toxin coupled to a specific targeting antibody or ligand were developed, with most of these hybrids being directed against tumor cells (Pastan et al., 2007).

First-generation immunotoxins were prepared by chemically conjugating antibodies to natural intact toxin units or to toxins with attenuated cell-binding capability. These constructs, however, were heterogeneous and nonspecific both because multiplicities of potential sites were available for chemical conjugation and since the presence of the cell-binding domain on the toxin led to an intoxication of nontumor cells as well. Immunotoxins of the second generation were also based on chemical conjugation between the targeting moiety and the toxin. Nevertheless, accumulated knowledge on the structure and function of the toxins enabled the removal of their native nonspecific cell-binding domain, thus generating immunotoxins that were much more target-specific when conjugated to monoclonal antibodies. Although more specific and thus better tolerated by animals, immunotoxins from this second generation were still chemically heterogeneous, and their large size hindered them from penetrating solid tumors. In order to avoid this heterogeneity, improve tumor penetration, and reduce production complexity and expense, recombinant-DNA techniques were applied in the production of the third-generation immunotoxins. In these constructs—mostly produced in the bacterium E. coli—the cell-binding domain of the toxin is genetically replaced with a ligand or with the Fv portion of an antibody in which the immunoglobilin light- and heavy-chain variable regions are either genetically linked or held together by a disulfide bond (Shapira & Benhar, 2010).

Among the bacterial toxins that were used for the construction of immunotoxins, the most common were the diphtheria toxin and the pseudomonas exotoxin A, which toxins are naturally produced by the Gram-positive, aerobic Corynebacterium diphtheria and by the Gram-negative, aerobic Pseudomonas aeruginosa, respectively. Clinical trials with different exotoxin A-immunotoxins have already been performed with positive results in leukemia and bladder cancer (Kreitman et al., 2001), (Kreitman et al., 2005), (Biggers & Scheinfeld, 2008).

In spite of the promise shown by bacterial toxin-based chimeric proteins, these hybrids still present several obstacles that limit their clinical application. The toxin part of the fusion proteins elicits a high degree of humoral response in humans. In addition, in developed countries, where people have become immunized against diphtheria, the patient's serum will have circulating antibodies against the diphtheria toxin that will result in a neutralization of diphtheria toxin-based immunotoxins (Hall *et al.*, 2001). Both the *Pseudomonas* exotoxin and the diphtheria toxin are large molecules and are difficult to humanize. At sufficiently high concentrations these fusion proteins lead to symptoms like the vascular-leak syndrome and thus exhibit a certain degree of nonspecific toxicity.

### HlyA as a possible candidate toxin for the synthesis of immunotoxins

Considering all the details of the structure and mode of action of HlyA discussed above, we can state that HlyA can be a good candidate for an effective immunotoxin. Although, certain additional details about the domains implicated in the binding of the toxin to target-cell receptors need to be clarified, we can consider that the domain that comprises amino acids 914–936 should be exchanged for the specific antibody (or ligand) chosen for interaction with the tumor cell of interest. Perhaps some amino acids within the repeat domain that might be involved in the interaction with the ß2-integrins should also be removed in the fusion protein.

### The reason for using HlyA in an immunotoxin

An observation deserving emphasis is that the more relevant effects that HlyA produces during an infection are sublytic rather than cytolytic. Among these effects we must bear in mind the one related to the modulation of the host-signaling cascades, where HlyA was found to produce an inactivation of the PKB (*cf.* **Section 2.4**) – a key protein involved in several pathways related to host-cell survival, inflammatory responses, proliferation, and metabolism (Manning & Cantley, 2007), (Fayard *et al.*, 2005). By inactivating PKB, HlyA is able to fine-tune host responses related to the inflammatory- and apoptosis-signaling cascades that are initiated during the course of an infection. PKB inactivation is produced by an extracellular calcium-dependent, potassium-independent process requiring HlyA insertion into the host plasma membrane and subsequent pore formation. Calcium influx induces the activation of host-protein phosphatases that dephosphorylate PKB, inactivating it and finally inducing host-cell apoptosis. Thus, if a ligand directed at a specific tumor-cell receptor is fused with HlyA, that immunotoxin might induce the apoptosis of the desired cell.

The advantage of using HlyA is that the translocation of the immunotoxin into a tumor cell is not necessary, only its binding to the membrane where HlyA can insert itself and form the pore needed to execute its apoptotic action. Of course, these hypotheses are only possibilities that would warrant further investigation.

### 4.2 Vaccines

In recent years, an increase in the development of vaccination technology has taken place, but the ideal vaccine has not yet been found. In general terms, there are certain criteria that a vaccine must satisfy: it must be capable of eliciting the appropriate immune response; and it should be safe, stable, and reproducible (Perrie *et al.*, 2008).

UTIs caused by UPEC still represent an enormous challenge for the development of vaccines targeted to induce an immunity that can either prevent the infectious agent from attaching to and colonizing the mucosal epithelium and/or can block the binding and action of microbial toxins, such as HlyA (Holmgren et al., 2003). Since the infections by these bacteria occur at, or take their departure from, a mucosal surface; a mucosal route of vaccination should be selected rather than a parenteral one.

A significant aspect of immune responses at mucosal surfaces is the production of a secretory IgA (S-IgA) and its transport across the epithelium. This S-IgA response represents the first line of defence against the invasion by bacterial pathogens. The mucosal immune system is an integrated network of tissues, cells, and effector molecules that functions to protect the host from those pathogens. Furthermore, mucosal lymphocytes exhibit unique homing receptors, the integrins, that recognize ligands expressed on mucosal endothelial cells so as to allow their retention within mucosal tissues for the delivery of cellular and humoral immune responses (Butcher & Picker, 1996). Because of the presence of specific interconnected mucosal induction and effector sites for eliciting the S-IgA antibody response, the mucosal immune system has been shown to be separated from the peripheral immune system. Thus, the induction of peripheral immune responses by parenteral immunization does not necessarily result in significant mucosal immunity; by contrast, mucosal immunization is capable of inducing immune protection in both the external secretions and the peripheral immune compartments (Kiyono et al., 1992), (McGhee et al., 1992).

The induction of immune responses following mucosal immunization is usually dependent upon the coadministration of the appropriate adjuvants that can initiate and support the transition from innate to adaptive immunity. While a number of substances of bacterial origin have been tested as mucosal adjuvants, the most widely used mucosal adjuvants in experimental animals are the cholera toxin (CT) and the closely related E. coli heat-labile enterotoxin (LT). Both CT and LT consist in homopentamers of cell-binding B subunits associated with a single toxically active A subunit. The A subunit enzymatically ADP-ribosylates the Gs protein of adenylate cyclase leading to an increased cAMP production in the affected cells (de Haan & Hirst, 2000). CT and LT can alter several steps in the induction of a mucosal immune response. These effects, which alone or in combination might explain their strong adjuvant action after oral immunization, include: (1) an increased permeability of the intestinal epithelium leading to an enhanced uptake of a coadministered antigen, (2) an enhanced antigen presentation by various antigen-presenting cells, (3) a promotion of isotype differentiation in B cells leading to increased IgA formation, and (4) complex stimulatory as well as inhibitory effects on T-cell proliferation and cytokine production. Finally, CT and LT have been shown not only to avoid an induction of oral tolerance but also to abrogate additional otherwise efficient regimens producing tolerance induction after oral antigen administration (Holmgren et al., 2005).

A number of studies have examined the activity of LT and CT as mucosal adjuvants in vaccines against a variety of bacterial, fungal, and viral pathogens. Representative examples include the tetanus toxoid (Xu-Amano et al., 1993), (Yamamoto et al., 1996), (Cheng et al., 1999), (Xu-Amano et al., 1994), the inactivated influenza virus (Hashigucci et al., 1996), (Tumpey et al., 2001), a recombinant urease from Helicobacter spp. (Lee et al., 1995), (Weltzin et al., 1997), (Lee, 2001), and the pneumococcal surface protein A from S. pneumoniae (Wu et al.,1997). Many other examples have been reported, and all these studies clearly indicate that both LT and CT have significant potential for use as adjuvants for mucosally administered antigens. Nevertheless, both LT and CT are potent enterotoxins, and this property has seriously limited the practical use of these molecules (Freytag & Clements 2005). To avoid such toxicity, a number of nontoxic mutant derivatives of CT or LT have been proposed (Douce G, 1997) (Douce et al. 1998). In particular, single-amino-acid-substitution mutants of LT (R7K, S63K and R192G) that lack ADP-ribosyltransferase activity have been shown to retain their adjuvant properties (Douce et al. 1995).

In contrast, because of their size, plasticity, and safety profile in humans, OMVs are attractive vehicles for vaccine delivery. OMV vaccines for serogroup-B meningococcal disease, consisting in vesicles from *Salmonella thyphimurium* and *Pseudomonas aeruginosa* that contain surface antigens native to the pathogens have been shown to exhibit immunogenic properties (Alaniz et al., 2007), (Bauman & Kuehn, 2006). Two vaccines for serogroup-B meningococcal disease currently exist that are formulations comprising bacterial surface antigens that have been naturally incorporated into OMVs (Oster et al., 2005), (Feiring et al. , 2006). These OMV-based vaccines represent a novel system where both the antigen and delivery vehicle are derived from the *Neisseria meningitidis* pathogen itself (Claassen et al., 1996), (Arigita et al., 2004). Moreover, genetically engineered OMVs offer an attractive possibility for use as easily purified vaccine-delivery systems capable of greatly enhancing the immunogenicity of low-immunogenicity protein antigens without the need for an added adjuvant.

With the development of controlled-release technologies, the engineering of OMVs emerged as a promising strategy for antigen delivery because these vesicles are similar in geometry to naturally occurring pathogens and are readily internalized by antigen-presenting cells, thus avoiding the complex manufacturing steps required to purify and encapsulate antigens into particulate delivery systems such as polymer particles (Singh et al., 2007), immune-stimulating complexes (Morein et al., 1984), liposomes, proteosomes, and related vesicles (Lowell et al., 1988), (Lowell et al., 1988), (Felnerova et al., 2004), (Copland et al., 2005) — all of which processes render these approaches economically unfeasible (Ulmer et al., 2006).

The genetic fusion of the green-fluorescent protein (GFP) as a model subunit antigen with the bacterial hemolysin ClyA resulted in a chimeric protein that elicited strong anti-GFP antibody titers in immunized mice, whereas immunization with GFP alone elicited no such titers. Similar to native unfused ClyA, the chimeric ClyA-fusion proteins were found localized in bacterial OMVs, where they retained the activity of the fusion partners, thus demonstrating for the first time that ClyA can be used to colocalize fully functional heterologous proteins directly in bacterial OMVs. The anti-GFP humoral response in mice immunized with the engineered OMV formulations was indistinguishable from the response to the purified ClyA-GFP fusion protein alone and was equal to the response to purified proteins adsorbed to aluminum hydroxide, a standard adjuvant. Engineered OMVs containing ClyA-GFP were easily isolated by ultracentrifugation, thus effectively eliminating the need for a laborious antigen purification from cell-culture expression systems (Chena et al., 2010). The retention of hemolytic-protein activity indicated that ClyA-antigen fusions maintained their conformations. Although no pathologic effects were observed in mice immunized with ClyA, a detoxification of the toxin through mutation, truncation, or chemical methods may attenuate any possible toxicity while still retaining the hybrid's immunomodulatory capabilities.

On the basis of all these data, HlyA presents many properties that can be considered when designing a vaccine.

**Anti- UPEC vaccine:** The urinary tract is one of the most common sites of bacterial infection. As mentioned above, over half (53%) of all women along with 14% of men experience at least one UTI in their lifetime (Griebling, 2005), (Griebling, 2005). *E. coli* is the infectious agent in more than 80% of the uncomplicated UTIs (Marrs *et al.*, 2005), (Foxman & Brown, 2003). In addition, the upper UTIs of young children can cause permanent kidney damage. An estimated 57% of children with acute pyelonephritis develop renal scarring (Lin *et al.*, 2003).

In recent years, an increase in the antibiotic resistance of UPEC isolates has been observed (Bours *et al.*, 2010) that imposed an urgent need for alternative treatment and prevention strategies to combat this serious and widespread human pathogen. With this aim, much research has been focussed on the development of vaccines to stimulate protective immunity against UPEC. In those studies, surface-exposed molecules such as P fimbriae, the lipopolysaccharide core, α-hemolysin, and the salmochelin receptor IroN have been utilized as antigens for subunit vaccines (Goluszko *et al.*, 2005), (Russo *et al.*,2003), (O'Hanley *et al.*, 1991); but the limited success of these strategies prevented any vaccine from being currently available. One consideration is that this vaccine has to generate immune responses at the level of mucosal surfaces.

Large-scale reverse-vaccinology approaches offer an alternative to the traditional vaccine design through applying genomic and bioinformatic methods to identify novel vaccine targets (Pizza *et al.*, 2000). Using this technique, Alteri *et al* identified a class of molecules involved in iron acquisition as vaccine candidates and reported that intranasal immunization with this UPEC outer-membrane iron receptor generated an antigen-specific humoral response to provide protection from UTI (Alteri *et al.*, 2009). The authors proposed that the targeting of an entire class of molecules instead a single protein would permit the identification of components of a more generally protective UTI vaccine and that this strategy could be used in the development of vaccines to prevent infections caused by other pathogenic bacteria. During this present year, these same authors, using the *in-vivo*-induced–antigen technology, identified a novel UPEC virulence factor (*tosA*, a gene encoding a predicted repeat-in-toxin family member) that could be useful as a potential vaccine target (Vigil *et al.*, 2011). Although this methodology did not identify HlyA as a potential candidate for this vaccine, the introduction of that toxin would be beneficial. First of all, HlyA has been recognized as one of the main virulence factors associated with the pathogenicity caused by UPEC (Wiles, Kulesus and Mulvey 2008); second, the toxin induced an immunity response in host organisms (O'Hanley *et al.*, 1991) and thus is immunogenic in its native state; third, it can also produce focal leaks in intestinal epithelia (Troeger *et al.*, 2007). Focal leaks are small openings within the epithelium where bacterial penetration occurs. HlyA induces such focal leaks in a proinflammatory environment—those being also induced by the secretion of the cytokines TNFα and IL-13. Of relevance to highlight is that HlyA can increase the permeability of the intestinal epithelium so as to lead to an enhanced uptake of a coadministered antigen, thus acting as both a coadjuvant and an antigen in its own right. The dose that induces this effect would naturally have to be extensively investigated.

HlyA can also be used as adjuvants in any other vaccine design against another pathogen. For example, the toxin can be included in any liposomal vaccine in order to facilitate uptake though epithelia for the induction of immunity.

**OMV vaccines**: Balsalobre *et al* (2006) demonstrated that physiologically active HlyA is associated with the OMVs produced from *E. coli* laboratory strains and also from natural and clinical isolates. In our laboratory, we found that the unacylated toxin (proHlyA) can also be associated with OMVs (Herlax *et al.*, 2010). On the basis of this finding, OMV vaccines can be designed by effecting a fusion of the desired antigen with ProHlyA. In this way, ProHlyA would direct the exposure of the antigen on the surface of the OMVs without inducing any cytotoxic response. An advantage of OMV vaccines is that, because of their size and lipopolysaccharide content, they are able to induce an adequate immune response.

Finally, mention must be made that these hypotheses are just speculative on the basis of what is known about the structure and function of HlyA, whose application in toxin-based therapy still has to be exhaustively investigated and especially the immune response the toxin might evoke.

## 5. Conclusion

*E. coli* is one of the predominant species of facultative anaerobes in the human gut and in the majority of the cases is harmless to the host. These strains are mostly commensals but also contain a group called the extraintestinal pathogenic *E. coli* (ExPEC). Usually the ExPEC are also harmless colonizers but under certain circumstances can translocate and cause infection. The main virulence factor responsible for this translocation is the HlyA toxin, which pathogen is mainly associated with severe UTI but in addition with bacteremia and extraintestinal infections. In this chapter an exhaustive description of the toxin has been delineated; including its synthesis, maturation, and export from the bacteria. Effects produced by HlyA in different target organs have also been discussed. The significance of the maturation process for the toxin cannot be understated. The acylation of the protein at two internal lysines gives the toxin its virulence, by exposing intrinsic disordered regions that are essential to different steps of the toxin's mechanism of action. The further exposure of regions involved in the protein-protein interaction within the oligomerization process is responsible for the permeability induced in all the target cells, despite the intracellular signal pathway the toxin induces in each specific organ. This activation is unique to prokaryotic proteins.

Based on the already known structural and functional characteristics of HlyA, we might speculate about its use in toxin-based therapy. Such therapy is a versatile and dynamic research area with a great potential application. Further investigation, however, is required in order to improve the efficiency and safety of toxin-based agents. Investments in the development of delivery and targeting techniques are definitely needed in order to achieve this goal, though the basic research on the structure and mechanism of natural toxins should nevertheless not be abandoned. Topics related to HlyA have still to be clarified concerning the existence of a toxin-specific receptor in target cells and the domains of the toxin involved in its interaction with those putative binding sites. The deeper our knowledge becomes about this unique family of secreted polypeptides, the more easily will we be able to harness their great potential for our own benefit.

## 6. Acknowledgements

We thank Prof. Norma Tedesco for revising the English grammar; Mario Ramos for the graphic designs; and Dr. Donald F. Haggerty, a retired career biochemist and native English speaker, for editing the final version of the manuscript.

This work was supported by grants from the Comisión de Investigaciones Científicas de la Provincia de Buenos Aires, and Agencia Nacional de Promoción Científica Grant PICT N° 647, Argentina.

## 7. References

Alaniz, R. C., B. L. Deatherage, J. C. Lara and B. T. Cookson (2007). "Membrane vesicles are immunogenic facsimiles of Salmonella typhimurium that potently activate dendritic cells, prime B and T cell responses, and stimulate protective immunity *in vivo*." *J Immunol* 179: 7692–7701.

Alteri, C. J., E. C. Hagan, K. E. Sivick, S. N. Smith and H. L. Mobley (2009). "Mucosal immunization with iron receptor antigens protects against urinary tract infection." *PLoS Pathog.* 5(9): e1000586

Arigita, C., W. Jiskoot, J. Westdijk, C. van Ingen, W. E. Hennink, D. J. Crommelin and G. F. Kersten (2004). "Stability of mono- and trivalent meningococcal outer membrane vesicle vaccines." *Vaccine* 22: 629–642.

Azami-El-Idrissi, M. E., C. Bauche, J. Loucka, R. Osicka, P. Sebo, D. Ladant and C. Leclerc (2003). " Interaction of *Bordetella pertussis* Adenylate Cyclase with CD11b/CD18 : Role of Toxin acylation and identification of the main Integrin interaction domain." *J. Biol. Chem* 278: 38514-38521

Bailey, M., V. Koronakis, T. Schmoll and C. Hughes (1992). "*Escherichia coli* HlyT protein, a transcriptional activator of haemolysin synthesis and secretion, is encoded by *rfaH* (*srfB*) locus requierd for expression of sex factor and lipopolisaccharide genes." *Mol Microbiol.* 6: 1003-1012

Bailey, M. J., C. Hughes and V. Koronakis (1996). "Increased distal gene transcription by the elongation factor RfaH, a specialized homologue of NusG." *Mol Microbiol* 22(4): 729-37

Bailey, M. J., C. Hughes and V. Koronakis (1997). "RfaH and the ops element, components of a novel system controlling bacterial transcription elongation." *Mol Microbiol* 26(5): 845-51

Bailey, R., A. Dunker, C. Brown, E. Garner and G. M (2001). "Clusterin, a Binding Protein with a Molten Globule-like Region." *Biochemistry* 40: 11818-11840

Bakas, L., A. Chanturiya, V. Herlax and J. Zimmerberg (2006). "Paradoxical lipid dependence of pores formed by the *Escherichia coli* alpha-hemolysin in planar phospholipid bilayer membranes." *Biophys J* 91(10): 3748-55

Bakás, L., H. Ostolaza, W. L. Vaz and F. M. Goñi (1996). "Reversible adsorption and nonreversible insertion of *Escherichia coli* alpha-hemolysin into lipid bilayers." *Biophys J.* 71(4): 1869-1876

Bakás, L., M. Veiga, A. Soloaga, H. Ostolaza and F. Goñi (1998). "Calcium-dependent conformation of E. *coli* alpha-haemolysin. Implications for the mechanism of membrane insertion and lysis." *Biochim Biophys Acta* 1368(2): 225-234

Balsalobre, C., J. M. Silvan, S. Berglund, Y. Mizunoe, B. E. Uhlin and S. N. Wai (2006). "Release of the type I secreted alpha-haemolysin via outer membrane vesicles from *Escherichia coli.*" *Mol Microbiol* 59(1): 99-112

Bauer, M. E. and R. A. Welch (1996). "Association of RTX toxins with erythrocytes." *Infec Immun* 64(11): 4665-4672

Bauman, S. J. and M. J. Kuehn (2006). "Purification of outer membrane vesicles from *Pseudomonas aeruginosa* and their activation of an IL-8 response." *Microbes Infect* 8: 2400–2408.

Bernardes, N., R. Seruca, A. Chakrabarty and A. M. Fialho (2010). "Microbial-based therapy of cancer: Current progress and future prospects" *Bioengineered Bugs* 1(3): 178-190

Beveridge, T. J. (1999). "Structures of Gram-negative cell walls and their derived membrane vesicles." *J.Bacteriol* 181: 4725-4733

Biggers, K. and N. Scheinfeld (2008). "VB4-845, a conjugated recombinant antibody and immunotoxin for head and neck cancer and bladder cancer." *Curr Opin Mol Ther* 10: 176-186

Boehm, D., R. Welch and I. Snyder (1990). "Domains of *Escherichia coli* hemolysin (HlyA) involved in binding of calcium and erythrocyte membranes." *Infect Immunity* 58(6): 1959-1964

Bours, P. H., R. Polak, A. I. Hoepelman, E. Delgado, A. Jarquin and A. J. Matute (2010). " Increasing resistance in community-acquired urinary tract infections in Latin America, five years after the implementation of national therapeutic guidelines." *Int. J. Infect. Dis.* 4: e770–e774

Brown, D. A. and E. London (1998). "Structure and origin of ordered lipid domains in biological membranes." *J Membr Biol* 164: 103-14

Butcher, E. C. and L. J. Picker (1996). "Lymphocyte homing and homeostasis." *Science* 272: 60-66.

Cavalieri, S., G. A. Bohach and I. Synder (1984). "*Escherichia coli* _-hemolysin characteristics and probable role in pathogenicity." *Microbiol. Rev* 48: 326-343

Cavalieri, S. J. and I. S. Snyder (1982). "Effect of *Escherichia coli* alpha-hemolysin on human peripheral leukocyte viability in vitro." *Infect Immun* 36(2): 455-61

Chen, M., R. Tofighi, W. Bao, O. Aspevall, T. Jahnukainen, L. E. Gustafsson, S. Ceccatelli and G. Celsi (2006). "Carbon monoxide prevents apoptosis induced by uropathogenic *Escherichia coli* toxins." *Pediatr Nephrol* 21(3): 382-9

Chena, D. J., N. Osterriederb, S. M. Metzgerb, E. Bucklesd, A. M. Doodye, M. P. DeLisaa and D. Putnam (2010). "Delivery of foreign antigens by engineered outer membrane vesicle vaccines." *PNAS* 107( 7): 3099–3104

Cheng, E., L. Cardenas-Freytag and J. D. Clements (1999). "The role of cAMP in mucosal adjuvanticity of *Escherichia coli* heat-labile enterotoxin (LT)." *Vaccine* 18: 38–49.

Cherukuri, A., T. Shoham, H. W. Sohn, S. Levy, S. Brooks, R. Carter and S. K. Pierce (2004). "The tetraspanin CD81 is necessary for partitioning of coligated CD19/CD21-B cell antigen receptor complexes into signaling-active lipid rafts." *J. Immunol.* 172: 370-380

Chow, M., C. J. Der and J. E. Buss (1992). "Structure and biological effects of lipid modifications on proteins." *Curr Opin Cell Biol* 4(4): 629-36

Claassen, I., J. Meylis, P. van der Ley, C. Peeters, H. Brons, J. Robert, D. Borsboom, A. van der Ark, I. van Straaten, P. Roholl, B. Kuipers and J. Poolman (1996). "Production, characterization and control of a *Neisseria meningitidis* hexavalent class 1 outer membrane protein containing vesicle vaccine." *Vaccine* 14: 1001–1008.

Copland, M. J., T. Rades, N. M. Davies and M. A. Baird (2005). "Lipid based particulate formulations for the delivery of antigen." *Immunol Cell Biol* 83: 97–105

Cortajarena, A., F. Goñi and H. Ostolaza (2001). "Glycophorin as a receptor for *Escherichia coli* α-hemolysin in erythrocytes." *J. Biol. Chem* 276(16): 12513-12519

Cortajarena, A. L., F. M. Goñi and H. Ostolaza (2003). "A receptor- binding region in *Escherichia coli* _-haemolysin." *J.Biol.Chem.* 278(21): 19159-19163

Cross, M. A., V. Koronakis, P. L. Stanley and C. Hughes (1990). "HlyB-dependent secretion of hemolysin by uropathogenic *Escherichia coli* requires conserved sequences flanking the chromosomal hly determinant." *J Bacteriol* 172(3): 1217-24

de Haan, L. and T. R. Hirst (2000). "Cholera toxin and related enterotoxins: a cell biological and immunological perspective." *J Nat. Toxins* 9: 281–97.

de Planque, M. R. and J. A. Killian (2003). "Protein-lipid interactions studied with designed transmembrane peptides: role of hydrophobic matching and interfacial anchoring." *Mol. Membr. Biol.* 20: 271–284

Dhakal, B. K., R. R. Kulesus and M. A. Mulvey (2008). "Mechanisms and consequences of bladder cell invasion by uropathogenic *Escherichia coli*." *Eur J Clin Invest* 38 Suppl 2: 2-11

Douce G, F. M., Pizza M, Rappuoli R, Dougan G. (1997). " Intranasal immunogenicity and adjuvanticity of site-directed mutant derivatives of cholera toxin." *Infect Immun* 65: 2821-2828.

Douce, G., M. M. Giuliani, V. Giannelli, M. Pizza, R. Rappuoli and G. Dougan (1998). "Mucosal immunogenicity of genetically detoxified derivatives of heat labile toxin from *Escherichia coli*." *Vaccine* 16: 1065-1073

Douce, G., C. Turcotte, I. Cropley and e. al. (1995). "Mutants of *Escherichia coli* heat-labile toxin lacking ADP-ribosyltransferase activity act as nontoxic, mucosal adjuvants." *Proc Natl Acad Sci USA* 92: 1644-1648.

Dunker, A. K., M. S. Cortese, P. Romero, L. M. Iakoucheva and V. Uversky (2005). "Flexible nets. The roles of intrinsic disorder in protein interaction networks." *FEBS J.* 272: 5129-5148.

Dunker, K., J. D. Lawson, R. Brown, P. Williams and J. Romero (2001). "Intrinsically disordered protein." *J. Mol. Graph. Model* 19: 26-59

Fayard, E., L. A. Tintignac, A. Baudy and B. A. Hemmings (2005). "Potein kinase B/Akt at a glance." *J.Cell Sci.* 118: 5675-5678

Feiring, B., J. Fuglesang, P. Oster, L. M. Naess, O. S. Helland, S. Tilman, E. Rosenqvist, M. A. Bergsaker, H. Nøkleby and I. S. Aaberge (2006). "Persisting immune responses indicating long-term protection after booster dose with meningococcal group B outer membrane vesicle vaccine." *Clin Vaccine Immunol* 13: 790–796

Felmlee, T., S. Pellet and R. Welch (1985). "Nucleotide sequence of en *Escherichia coli* chromosomal hemolysin." *J.Bacteriol.* 163(1): 94-105

Felmlee, T., S. Pellett, E. Y. Lee and R. A. Welch (1985). "*Escherichia coli* hemolysin is released extracellularly without cleavage of a signal peptide." *J Bacteriol* 163(1): 88-93

Felmlee, T. and R. A. Welch (1988). "Alterations of amino acid repeats in the *Escherichia coli* hemolysin affect cytolytic activity and secretion." *Proc Natl Acad Sci U S A* 85(14): 5269-73

Felnerova, D., J. F. Viret, R. Glück and C. Moser (2004). "Liposomes and virosomes as delivery systems for antigens, nucleic acids and drugs." *Curr Opin Biotechnol* 15: 518–529.

Forster, T. (1959). "Transfer mechanisms of electronic excitation." *Discuss. Faraday Soc.* 27: 7-17

Foxman, B. (2002.). "Epidemiology of urinary tract infections: incidence, morbidity, and economic costs." *Am. J. Med.* 113: 5S-13S

Foxman, B. and P. Brown (2003). "Epidemiology of urinary tract infections: transmission and risk factors, incidence, and costs." *Infect Dis Clin North Am* 17(2): 227-41

Foxman, B., S. D. Manning, P. Tallman, R. Bauer, L. Zhang, J. S. Koopman, B. Gillespie, J. D. Sobel and C. F. Marrs ( 2002). "Uropathogenic *Escherichia coli* are more likely than commensal *E. coli* to be shared between heterosexual sex partners." *Am. J. Epidemiol.* 156: 1133-1140.

Fragoso, R., D. Ren, X. Zhang, M. W. Su, S. J. Burakoff and Y. J. Jin (2003). "Lipid raft distribution of CD4 depends on its palmitoylation and association with Lck, and evidence for CD4-induced lipid raft aggregation as an additional mechanism to enhance CD3 signaling." *J. Immunol.* 170: 913-921

Freytag, L. C. and J. D. Clements (2005). "Mucosal adjuvants." *Vaccine.* 23: 1804-1813.

Gentschev, I., G. Dietrich and W. Goebel (2002). "The *E. coli* _-hemolysin secretion system and its use in vaccine development." *Trends Microbiol.* 10(1): 39-45

Gentschev, I., H. Mollenkopf, Z. Sokolovic, J. Hess, S. H. E. Kaufmann and W. Goebel (1996). "Development of antigen-delivery systems, based on the *Escherichia coli* haemolysin secretion pathway." *Genes* 179: 133-140

Goebel, W. and J. Hedgpeth (1982). "Cloning and functional characterization of the plasmid-encoded hemolysin determinant of Escherichia coli." *J Bacteriol* 151(3): 1290-8

Goluszko, P., E. Goluszko, B. Nowicki, S. Nowicki, V. Popov and e. al. (2005). "Vaccination with purified Dr Fimbriae reduces mortality associated with chronic urinary tract infection due to *Escherichia coli* bearing Dr adhesin." *Infect Immun* 73: 627-631.

Griebling, T. L. (2005). "Urologic diseases in america project: trends in resource use for urinary tract infections in men." *J Urol* 173: 1288-1294.

Griebling, T. L. (2005). "Urologic diseases in America project: trends in resource use for urinary tract infections in women." *J Urol* 173: 1281-1287.

Guyer, D. M., S. Radulovic, F. E. Jones and H. L. Mobley (2002). "Sat, the secreted autotransporter toxin of uropathogenic *Escherichia coli*, is a vacuolating cytotoxin for bladder and kidney epithelial cells." *Infect Immun* 70(8): 4539-46

Hackett, M., C. Walker, L. Guo, M. C. Gray, S. V. Cuyk, U. Ullmann, J. Shabanowitz, D. F. Hunt, E. L. Hewlett and P. Sebo (1995). "Hemolytic, but not cell-invasive activity of adenylate cyclase toxin is selectively affected by differential fatty acylation in *Escherichia coli*." *J Biol Chem* 270: 20250-20253

Hall, P. D., G. Virella, T. Willoughby, D. H. Atchley, R. J. Kreitman and A. E. Frankel (2001). "Antibody response to DT-GM, a novel fusion toxin consisting of truncated diphtheria toxin (DT) linked to human granulocyte-macrophage colonystimulating factor (GM), during a phase I trial of patients with relapsed or refractory acute myeloid leukemia." *Clin Immunol* 100: 191-197.

Hamon, M. A., E. Batsche, B. Regnault, T. N. Tham, S. Seveau, C. Muchardt and P. Cossart (2007). "Histone modifications induced by a family of bacterial toxins." *Proc Natl Acad Sci U S A* 104(33): 13467-72

Hashigucci, K., H. Ogawa, T. Ishidate and et. al. (1996). "Antibody responses in volunteers induced by nasal influenza vaccine combined with Escherichia coli heat-labile enterotoxin B subunit containing a trace amount of the holotoxin." *Vaccine* 14: 113-119

Herlax, V. and L. Bakas (2003). "Acyl chains are responsible for the irreversibility in the *Escherichia coli* alpha-hemolysin binding to membranes." *Chem Phys Lipids.* 122: 185-190

Herlax, V. and L. Bakas (2007). "Fatty acids covalently bound to alpha-hemolysin of Escherichia coli are involved in the molten globule conformation: implication of disordered regions in binding promiscuity." *Biochemistry* 46(17): 5177-84

Herlax, V., M. F. Henning, A. M. Bernasconi, F. M. Goni and L. Bakas (2010). "The lytic mechanism of *Escherichia coli* _-hemolysin associated to outer membrane vesicles." *Health.* 2: 484-492

Herlax, V., S. Mate, O. Rimoldi and L. Bakas (2009). "Relevance of fatty acid covalently bound to *Escherichia coli* alpha-hemolysin and membrane microdomains in the oligomerization process." *J Biol Chem* 284(37): 25199-210

Heyduk, T. (2002). "Measuring protein conformational changes by FRET/LRET." *Curr. Opin. Biotechnol.* 13: 292-296

Holmgren, J., J. Adamsson, F. Anjuère, J. Clemens, C. Czerkinsky, K. Eriksson, C. F. Flach, A. George-Chandy, A. M. Harandi, M. Lebens, T. Lehner, M. Lindblad, E. Nygren, S. Raghavan, J. Sanchez, M. Stanford, J. B. Sun, A. M. Svennerholm and S. Tengvall (2005). "Mucosal adjuvants and anti-infection and anti-immunopathology vaccines based on cholera toxin, cholera toxin B subunit and CpG DNA." *Immunol Lett.* 97(2): 181-188.

Holmgren, J., C. Czerkinsky, K. Eriksson and A. Mharandi (2003). "Mucosal immunisation and adjuvants: a brief overview of recent advances and challenges." *Vaccine.* 21(2): S89-95.

Hooton, T. M., D. Scholes, A. E. Stapleton, P. L. Roberts, C. Winter, K. Gupta, M. Samadpour and W. E. Stamm (2000.). "A prospective study of asymptomatic bacteriuria in sexually active young women." *N. Engl. J. Med* 343: 992-997

Hui, D., C. Morden, F. Zhang and V. Ling (2000). "Combinatorial analysis of the structural requirements of the *Escherichia coli* hemolysin signal sequence." *J.Biol.Chem.* 275: 2713-2720

Hyland, C., L. Vuillard, C. Hughes and V. Koronakis (2001). "Membrane interaction of *Escherichia coli* Hemolysin: Flotation and Insertion-Dependent Labeling by Phospholipid Vesicles." *J. Bacteriol* 183: 5364-5370

Issartel, J., V. Koronakis and C. Hughes (1991). "Activation of *Escherichia coli* prohaemolysin to the mature toxin by acyl carrier protein-dependent fatty acylation." *Nature* 351: 759-761

Jarchau, T., T. Chakraborty, F. Garcia and W. Goebel (1994). "Selection for transport competence of C-terminal polypeptides derived from *Escherichia coli* hemolysin: the shortest peptide capable of autonomous HlyB/HlyD-dependent secretion comprises the C-terminal 62 amino acids of HlyA." *Mol Gen Genet* 245(1): 53-60

Jonas, D., B. Schultheis, C. Klas, P. Krammer and S. Bhadki (1993). "Cytocidal effects of *Escherichia coli* hemolysin on human T lymphocytes." *Infect Immun* 61: 1715-1721

Kiyono, H., J. Bienenstock, J. R. McGhee and P. B. Ernst (1992). "The mucosal immune system: features of inductive and effector sites to consider in mucosal immunization and vaccine development." *Reg Immunol* 4: 54-62.

Koronakis, E., C. Hughes, I. Milisav and V. Koronakis (1995). "Protein exporter function and *in vivo* ATPase activity are correleted in ABC-domain mutants of HlyB." *Mol.Microbiol* 16: 87-96

Koronakis, V., C. Hughes and E. Koronakis (1991). "Energetically distinct early and late stages of HlyB/HlyD-dependent secretion across both *Escherichia coli* membranes." *EMBO J.* 10: 3263-3272

Koronakis, V., E. Koronakis and C. Hughes (1989). "Isolation and analysis of the C-terminal signal directing export of *Escherichia coli* hemolysin protein across both bacterial membranes." *EMBO J* 8(2): 595-605

Koronakis, V., J. Li, E. Koronakis and K. Stauffer (1997). "Structure of TolC, the outer membrane component of the bacterial type I efflux system, derived from two-dimensional crystals." *Mol Microbiol.* 23: 617-626

Koronakis, V., A. Sharff, E. Koronakis, B. Luisi and C. Hughes (2000). "Crystal structure of the bacterial membrane protein TolC central to multidrug efflux and protein export." *Nature* 405: 914-919

Koschinski, A., H. Repp, H. Unver, F. Dreyer, D. Brockmeier, A. Valeva, S. Bhakdi and I. Walev (2006). " Why Escherichia coli _-hemolysin induces calcium oscillations in mammalian cells–the pore is on its own." *FASEB J* E80-E87

Kreitman, R. J., D. R. Squires, M. Stetler-Stevenson, P. Noel, D. J. FitzGerald, W. H. Wilson and et. al (2005). " Phase I trial of recombinant immunotoxin RFB4 (dsFv)-PE38 (BL22) in patients with B-cell malignancies." *J Clin Oncol* 23: 6719-6729.

Kreitman, R. J., W. H. Wilson, K. Bergeron, M. Raggio, M. Stetler-Stevenson and D. J. FitzGerald (2001). "Efficacy of the anti-CD22 recombinant immunotoxin BL22 in chemotherapy-resistant hairy-cell leukemia." *N Engl J Med* 345: 241-247

Kuehn, M. J. and N. C. Kesty (2005). "Bacterial outer membrane vesicles and host-pathogen interaction." *Genes and Development* 19: 2645-2655

Lakowicz, J. R., I. Gryczynski, W. Wiczk, G. Laczko, F. C. Prendergast and M. L. Johnson (1990). "Conformational distributions of melittin in water/methanol mixtures from frequency-domain measurements of nonradiative energy transfer." *Biophys. Chem.* 36: 99-115

Lally, E. T., I. R. Kieba, A. Sato, C. L. Green, J. Rosenbloom, J. Korostoff, J. F. Wang, B. J. Shenker, S. Ortlepp, M. K. Robinson and P. C. Billings (1997). "RTX toxins recognize a beta2 integrin on the surface of human target cells." *J Biol Chem.* 272(48): 30463-9

Langston, K. G., L. M. Worsham, L. Earls and M. L. Ernst-Fonberg (2004). "Activation of hemolysin toxin: relationship between two internal protein sites of acylation." *Biochemistry* 43(14): 4338-46

Lee, C. K. (2001). "Vaccination against Helicobacter pylori in non-human primate models and humans." *Scand J Immunol* 53: 437-442.

Lee, C. K., R. Weltzin, W. D. Thomas and et. al. (1995). "Oral immunization with recombinant *Helicobacter pylori* urease induces secretory IgA antibodies and protects mice from challenge with Helicobacter felis." *J Infect Dis* 172: 161-171.

Lichtenberg, D., F. M. Goñi and H. Heerklotz (2005). "Detergent-resistant membranes should not be identified with membrane rafts." *Trends Biochem Sci* 30(8): 430-436

Lim, K. B., C. R. Bazemore Walker, L. Guo, S. Pellett, J. Shabanowitz, D. Hunt, E. L. Hewlett, A. Ludwig, W. Goebeli, R. A. Welch and Hackett.M. (2000). "*Escherichia coli* _-Hemolysin (HlyA) Is Heterogeneously Acylated in Vivo with 14-, 15-, and 17-Carbon Fatty Acids." *J Biol Chem* 275: 36698-36702

Lin, K. Y., N. T. Chiu, M. J. Chen, C. H. Lai, J. J. Huang and e. al. (2003). "Acute pyelonephritis and sequelae of renal scar in pediatric first febrile urinary tract infection." *Pediatr Nephrol* 18: 362–365.

Linhartova, I., L. Bumba, J. Masin, M. Basler, R. Osicka, J. Kamanova, K. Prochazkova, I. Adkins, J. Hejnova-Holubova, L. Sadilkova, J. Morova and P. Sebo (2010) "RTX proteins: a highly diverse family secreted by a common mechanism." *FEMS Microbiol Rev* 34(6): 1076-112

Lowell, G. H., W. R. Ballou, L. F. Smith, R. A. Wirtz, W. D. Zollinger and W. T. Hockmeyer (1988). "Proteosome-lipopeptide vaccines: Enhancement of immunogenicity for malaria CS peptides." *Science* 240: 800–802.

Lowell, G. H., L. F. Smith, R. C. Seid and W. D. Zollinger (1988). "Peptides bound to proteosomes via hydrophobic feet become highly immunogenic without adjuvants." *J Exp Med* 167: 658–663

Ludwig, A., M. Vogel and W. Goebel (1987). "Mutations affecting activity and transport of haemolysin in *Escherichia coli*." *Mol Gen Genet.* 206: 238-254

Manning, B. D. and L. C. Cantley (2007). "Akt/PKB signaling: navigating downstream." *Cell* 129: 1261-1274

Mansson, L. E., P. Kjall, S. Pellett, G. Nagy, R. A. Welch, F. Backhed, T. Frisan and A. Richter-Dahlfors (2007). "Role of the lipopolisacharide -CD14 complex for the activity of hemolysin from uropathogenic Escherichia coli." *Infec Immun* 75: 997-1004

Marrs, C. F., L. Zhang and B. Foxman (2005). "*Escherichia coli* mediated urinary tract infections: are there distinct uropathogenic *E. coli* (UPEC) pathotypes?" *FEMS Microbiol Lett* 252: 183–190

Mashburn-Warren, L. and M. Whiteley (2006). "Special delivery: Vesicle trafficking in prokaryotes." *Molecular Microbiology* 61(4): 839-846.

McGhee, J. R., J. Mestecky, M. T. Dertzbaugh, J. H. Eldridge, M. Hirasawa and H. Kiyono (1992). "The mucosal immune system: from fundamental concepts to vaccine development." *Vaccine* 10: 75-88

Menestrina, G., N. Mackman, I. Holland, B. and S. Bhadki (1987). "*Escherichia coli* haemolysin forms voltage-dependent ion channels in lipid membranes." *Biochem Biophys Acta* 905: 109-117

Moayeri, M. and R. Welch (1994). "Effects of temperature, time, and toxin concentration on lesion formation by the *Escherichia coli* hemolysin." *Infection and immunity* 62(10): 4124-4134

Moayeri, M. and R. Welch (1997). "Prelytic and lytic Conformation of erythrocyte- associated *Escherichia coli* hemolysin." *Infection and immunity* 65(6): 2233-2239

Mobley, H. L. T., M. S. Donnenberg and E. C. Hagan (2009.) *Uropathogenic Escherichia coli.* Washington, DC, ASM Press.

Moolten, F., S. Zajdel and S. Cooperband (1976). "Immunotherapy of experimental animal tumors with antitumor antibodies conjugated to diphtheria toxin or ricin." *Ann. NY Acad. Sci.* 277: 690–699.

Moolten, F. L. and S. R. Cooperband (1970). "Selective destruction of target cells by diphtheria toxin conjugated to antibody directed against antigens on the cells." *Science* 169: 68–70

Morein, B., B. Sundquist, S. Höglund, K. Dalsgaard and A. Osterhaus (1984). " Iscom, a novel structure for antigenic presentation of membrane proteins from enveloped viruses." *Nature* 308: 457–460

Morova, J., R. Osicka, J. Masin and P. Sebo (2008). "RTX cytotoxins recognize beta2 integrin receptors through N-linked oligosaccharides." *P Natl Acad Sci USA* 105: 5355–5360.

Muller, D., C. Hughes and W. Goebel (1983). "Relationship between plasmid and chromosomal hemolysin determinants of *Escherichia coli*." *J .Bacteriol.* 153: 846-851

Murphy, J. R. (1996). *Corynebacterium Diphtheriae.* University of Texas Medical Branch at Galveston.

Nicaud, J. M., N. Mackman, L. Gray and I. B. Holland (1986). "The C-terminal, 23 kDa peptide of *E. coli* haemolysin 2001 contains all the information necessary for its secretion by the haemolysin (Hly) export machinery." *FEBS Lett* 204(2): 331-5

Nieto, J., C. Hughes, M. Bailey and V. Koronakis (1996). "Suppression of transcription polarity in the *E.coli* hemolysin operon by a short upstream element shared by polysaccharide and DNA transfer determinants." *Mol.Microbiol* 19: 705-714

O'Hanley, P., G. Lalonde and G. Ji (1991). "Alpha-hemolysin contributes to the pathogenicity of piliated digalactoside-binding *Escherichia coli* in the kidney: efficacy of an alpha-hemolysin vaccine in preventing renal injury in the BALB/c mouse model of pyelonephritis." *Infect Immun* 59 1153–1161.

Oster, P., D. Lennon, J. O'Hallahan, K. Mulholland, S. Reid and D. Martin (2005). "MeNZB: A safe and highly immunogenic tailor-made vaccine against the New Zealand *Neisseria meningitidis* serogroup B disease epidemic strain." *Vaccine* 23: 2191–2196.

Ostolaza, H., L. Bakas and F. Goñi (1997). "Balance of electrostatic and hydrophobic interactions in the lysis of model membranes by *E. coli* alpha-haemolysin." *J Membr Biol.* 158(2): 137-145

Ostolaza, H., B. Bartolome, I. Ortiz de Zarate, F. de la Cruz and F. M. Goñi (1993). "Release of lipid vesicle contents by the bacterial protein toxin alpha-haemolysin." *Biochim Biophys Acta* 1147(1): 81-88

Ostolaza, H. and F. M. Goñi (1995). "Interaction of the bacterial protein toxin alpha-haemolysin with model membranes: protein binding does not always lead to lytic activity." *FEBS Lett.* 371(3): 303-306

Parsons, M., B. Vojnovic and S. Ameer-Beg (2004). "Imaging proteinprotein interactions in cell motility using fluorescence resonance energy transfer (FRET)." *Biochem. Soc. Trans.* 32: 431-433

Pastan, I., R. Hassan, D. J. FitzGerald and R. J. Kreitman (2007). "Immunotoxin treatment of cancer." *Annu.Rev. Med.* 58: 221-237.

Pellett, S., D. Boehm, I. Snyder, G. Rowe and R. Welch (1990). "Characterization of monoclonal anibodies against the *Escherichia coli* hemolysin." *Infection and immunity* 58(3): 822-827

Perrie, Y., A. R. Mohammed, D. J. Kirby, S. E. McNeil and V. W. Bramwell (2008). "Vaccine adjuvant systems: enhancing the efficacy of sub-unit protein antigens." *International Journal of Pharmaceutics* 364: 272–280

Pike, L. J. (2003). "Lipid rafts: bringing order to chaos." *J Lipid Res* 44(4): 655-67

Pimenta, A. L., K. Racher, L. Jamieson, M. A. Blight and I. B. Holland (2005). "Mutations in HlyD, part of the type 1 translocator for hemolysin secretion, affect the folding of the secreted toxin." *J Bacteriol* 187: 7471–7480.

Pizza, M., V. Scarlato, V. Masignani, M. M. Giuliani, B. Arico and et. al. (2000). "Identification of vaccine candidates against serogroup B meningococcus bywhole-genome sequencing." *Science* 287: 1816–1820

Ratner, A. J., K. R. Hippe, J. L. Aguilar, M. H. Bender, A. L. Nelson and J. N. Weiser (2006). "Epithelial cells are sensitive detectors of bacterial pore-forming toxins." *J Biol Chem* 281(18): 12994-8

Rose, T., P. Sebo, J. Bellalou and D. Ladant (1995). "Interaction of calcium with Bordetella pertussis adenylate cyclase toxin. Characterization of multiple calcium-binding sites and calcium-induced conformational changes." *J Biol Chem* 270(44): 26370-6

Rowe, G., S. Pellet and R. Welch (1994). "Analysis of toxinogenic functions associated with the RTX repeat region and monoclonal antibody D12 epitope of *Escherichia coli* hemolysin (HlyA)." *Infect. Immun.* 62: 579-588

Russo, T. A., C. D. McFadden, U. B. arlino-MacDonald, J. M. Beanan, R. Olson and e. al. (2003). "The Siderophore receptor IroN of extraintestinal pathogenic Escherichia coli is a potential vaccine candidate." *Infect Immun* 71: 7164–7169.

Sanchez-Magraner, L., A. Cortajarena, F. Goni and H. Ostolaza (2006). "Membrane insertion of *Escherichia coli* alpha-hemolysin is independent from membrane lysis." *J. Biol. Chem* 281: 5461-5467

Sanchez-Magraner, L., A. R. Viguera, M. Garcia-Pacios, M. P. Garcillan, J. L. Arrondo, F. de la Cruz, F. M. Goni and H. Ostolaza (2007). "The calcium-binding C-terminal domain of *Escherichia coli* alpha-hemolysin is a major determinant in the surface-active properties of the protein." *J Biol Chem* 282(16): 11827-35

Sanchez, J. and J. Holmgren (2011). "Cholera toxin - a foe & a friend." *Indian J Med Res.* 133(2): 153-163

Schlessinger, D. and M. Schaechter (1993). *Bacterial toxins.* Baltimore, Williams and Wilkins.

Schulein, R., I. Gentschev, H. J. Mollenkopf and W. Goebel (1992). "A topological model for the haemolysin translocator protein HlyD." *Mol Gen Genet* 234(1): 155-63

Shapira, A. and I. Benhar (2010). "Toxin-Based Therapeutic Approaches." *Toxins* 2: 2519-2583

Shogomori, H., A. T. Hammond, A. G. Ostermeyer-Fay, D. J. Barr, G. W. Feigenson, E. London and D. A. Brown (2005). "Palmitoylation and intracellular domain interactions both contribute to raft targeting of linker for activation of T cells." *J Biol Chem.* 280: 18931-42

Singh, M., A. Chakrapani and D. O'Hagan (2007). "Nanoparticles and microparticles as vaccine-delivery systems." *Expert Rev Vaccines* 6: 797–808

Smith, Y. C., S. B. Rasmussen, K. K. Grande, R. M. Conran and A. D. O'Brien (2008). "Hemolysin of uropathogenic *Escherichia coli* evokes extensive shedding of the uroepithelium and hemorrhage in bladder tissue within the first 24 hours after intraurethral inoculation of mice." *Infect Immun* 76(7): 2978-90

Soloaga, A., H. Ostolaza, F. Goñi and F. De la Cruz (1996). "Purification of *Escherichia coli* pro-haemolysin, and a comparison with the properties of mature _-haemolysin." *Eur. J. Biochem.* 238: 418-422

Soloaga, A., P. Veiga, L. García Segura, H. Ostolaza, R. Brasseur and G. F (1999). "Insertion of *Escherichia coli* _-haemolysin in lipid bilayer as a non-transmembrane integral protein:prediction and experiment." *Molecular Microbiology* 31: 1013-1024

Stanley, P., V. Koronakis, K. Hardie and C. Hughes (1996). "Independent interaction of the acyltransferase HlyC with two maturation domains of the *Escherichia coli* toxin HlyA." *Mol Microbiol* 20(4): 813-22

Stanley, P., V. Koronakis and C. Hughes (1991). "Mutational analysis supports a role for multiple structural features in the C-terminal secretion signal of *Escherichia coli* haemolysin." *Mol Microbiol* 5(10): 2391-403

Stanley, P., V. Koronakis and C. Hughes (1998). "Acylation of *Escherichia coli* hemolysin: A unique protein lipidation mechanism underlying toxin fuction." *Microbiology and Molecular Biology Reviews* 62: 309-333

Stanley, P., L. Packman, V. Koronakis and C. Hughes (1994). "Fatty acylation of two internal lysine residues required for the toxic activity of *Escherichia coli* hemolysin." *Science* 266: 1992-1996

Thanabalu, T., E. Koronakis, C. Hughes and V. Koronakis (1998). "Substrate-induced assembly of a contiguous channel for protein export from *E.coli* : reversible bridging of an inner-membrane translocase to an outer membrane exit pore." *The EMBO Journal* 17(22): 6487-6496

Tran Van Nhieu, G., C. Clair, G. Grompone and P. Sansonetti (2004). "Calcium signalling during cell interaction with bacterial pathogens." *Biology of the cell* 96: 93-101

Trent, M. S., L. M. Worsham and M. L. Ernst-Fonberg (1998). "The biochemistry of hemolysin toxin activation: characterization of HlyC, an internal protein acyltransferase." *Biochemistry* 37(13): 4644-52

Troeger, H., J. F. Richter, L. Beutin, D. Gunzel, U. Dobrindt, H. J. Epple, A. H. Gitter, M. Zeitz, M. Fromm and J. D. Schulzke (2007). "*Escherichia coli* alpha-haemolysin induces focal leaks in colonic epithelium: a novel mechanism of bacterial translocation." *Cell Microbiol* 9(10): 2530-40

Tumpey, T. M., M. Renshaw, J. D. Clements and J. M. Katz (2001). "Mucosal delivery of inactivated influenza vaccine induces B-cell-dependent heterosubtypic cross-protection against lethal influenza A H5N1 virus infection." *J Virol* 75: 5141-5150

Uhlen, P., A. Laestadius, T. Jahnukainen, T. Soderblom, F. Backhed, G. Celsi, H. Brisman, S. Normark, A. Aperia and A. Richter- Dahlfors (2000). "Alpha-haemolysin of uropathogenic *E. coli* induces $Ca^{+2}$ oscillations in renal epithelial cells." *Nature* 405: 694-697

Ulmer, J. B., U. Valley and R. Rappuoli (2006). "Vaccine manufacturing: Challenges and solutions." *Nat Biotechnol* 24: 1377-1383

Valeva, A., I. Walev, H. Kemmer, S. Weis, I. Siegel, F. Boukhallouk, T. Wassenaar, T. Chavakis and S. Bhakdi (2005). "Binding of *Escherichia coli* Hemolysin and Activation of the Target Cells is Not Receptor-dependent." *J Biol Chem* 280: 36657-36663 van der Goot, G. and J. A. Young (2009). "Receptors of anthrax toxin and cell entry." *Mol Aspects Med.* 30(6): 406-412

Vigil, P. D., C. Alteri and H. L. Mobley (2011). "Identification of In Vivo-Induced Antigens Including an RTX Family Exoprotein Required for Uropathogenic *Escherichia coli* Virulence." *Infection and Immunity* 79: 2335-2344

Wandersman, C. and P. Delepelaire (1990). "TolC, an *Escherichia coli* outer membrane protein required for hemolysin secretion." *Proc Natl Acad Sci U S A.* 87: 4776-4780

Wang, R., S. Seror, M. Blight, J. Pratt, J. Broome-Smith and I. Holland (1991). "Analysis of the membrane organization of an *Escherichia coli* protein traslocator HlyB, a member of a large family of prokaryote and eukaryote surface transport proteins." *J.Mol.Biol* 217: 441-454

Welch, R. (2001). "RTX Toxin Structure and Function: A Story of Numerous Anomalies and Few Analogies in Toxin Biology." *Current Top Microbiol Immunol* 257: 85-111

Welch, R. A. (1991). "Pore-forming cytolisins of Gram-negative bacteria." *Mol. Microbiol* 5: 521-528

Weltzin, R., H. Kleanthous, F. Guirakhoo, T. P. Monath and C. K. Lee (1997). "Novel intranasal immunization techniques for antibody induction and protection of mice against gastric Helicobacter felis infection." *Vaccine* 4: 370-376.

Wiles, T. J., J. M. Bower, M. J. Redd and M. A. Mulvey (2009). "Use of zebrafish to probe the divergent virulence potentials and toxin requirements of extraintestinal pathogenic *Escherichia coli*." *PLoS Pathog* 5(12): e1000697

Wiles, T. J., B. K.-. Dhakal, D. S. Eto and M. A. Mulvey (2008). "Inactivation of host Akt/protein kinase B signalling by bacterial pore- foming toxins." *Moleculas Biology of the cell* 19: 1427-1438

Wiles, T. J., R. R. Kulesus and M. A. Mulvey (2008). "Origins and virulence mechanisms of uropathogenic *Escherichia coli*." *Exp Mol Pathol* 85(1): 11-9

Worsham, L. M., K. G. Langston and M. L. Ernst-Fonberg (2005). "Thermodynamics of a protein acylation: activation of *Escherichia coli* hemolysin toxin." *Biochemistry* 44: 1329-1337

Worsham, L. M., M. S. Trent, L. Earls, C. Jolly and M. L. Ernst-Fonberg (2001). "Insights into the catalytic mechanism of HlyC, the internal protein acyltransferase that activates *Escherichia coli* hemolysin toxin." *Biochemistry* 40(45): 13607-16

Wu, Y.-Y., M. H. Nahm, Y. Guo, M. W. Russell and D. E. Briles (1997). "Intranasal immunization of mice with PspA (pneumococcal surface protein A) can prevent intranasal carriage, pulmonary infection, and sepsis with *Streptococcus pneumoniae*." *J Infect Dis* 175: 839-846.

Xu-Amano, J., R. J. Jackson, K. Fujihashi, H. Kiyono, H. F. Staats and J. R. McGhee (1994). "Helper Th1 and Th2 cell responses following mucosal or systemic immunization with cholera toxin." *Vaccine* 12: 903-911

Xu-Amano, J., H. Kiyono and R. J. Jackson (1993). "Helper T cell subsets for immunoglobulin A responses: oral immunization with tetanus toxoid and cholera toxin as adjuvant selectively induces Th2 cells in mucosa associated tissues." *J Exp Med* 178(4): 1309-1320.

Yamamoto, M., J. L. Vancott, N. Okahashi and e. al. (1996). "The role of Th1 and Th2 cells for mucosal IgA responses." *Ann NY Acad Sci* 778: 64-71.

Yamamoto, S., T. Tsukamoto, A. Terai, H. Kurazono, Y. Takeda and O. Yoshida (1997). "Genetic evidence supporting the fecal-perinealurethral hypothesis in cystitis caused by *Escherichia coli.*" *J. Urol.* 157: 1127–1129.

Young, J. and I. B. Holland (1999). "ABC transporters: bacterial exporters-revisited five years on." *Biochim Biophys Acta* 1461: 177–200.

# 6

# GPCRs and G Protein Activation

Waelbroeck Magali
*Université Libre de Bruxelles*
*Belgium*

## 1. Introduction

An efficient intercellular communication system is essential to allow the correct functioning of multicellular organisms. This necessitates extracellular messengers (hormones, or neurotransmitters) as well as receptors , that is, proteins capable of recognizing these extracellular messengers and transducing a signal inside the cell. Each cell expresses several different types of receptors: signal transduction is both temporally and spatially integrated in order to generate the appropriate cellular response to each physiological situation.

Hydrophobic ligands are able to penetrate inside the cell: they recognize intracellular receptors that migrate to the nucleus and regulate protein transcription. Hydrophilic ligands, in contrast, are unable to cross the plasma membrane: these extracellular ligands recognize transmembrane receptors that then produce intracellular messengers to affect the target cell function. Several families of transmembrane receptors are known: some are ligand-gated ion channels, and regulate the transmembrane voltage (depolarization or hyperpolarisation) or the intracellular $Ca^{++}$ concentration; others are ligand-activated enzymes: some synthetize cGMP, others phosphorylate specific target proteins upon ligand recognition; and yet other receptors (known as "G Protein Coupled Receptors" or "GPCRs") activate intracellular trimeric G proteins in response to extracellular signals. These receptor-activated G proteins in turn activate enzymes responsible for "second messenger" synthesis (adenylate cyclase $\rightarrow$ cAMP, or phospholipase C $\rightarrow$ Inositol trisphosphate (Inositol(1,4,5)P$_3$ or "IP3") and diacylglycerol), regulate ion channels, or activate other ("small") G proteins.

## 2. G protein coupled receptors

### 2.1 A few examples

The human genome contains at least 800 GPCRs, grouped in five main families (Fredriksson *et al.*, 2003). One of the best characterized GPCR, rhodopsin, is responsible for vision in the dark: it captures photons thanks to its prosthetic group (11-cis retinal), and leads to phosphodiesterase activation in retina rod cells. It is extremely abundant in the rod cell disks, comparatively easy to purify, and therefore has been very extensively studied for many years by biochemists. Other GPCRs allow us to taste and smell, control our appetite, fertility, stress, heart rate and breathing, etc. Adrenaline (the stress hormone), histamine (allergic reactions), glucagon (glycemia control), but also taste and odorant receptors, luteotropic and follicular stimulating hormone receptors (ovule and spermatozoid development), etc. recognize GPCRs and induce G protein activation.

## 2.2 GPCR families

All G protein coupled receptors possess a glycosylated extracellular amino-terminal (N-term) and an intracellular carboxyl-terminal (C-term) domain, separated by 7 transmembrane helices (TM1 to TM7) joined by three intracellular (IC1 to IC3) and three extracellular (EC1 to EC3) loops (Figure 1).

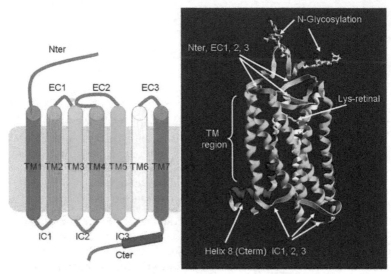

Fig. 1. Ribbon representation of the X-ray structure of rhodopsin. Left: schematic representation of a GPCR, showing the TM helices, intracellular (IC) and extracellular (EC) loops. Right: ribbon representation of the crystal structure of rhodopsin (1GZM). The prosthetic group, retinal, is covalently bound to a lysine side chain in TM7 (sticks).

The first and second EC loops are joined by a conserved disulfide bridge. The C-term region begins by an intracellular $\alpha$-helix, H8, which lies horizontally on the plasma membrane: it forms an aromatic cluster with a tyrosine side chains from TM7 and interacts with the phospholipid head groups through lysine and arginine side chains. The 7 helices are arranged in a bundle (Figure 1). GPCRs have been identified in animals, yeast, plants. They probably arise from a common ancestor (Fredriksson *et al.*, 2003). Several hundred putative GPCRs have been identified in the human genome where they represent 1-3% of the genes (Fredriksson *et al.*, 2003). The vast majority of GPCRs (including most odorant receptors) share "signature" amino acids with rhodopsin (see below). They have been grouped in "Family A" (Kolakowski, Jr., 1994) or "rhodopsin-like receptor family" (Fredriksson *et al.*, 2003). Other GPCR families do not possess these highly conserved amino acids, but share other signature amino acids. For instance, all "family B" (secretin-receptor like) receptors possess a typical N-terminal "sushi" domain with three conserved disulfide bridges and have very strong sequence homologies in the transmembrane domain. Fifteen of these receptors with a comparatively short N-term domain ("sushi" domain only) are specialized in recognition of peptide hormones and neurotransmitters (glucagon, Growth Hormone Releasing Hormone (GHRH or GRF), parathyroid hormone (PTH) and others).

The majority of family B receptors possess several additional modules (EGF-like, Immunoglobulin-like, etc) before the sushi domain, suggesting that they might function as adhesion proteins (Mizuno and Itoh, 2011). Most of these are "orphan" receptors (that is: their ligand is unknown) and their ability to activate G proteins has not been proven yet. "Family C" receptors recognize amino acids (metabotropic receptors for glutamate and $GABA_B$ receptors for GABA) or calcium ions, through N-terminal "venus flytrap" domains (Jensen et al., 2002; Wellendorph and Brauner-Osborne, 2009). Some of the taste receptors also are GPCRs: sweet and "umami" tasting molecules are recognized by "family C" GPCRs, and the "bitter" taste, by GPCRs that present very little homology with the other GPCRs, and form an additional receptor family (Fredriksson et al., 2003).

## 2.3 Conserved residues in "family A" receptors

Rhodopsin and related receptors possess a few extremely conserved residues in each TM helix. In the Ballesteros and Weinstein nomenclature, the most conserved amino acid in each TM helix is numbered $X^{h.50}$ (where "h" is the helix number): for instance, $R^{3.50}$ is the most conserved amino acid in TM3; $D^{3.49}$ and $Y^{3.51}$ are the two conserved amino acids immediately preceding and following this arginine.

Some of these very conserved side chains are involved in structural features, like the prolines in helices 5, 6 and 7 that induce kinks in the TM helices. In the different family A receptors crystal structures (rhodopsin but also β-adrenergic, adenosine, histamine H3 receptors), the conserved asparagine of the TM7 $NPxxY(x)_{5-6}F$ motif is part of a hydrogen bond network involving TM1, TM2 ($D^{2.50}$) and TM7, while the tyrosine in this motif constrains TM7 in contact with aromatic side chains in the C term helix 8. Other conserved side chains play a role in the resting and/or active receptor conformation For instance, the arginine of the TM3 "DRY motif" ($E/DR^{3.50}Y$) at the intracellular end of the third transmembrane helix forms in rhodopsin a H bond network with $E^{3.49}$, $E^{6.30}$ and $T^{6.34}$. The "ionic lock" $R^{3.50}$-$E^{6.30}$ stabilizes the resting state: it is broken up in metarhodopsin II (the active rhodopsin conformation). In that structure, $R^{3.50}$ folds back inside the G protein to interact with $Y^{5.58}$, thereby creating an intracellular binding pocket, able to accommodate the G-protein. The ionic lock is less stable in the β-adrenergic receptors compared to rhodopsin, and this is perhaps responsible for their detectable constitutive activity (ability to activate G proteins in the absence of agonist) (Moukhametzianov et al., 2011). $W^{6.48}$ of the $CWxP^{6.50}$ motif is in very close contact with the agonist ligands, and was thought to trip the switch of receptor activation by toggling between different rotamer conformations and thereby affecting the position of neighbouring aromatic side chains. Although this hypothesis is supported by computational mapping (Bhattacharya and Vaidehi, 2010), the toggle is not evident in the metarhodopsin II (Standfuss et al., 2011; Choe et al., 2011) or $\beta_2$-adrenergic receptor crystal structures (Rasmussen et al., 2011b; Rasmussen et al., 2011a).

## 3. Trimeric G proteins

### 3.1 G protein subtypes and GPCR effectors

The G proteins that transduce the signal from GPCRs are heterotrimeric (Gαβγ) proteins. Some of the mRNAs encoding the Gα subunits are subject to alternative splicing so that sixteen genes encode 23 known G$\alpha$ proteins: (Birnbaumer, 2007). The G$\alpha$ proteins are anchored to the plasma membrane by N-terminal myristoylation or palmitoylation. They can be grouped into four families based upon sequence homologies, and each GPCR has a preference for a single G$\alpha$ or for a single family of G$\alpha$ subunits. Each G$\alpha$ subunit regulates one or a few effectors (Birnbaumer, 2007):

- G proteins in the $G_s$ ($G_{s/olf}$) G protein family stimulate adenylate cyclase,
- G proteins in the $G_i$ ($G_{i/o/t/gust/z}$) G protein family inhibit adenylate cyclase and/or regulate ion channels,
- G proteins in the $G_{q/11}$ ($G_{q/11/14/15/16}$) G protein family activate phospholipase C,
- G proteins in the G12/13 G protein family activate "Guanyl nucleotide Exchange Factors" (GEFs) that in turn activate another group of "small" (monomeric) G proteins, the Rho G proteins

The carboxyl-terminal G$\alpha$ sequence is the major determinant for receptor recognition: exchanging this sequence allows the construction of promiscuous chimeric G proteins that can be used to drive GPCR coupling to a non-physiological effector (Kostenis $et\ al.$, 2005).

There are five known human G$\beta$ and 12 G$\gamma$ genes (Birnbaumer, 2007). Most but not all of the G$\beta\gamma$ and G$\alpha$- G$\beta\gamma$ combinations are allowed. All G$\gamma$ subunits are C-terminally prenylated (some with geranyl-geranyl, others with farnesyl groups) and carboxymethylated: this helps to anchor the G$\beta\gamma$ subunits to the plasma membrane. The C-terminal sequence determines the nature of the prenyl group (farnesyl or geranyl-geranyl) modifying the G$\gamma$ subunit; both the C-terminal sequence and the prenyl group play an active role in the recognition of both rhodopsin and phospholipids (Katadae $et\ al.$, 2008). Although the literature on this subject is sparse, there is some evidence that other GPCRs also recognize preferentially specific G$\beta\gamma$ subunits (Jian $et\ al.$, 2001; Kisselev and Downs, 2003; Birnbaumer, 2007). The G$\beta\gamma$ subunits recognize and regulate a growing list of effectors, including ion channels, phospholipase C (PLC), phosphoinositide-3' kinase-$\gamma$ (PI3K$\gamma$), various adenylate cyclase isoforms, etc. Different PLC isoforms respond differently to different G$\beta\gamma$ isoforms; and the cardiac ATP-inhibited inwardly rectifying $K^+$ channel (KirATP) is either inhibited or activated by G$\beta\gamma$ depending on the nature of the G$\beta$ subunit (Birnbaumer, 2007).

### 3.2 G proteins as (inefficient) GTPases

G proteins are (poor) GTPases (Birnbaumer, 2007): they hydrolyze GTP slowly to GDP + inorganic phosphate, then release GDP extremely slowly. The GDP release and the GTP hydrolysis reactions are highly regulated, accompanied by conformation changes, and used as molecular clocks.

Trimeric G proteins are no exception to this rule: GTP binding is necessary to allow transient effectors activation (Oldham and Hamm, 2006; Birnbaumer, 2007). As summarized in Figure 2, the GDP release from trimeric G proteins is accelerated by G Protein Coupled Receptors (GPCRs) that function as "Guanyl nucleotides Exchange Factors" (GEFs): they allow GDP release, and this is rapidly followed by GTP recognition and dissociation of the two G protein subunits. Both subunits can then transiently recognize their respective effectors. The GTP hydrolysis reaction (leading to signal interruption) is accelerated by "Regulators of G protein Signaling" (RGS) proteins, that function as "GTPase Activator Proteins" (GAPs) and accelerate signal interruption.

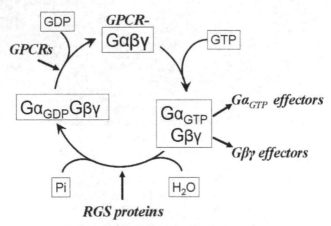

Fig. 2. the G protein activation cycle. In the resting state, the G protein is trimeric ($Ga_{GDP}G\beta\gamma$) and occupied by GDP. GPCRs interact with resting (GDP-bound) G proteins, facilitate the GDP release and stabilize an empty G protein conformation. GTP induces the dissociation of the two G protein subunits, $Ga_{GTP}$ and $G\beta\gamma$: this allows both subunits to recognize and regulate their respective effectors (enzymes, channels or regulators of G protein signaling). $Ga_{GTP}$ hydrolyses GTP to GDP and the $Ga_{GDP}$ complex recognizes $G\beta\gamma$ with a very high affinity: the resting trimeric complex reforms spontaneously. GTP hydrolysis can be accelerated by "Regulators of G protein Signaling" (RGS) molecules.

## 3.3 G protein structures

The G proteins regulated by GPCRs are heterotrimeric (Birnbaumer, 2007). The three polypeptide chains form two independent subunits: the $Ga$ and $G\beta\gamma$ subunits (Figure 3). The $Ga$ protein structure can be divided into two domains held together by mutual interactions with the guanyl nucleotide: a N-terminal "Ras-like domain" (with strong structural homology with the small GTPases, Ras) and a C-terminal α-helical domain (Figure 3). In the agonist-receptor-G protein complex, the guanyl nucleotide has dissociated, and the helical domain "floats away" from the Ras –like domain (Rasmussen et al., 2011b). The $G\beta$ protein "WD repeats" (blue) forms a 7 blades beta-propeller domain, and the $G\gamma$ protein (green) wraps around $G\beta$, one of the two small α-helices forming a coiled-coil with the $G\beta$ protein α-helix (Figure 3). $G\beta\gamma$ forms a stable complex that cannot be dissociated without denaturation but $Ga$ can dissociate from $G\beta\gamma$ upon GTP binding.

As shown in Figure 4, the conformation of three segments ("switch regions") of $Ga$ changes during the GTPase catalytic cycle (Oldham and Hamm, 2006; Rasmussen et al., 2011b). This regulates the interaction of $Ga$ with $G\beta\gamma$ and with its effectors. Switch 2 together with either switch 1 or 3 indeed forms part of the $Ga$ – protein binding interface (Figure 5). It participates to the recognition of $G\beta\gamma$, but also of effectors and "Regulators or G protein Signaling" (RGS) proteins (Figure 5): the $Ga$- $G\beta\gamma$ dissociation is essential to allow effectors activation by $Ga_{GTP}$.

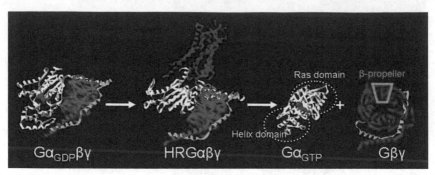

Fig. 3. Ribbon representations of G protein structures. The $G\alpha$ subunits are presented in yellow, the $\beta$ subunits in blue and the $\gamma$ subunits in green and the $\beta2$-adrenergic receptor, in red. GDP and GTP are shown in fuchsia as space-filling structures. From left to right: the $G_i$ protein (1GG2), the $\beta2$ adrenergic receptor-agonist-$G_s$complex (3SN6), the activated $Ga_{GTP}$ (1GIL) and $G\beta1\gamma2$ (1TBG) structures. The latter two structures have been rotated separately by approximately 90° compared to $Ga_{GDP}\beta\gamma$, to show the different domains (center right: the ras-like and $\alpha$-helical domains of the $G\alpha$ subunit; far right: a $\beta$-propeller structure in the $G\beta$ subunit).

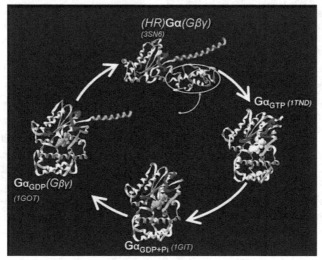

Fig. 4. Effect of guanyl nucleotide binding on the $G\alpha$ subunit conformation. Ribbon representation of $G\alpha$ in different crystallized complexes : the $G\alpha$ subunit only is shown for simplicity; the $G\beta\gamma$ subunit when present would be in front and to the right of the $G\alpha$ subunit. The N-terminal region, when structurally defined (stabilized by interactions with $G\beta\gamma$) is represented by a blue ribbon. Three regions change conformation during the GTPase catalytic cycle: "switch 1" is shown in green, "switch 2" in gold, and "switch 3" in red. GDP (pink), GTP (yellow) and phosphate (pink) are shown as space filling structures. Left: the $Ga_t$ subunit in the $Ga_{GDP}\beta\gamma$ complex (PDB 1GOT); top center: the structure of $Ga_S$ in the ternary complex, HRG (PDB 3SN6), stabilized by a nanobody; right: the structure of the GTP analogue GTP$\gamma$S-activated $Ga_t$ (PDB 1TND) and bottom center: the structure of $Ga_i$ bound to GDP and inorganic phosphate during GTP hydrolysis (PDB 1GIT).

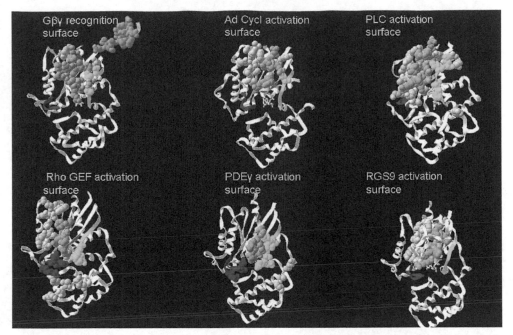

Fig. 5. Ribbon representation of Gα subunits in complex with Gβγ or with their effector and regulator proteins, showing the side chains that belong to the protein binding site. The Gα subunit only is shown for simplicity. The amino acids that belong to the protein binding sites of the different Gα structures are shown as space filling. When structurally defined, the ribbon (and side chains) that belong to the N-terminal helix are shown in blue, those that belong to switch 1 in green, to switch 2 in orange, to switch 3 in red, and to the rest of the protein in light grey. Top left: the GDP-bound transducin-$G_i$ chimera showing the interaction surface with Gβγ (1GOT); top center: GTP*-bound $Gα_S$ showing the interaction surface with adenylate-cyclase (1CUL); top right: GTP-bound $Gα_q$ showing the interaction surface with phospholipase-C (3OHM); bottom left:GTP-bound $Gα_{13}$ showing the interaction surface with the Rho GEF, p115 (1SHZ); bottom center: GTP-bound transducin-$G_i$ chimera showing the interaction surface with the phosphodiesterase inhibitor subunit, PDEγ (1FQJ) and bottom right, GTP-bound transducin-$G_i$ chimera showing the interaction surface with RGS9 in a complex with PDEγ and RGS9 (1FQK).

Gβγ is also capable of activating certain effectors. Its binding site for Gα overlaps in part the Gβγ-effector and Gβγ- regulator binding sites: the dissociation of $Gα_{GTP}$ from Gβγ is essential to allow Gβγ to recognize its effectors (Figure 6).

GTP hydrolysis, rapidly followed by the release of the phosphate ion, modifies the switch regions conformation (Figure 4). The conformation change does not only inhibit the Gα-effector interaction (Figure 5) but also favors Gβγ recognition by $Gα_{GDP}$, thereby also inactivating Gβγ (Figure 6). Agonist-bound receptors interact with both G protein subunits (Figure 3): the formation of the "ternary complex" is an essential step for G protein activation, but the ternary complex Gα subunit is not in the right conformation to activate G protein effectors (Figure 3).

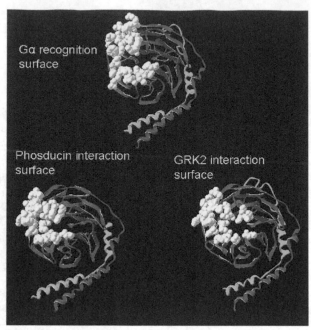

Fig. 6. Ribbon representation of Gβγ showing the interaction surface with Gα or with effectors. Ribbon representation of Gβγ in different crystallized complexes – the Gβ (blue) and Gγ (green) subunits only are shown for simplicity. The amino acids that belong to the Gα recognition surface (PDB ref 2BCG: top), to the phosducin binding site (1GP2: bottom left) or to the kinase, GRK2 binding surface (bottom right : 1OMW) are shown as space filling structures, in yellow.

## 4. G protein activation kinetics

Rhodopsin and related receptors **catalyze** G protein activation (Hamm, 1998): this means that each receptor sequentially activates several G proteins by facilitating the release of GDP, thereby allowing GTP binding. Rhodopsin does not enter in one of the Enzyme Commission "E.C." subclasses, as it does not catalyze the rupture or formation of covalent bond(s). The equations describing the reaction kinetics are nevertheless identical to those describing "ping pong" (double displacement) enzyme reaction (Waelbroeck et al., 1997; Heck and Hofmann, 2001; Ernst et al., 2007): G protein binding (substrate 1) is followed by GDP release (product 1), and GTP binding (substrate 2) is followed by the release of the activated G protein (product 2).

The kinetics of transducin activation by rhodopsin have been analyzed in detail (Heck and Hofmann, 2001; Ernst et al., 2007). Rhodopsin recognizes transiently the GDP-bound trimeric G protein, transducin, and activates transducin at the diffusion limit (Ernst et al., 2007). The physiological concentrations of the two "substrates" (GDP-bound transducin and GTP) are close to their respective Michaelis constants, $K_M$. In the case of double displacement reactions, it is unfortunately impossible to the individual rate constants of each reaction from the kinetic data ($K_M$, $V_{max}$). The [Substrate]/$K_M$ ratios at physiological concentrations nevertheless strongly suggest that the concentrations of the four reaction intermediates, $Rh^*$, $Rh^*-G_{GDP}$, $Rh^*-G$ and $Rh^*-G_{GTP}$ (where $Rh^*$ is the light activated rhodopsin) are similar (Roberts and Waelbroeck, 2004): none of the reaction intermediates accumulates. These characteristics are reminiscent of the properties of triose phosphate isomerase and other "kinetically perfect enzymes" (Albery and Knowles, 1976): all the reaction intermediates have very similar free energies at physiological substrate concentrations and the energy barriers separating the different enzyme states are very low, thereby allowing the reaction to proceed at the diffusion limit.

Enzymes accelerate reactions by stabilizing the "transition state", that is, the state with the highest energy along the reaction coordinates. Trimeric G proteins cannot be purified in the absence of guanyl nucleotides: they are unstable when empty. As explained below, agonists stabilize the agonist-receptor-G protein ternary complex (that includes an empty G protein): like enzymes, active GPCRs catalyze G protein activation by decreasing the free energy of the transition state, that is, the empty G protein (Waelbroeck, 1999). GTP recognition by the G protein destabilizes the ternary complex: this induces activated G protein release - and allows the catalytic activation of several G proteins by a single receptor.

## 5. Ligand binding studies and the ternary complex model

Ligands that induce G protein activation are termed "agonists", and ligands that do not affect the receptor activity, "antagonists". Even at 100% receptor occupancy, some agonists have a larger effect than others on G protein activation: the more effective agonists are called "full agonists" and the less efficient compounds, "partial agonists". More recently, it has been demonstrated that most GPCRs have the ability to activate (inefficiently) their cognate G proteins in the absence of any ligand: this is called "constitutive activity". Compounds that counteract the receptors' constitutive activity are called "inverse agonists".

Agonist binding to GPCRs in the absence of either GDP or GTP facilitates the formation of the ternary complex involving the receptor, an agonist, and an empty G protein (Figure 3) (Lefkowitz et al., 1976; De Lean et al., 1980; Rasmussen et al., 2011b). This is evident from the formation of a high molecular weight "ternary complex" (ligand-receptor-G protein, LRG) with a much higher affinity for agonists compared to isolated receptors. Guanyl nucleotides (GTP, GTP analogues or GDP) destabilize the G protein interaction with agonist-bound receptors by markedly decreasing the G-protein affinity for the receptor. When recognizing the high affinity ternary complex, guanyl nucleotides dramatically increase the agonists' dissociation rate, and decrease the receptor affinity for agonists while increasing their affinity for inverse agonists (see for instance (Lefkowitz et al., 1976; Berrie et al., 1979)); GDP is typically needed in larger concentrations than GTP or GTP analogues.

The effect of GTP on ligand binding can be used as a measure of the relative stability of the ternary complex compared to the binary ligand-receptor complex (Lefkowitz et al., 1976). Full agonists have a higher affinity in the absence of GTP and inverse agonists have a higher affinity in its presence: the effect of GTP on ligand recognition is correlated with the ligands' ability to induce or inhibit G protein activation by the receptor (Lefkowitz et al., 1976; De Lean et al., 1980).

The original ternary complex model (Figure 7: top left) (Lefkowitz et al., 1976) was designed to describe ligand binding to GPCRs. It describes the allosteric interactions between the ligand (L) and the G protein (G) recognizing different binding sites on the same receptor (R). Guanyl nucleotides were assumed to "prevent" G protein interaction with the receptor. The ternary complex model was later completed to the cubic ternary complex model (Figure 7: bottom right) (Weiss et al., 1996a; Weiss et al., 1996c; Weiss et al., 1996b). Two receptor conformations (R and R*) without and with the ability to activate G proteins respectively are assumed to coexist at equilibrium (R←→R*) in the absence and presence of ligands or G proteins. Agonists and G proteins favor the active (R*) receptor conformation while inverse agonists stabilize the inactive (R) conformation. As in the ternary complex model, the cubic model describes the binding (as opposed to functional) properties of GPCRs; guanyl nucleotides are assumed to "prevent" the receptor-G protein interaction.

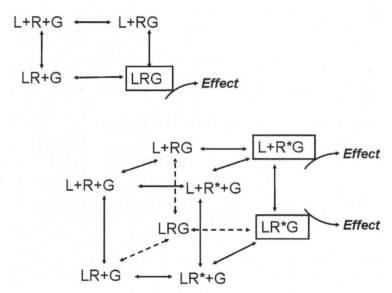

Fig. 7. The ternary complex model. Top left: the ternary complex model assumes that the receptor (R) can interact simultaneously with a ligand (L) and the G protein (G). Agonists facilitate and antagonists inhibit the receptor-G protein interaction; the ternary complex (LRG) is somehow responsible for transduction of the effect. Bottom right: the "cubic ternary complex model" assumes that the receptor can be found in a resting (R) or in an active (R*) conformation. G proteins (G) and agonist ligands stabilize R* while inverse agonists stabilize the R conformation. R*G and LR*G complexes are responsible for the biological effects of the active receptor.

Both ternary complex models have had a tremendous impact on our vision of GPCR function: the agonist-receptor-G protein complex is more and more often considered as "the active receptor". It should be remembered, however, that the ternary complex is certainly not "biologically active": it accumulates only under conditions where the G protein is unable to activate its' effectors (in the absence of GTP), and the G protein conformation in the β-adrenergic receptor-G protein complex (Figure 3) is not compatible with $G_S$-adenylate cyclase interaction (Figure 5)! The ternary complex model was designed to describe ligand binding to the receptors, as opposed to effectors activation.

## 6. "Resting" and "active" receptor structures

When rhodopsin is illuminated, its conformation passes through a number of intermediates (bathorhodopsin ($t_{1/2}$ 50ns), lumirhodopsin ($t_{1/2}$ 50μs), followed by metarhodopsin I and II) before releasing the all-trans retinal. Metarhodopsin II is biologically active: it catalyses G protein (transducin) activation.

The three dimensional structure of rhodopsin and several "family A" GPCRs has been elucidated by X-ray crystallography in the absence and presence of antagonists, agonists, G protein surrogates, or of the trimeric-$G_s$ protein (Choe *et al.*, 2011; Lebon *et al.*, 2011; Rasmussen *et al.*, 2011a; Rasmussen *et al.*, 2011b; Rosenbaum *et al.*, 2011; Standfuss *et al.*, 2011; Warne *et al.*, 2011; Xu *et al.*, 2011). A conserved ionic bond between rhodopsin arginine 135 ($R^{3.50}$) (in the conserved E/DRY motif at the intracellular end of TM3) and glutamate 247 ($E^{6.30}$) , at the end of the third intracellular loop-TM6 junction tethers rhodopsin TM3 to TM6. In antagonist-bound β2-adrenergic receptors this hydrogen bond between the conserved arginine and glutamate is not visible in the crystal and β2-adrenergic receptors are known to activate slightly their cognate $G_S$ G protein even in the absence of agonist; the ionic bridge is less stable than in rhodopsin (Moukhametzianov *et al.*, 2011).

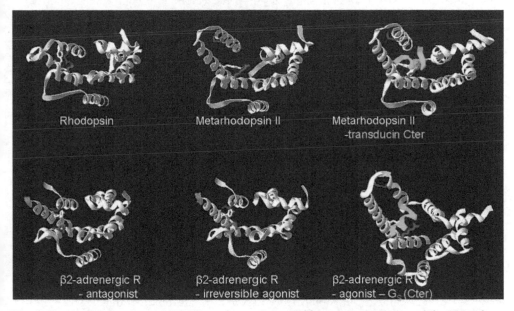

Fig. 8. Resting and activated GPCR conformations. Ribbon representation of the IC and TM regions of rhodopsin (1GZM), metarhodopsin II (3PXO), metarhodopsin II – Gt C-term complex (3PQR), antagonist-bound β2-adrenergic receptor (2RH1), agonist-bound β2-adrenergic receptor (3PDS) and agonist-$G_S$ (Cterm) bound β2-adrenergic receptor (3SN6), seen from the cytosol. The conserved TM3 arginine ($R^{3.50}$), TM5 tyrosine ($Y^{5.58}$) and TM6 glutamate ($E^{6.30}$) side chains are shown in red, yellow and blue respectively. The C-terminal transducin peptide (top right) and the C-terminal region of Gα$_S$ (bottom right) are shown as green ribbons.

In contrast with (dark adapted) rhodopsin, a large intracellular binding pocket is present between the TM helices of metarhodopsin II (Choe *et al.*, 2011): the distance between the conserved arginine ($R^{3.50}$) at the intracellular end of TM3 (E/DRY motif) and the conserved glutamate at the junction between the third intracellular loop and TM6 ($E^{6.30}$) increases from less than 3.3Å in rhodopsin (PDB 1GZM), bathorhodopsin (PDB 2G87) and lumirhodopsin (PDB 2HPY) to >15 Å in opsin (PDB 3CAP, 3DBQ) and metarhodopsin II (PDB 3PQR, 3PXO) (Figure 8). Arginine $R^{3.50}$ forms in opsin and metarhodopsin II a strong ion-dipole interaction with the conserved tyrosine, $Y^{5.58}$ in TM5. The G-protein binding pocket is created by the rotation of TM 5 and 6. It is large enough to accommodate the C-terminal helix of the transducin Gα subunit or of Gα$_S$, at almost 40° from the membrane surface (Park *et al.*, 2008; Scheerer *et al.*, 2008; Choe *et al.*, 2011) (Figure 8): this movement forces the opening of the GDP binding pocket and release GDP from the G protein (Rasmussen *et al.*, 2011b).

The sixth transmembrane helix (TM6) of the crystallized β1-, β2-adrenergic and adenosine A$_{2A}$ receptors remains very close to TM3 even in the presence of agonists (Rasmussen *et al.*, 2011a; Rosenbaum *et al.*, 2011; Warne *et al.*, 2011); and the conserved E/DRY motif arginine folds towards the cytoplasm, in the direction of the conserved TM6 glutamate: the G protein binding pocket is unavailable (Figure 8). An open, "metarhodopsin II-like" structure is achieved by β2-adrenergic receptors only in the presence of a G protein surrogate or of G$_S$ (Rasmussen *et al.*, 2011a; Rosenbaum *et al.*, 2011; Rasmussen *et al.*, 2011b): TM5 and TM6 rotate away from TM3, and the arginine side chain $R^{3.50}$ toggles away from the IC loop $E^{6.30}$ towards the conserved tyrosine, $Y^{5.38}$.

## 6.1 Is the "active receptor" mobile or rigid?

It is well known that crystallization rigidifies proteins and can lead to selection of an unusual conformation stabilized by "within the crystal" (non physiological) protein-protein interactions. For instance, opsin and metarhodopsin II conserve the same "opened" conformation in the crystal in the absence and presence of a transducin surrogate (Figure 8) (Altenbach *et al.*, 2008). Nevertheless, opsin – in contrast to metarhopsin II – has a very low ability to activate transducin: it is thus clear that crystallization of opsin "selected" a protein conformation with a low probability in intact membranes, stabilized through its contacts with other opsin molecules in the crystal.

X-ray diffraction studies have a tremendous impact on our perception of protein structure: they enhance the impression that proteins are rigid molecules with a well defined, stable conformation. In addition, the activity of most allosteric enzymes can explained in terms of two conformations with very different enzyme activities, stabilized by allosteric enhancers and inhibitors, respectively (Monod *et al.*, 1965). This is not an absolute rule, however: some enzymes change markedly in conformation upon substrate binding and dissociation – a phenomenon known as "induced fit" (Ma and Nussinov, 2010) and this can play an important role in enzyme regulation (Heredia *et al.*, 2006; Molnes *et al.*, 2011).

As suggested above, the ternary complex model had a tremendous impact on the way we understand GPCR activation, to the extent that the agonist-receptor-(empty) G protein complex is now described as "the" (one and only?) active receptor conformation – despite the fact that the Gα subunit conformation in the ternary complex (Figure 3) is not compatible with effectors activation. This interpretation was further supported by the initial computational mapping of conformational energy landscape of the β2-adrenergic receptor:

preliminary results (Bhattacharya and Vaidehi, 2010) indeed suggested that full agonists-bound receptors switch spontaneously to a more stable, active conformation very similar to metarhodopsin II. Detailed computational mapping in the presence of water and lipid molecules of rhodopsin (Provasi and Filizola, 2010) and of the agonist-bound $\beta$2-adrenergic receptor (Niesen et al., 2011) however indicate that both proteins are flexible and able to sample a large number of conformations. In the case of the $\beta$2-adrenergic receptor, reversible "shearing" movements of TM5-6 relative to TM 1-4 and 7 and "breathing" movements (opening and closing of the ligand binding pocket) have been predicted.

In the case of traditional receptors (including the $\beta$2-adrenergic receptor), GTP has a tremendous effect on agonists' recognition: agonists have a significantly lower affinity and much greater dissociation rate in the presence of GTP. This suggests that the predominant receptor conformation under "functional" conditions (in the presence of GTP) is different from the ternary complex conformation (that accumulates in the absence of GTP). I should like to suggest that most GPCRs are able to recruit G proteins while in the "closed" (low agonist affinity) conformation (Hu et al., 2010), open a G protein binding pocket and force GDP release to achieve the "high affinity" (ternary complex) conformation, then return to the "closed" conformation upon GTP recognition and activated G protein release. Agonists do not only facilitate the transition between the "closed" and "opened" conformations described by X-ray diffraction, but also decrease the free energy difference between the ternary complex and uncoupled receptors, thereby stabilizing the empty G protein conformation and facilitating GDP release, GTP binding. G protein dissociation from the receptor is then necessary to complete G protein activation.

### 6.2 Partial GPCR activation: Agonist efficacy

Some compounds seem less efficient than others to activate G protein coupled receptors: the rate of G protein activation by agonist-bound receptors varies depending on the ligand. Two explanations are usually put forward to account for this very common observation: partial agonists might stabilize the same "active" receptor conformation as full agonists but to a lesser extent; alternatively, they might stabilize an alternative receptor conformation, not quite as appropriate as the conformation induced by full agonists for G protein activation. These two explanations are non-exclusive and both explanations might in fact be correct at least where $\beta$2-adrenergic agonists are concerned (Bhattacharya and Vaidehi, 2010). Indeed, while dopamine was predicted to stabilize (less efficiently) the same "opened" receptor conformation as norepinephrine, salbutamol was predicted to stabilize a slightly different, less opened, receptor conformation. Yet a third explanation has been suggested for muscarinic receptors: agonists dissociate from muscarinic receptors with a rate constant comparable to the G protein exchange reaction rate. The efficacy of agonists activating $M_3$ muscarinic receptors was correlated with their dissociation rate constant, suggesting that the G protein activation reaction can be aborted prematurely if the agonist dissociates too early in the reaction cycle (Sykes et al., 2009).

### 7. Do GPCRs function as monomers or dimers?

$GABA_B$ receptors (a "family C" GPCR) function as obligate dimers (Jones et al., 1998; White et al., 2002). One of the two subunits is trapped intracellularly by an endoplasmic reticulum retention signal; the second forms non-functional homodimers. Upon coexpression, formation of a heterodimer is driven by dimerization of the N-terminal region, and by formation of a coiled coil by the C-terminal regions α-helices. This masks the E.R. retention signal of the first subunit, and allows the expression of the functional heterodimer at the plasma membrane, (Jones *et al.*, 1998; White *et al.*, 2002). Likewise, all other family C receptors form heterodimers.

Rhodopsin, the best known "family A" GPCR, forms quasi crystalline arrays in rod cells disk membranes: this led to the suggestion that not only family C receptors but all GPCRs might function as homo- or heterodimers. Non-radiative energy transfer between two fluorophores ("FRET") or from a luminescent protein to a fluorophore ("BRET") can be easily demonstrated if the "donor" and "acceptor" molecules are close enough (typically less than 50 Å from each other). Chimeric constructs including "donor" and "acceptor" proteins (luciferase, fluorescent proteins from jellyfish, etc.) and the protein of interest can be built by molecular biology techniques; alternatively, the donor and acceptor fluorophores can be tagged chemically to the protein of interest, or to an antibody raised against this protein. BRET and FRET have been used to demonstrate not only protein-protein interaction, but also conformational changes of a single protein (by tagging for instance the N- and C-terminal of the protein of interest). In analogy with "family C" receptors, the vast majority of family A and several family B receptors have been shown by BRET or FRET experiments to either dimerize or oligomerize. This idea raised a lot of interest, because the potential consequences of dimerization are so multiple and important (Milligan, 2009; Milligan, 2010; Birdsall, 2010):

- Dimerization is essential for "family C" receptor expression at the plasma membrane (Temussi, 2009)) and might play a role in several other systems;
- Dimerization affects the "pharmacology" of some receptors. For instance: the sweet taste is sensed by a T1R2-T1R3 heterodimer, while "umami" is detected by a T1R1-T1R3 heterodimer (Temussi, 2009) (NB. The bitter taste is sensed by a non-family C GPCR, the T2R receptor; and salt and acid are recognized by "ligand gated channels" receptors (Temussi, 2009)). Likewise, dimerization of some GPCRs (i.e. dopamine, opiate or taste receptors) has been shown to alter their pharmacological properties - suggesting that their interaction is stable enough to affect the receptor conformation (Milligan, 2010; Milligan, 2009). Negative cooperativity has been observed between agonists binding to TSH and chemokine receptor dimers (Springael *et al.*, 2005; Urizar *et al.*, 2005)): binding of one agonist ligand to the dimer decreased the affinity of the second agonist by increasing its dissociation rate.
- Dimerization may have important functional consequences: a single agonist is sufficient for activation of $G_S$ by the TSH receptor, but (low affinity) double occupancy of the dimer is necessary to support the activation of $G_{q/11}$ proteins and of phospholipase C (see below).
- Most if not all GPCRs do not only interact with G proteins, but also with other associated proteins, often in an agonist-modulated manner (Magalhaes *et al.*, 2011). Two receptors rather than one might be necessary to form optimal interactions with the receptor-associated proteins; alternatively, interaction of one subunit in the dimer with an associated protein might hinder or prevent the recognition of the second receptor subunit by steric hindrance, leading to "half of the sites reactivity".
- Several of the receptor-associated proteins act as scaffolds, recruiting in their turn other proteins in the vicinity of the receptor and of each other (Magalhaes *et al.*, 2011).

Dimerization of the receptors might be necessary to bring together some of the different accessory proteins recruited by each monomer.

It is unfortunately necessary to reassess the presence and consequences of dimerization for each receptor of interest: no generalization can be made in this respect. Indeed:

- Monomeric rhodopsin and β-adrenergic receptors, isolated and reconstituted in high density lipoprotein particles, function normally (Whorton *et al.*, 2007; Whorton *et al.*, 2008); and the receptor associated proteins arrestins, like GRKs, are able to recognize one receptor per protein (Hanson *et al.*, 2007; Bayburt *et al.*, 2011).
- Muscarinic M1 receptors dimerize only transiently: monomeric and dimeric forms are present at comparable concentrations in the plasma membrane at equilibrium and dimers dissociate rapidly ((Hern et al., 2010); see also (Johnston et al., 2011)). Cross-talk between two or more receptors is likely to necessitate strong interactions between the different monomers.
- Muscarinic M3 receptors (McMillin *et al.*, 2011) (and perhaps other receptors: (Johnston *et al.*, 2011)) are able to use several dimerization interfaces: this might explain why so many different dimerization interfaces have been observed when studying different receptors. This does not support the hypothesis that proteins like arrestin, G proteins or receptor kinases **need** a dimer for receptor recognition: the relative position of the two receptor monomers would be very important in that case.

## 8. Receptor promiscuity and biased signaling

GPCR "promiscuity" is defined as the ability of a given receptor to activate several different effectors (for review: (Hermans, 2003). While most receptors can probably induce parallel signaling by the $G\alpha_{GTP}$ and $G\beta\gamma$ subunits (see above), some are capable of activating different G proteins; and some use both G protein dependent and G protein independent signaling pathways.

"Biased signaling" refers to the observation that when receptors two or more signaling pathways, a few agonists preferentially use only one of the signaling pathways available to the other agonists – an observation that suggests that the activated receptor takes different conformations, depending on the agonist occupying its binding site.

### 8.1 Activation of several G protein subtypes by the same receptors

Each cell expresses several G proteins, belonging or not to the same family: all these G proteins will compete for recognition of each activated GPCR. Most $G_i$-coupled receptors activate several $G_i$ isoforms with variable efficiency; some $G_i$ and $G_S$ coupled receptors activate in addition $G_{q/11}$ G proteins - less efficiently, or only at much higher agonist concentrations... Does this reflect a lower (but measurable) affinity of the non-cognate G protein, or less efficient activation?

GPCRs catalyze G protein activation: they should be considered like honorary enzymes. If several substrates compete for transformation by the same enzyme, the proportion of substrates transformed by the enzyme per minute, at steady state, is proportional to their relative substrate concentration over specificity constant ratios, $[S]/K_S$ :

$$\frac{v^A}{v^B} = \frac{[A]/K_S^A}{[B]/K_S^B} \tag{1}$$

(where A and B represent the two substrates (G proteins), respectively, and $K_S^A$ and $K_S^B$ are their respective specificity constants : $K_s = \frac{k_{cat}}{K_M}$ ).

The equation is extremely similar to the equation describing the competition of several ligands for the same receptor: the proportion of receptor occupied by each ligand ([RA] and [RB]) is proportional to their relative ligand concentration over dissociation constant ratios:

$$\frac{[RA]}{[RB]} = \frac{[A]/K_D^A}{[B]/K_D^B} \tag{2}$$

The meaning of "$K_D$" and "$K_S$" is however very different: the dissociation constant, $K_D$ = $1/K_{affinity}$, is a concentration. It measures the ligand concentration necessary to occupy, at equilibrium and in the absence of competitors, 50% of the receptors. The specificity constant $K_S$, in contrast is a bimolecular reaction rate constant and measured in $M^{-1}sec^{-1}$. It measures the rate of formation of the "productive complex", ES† in the absence of alternative substrates of inhibitors.

Multiple G protein signaling has more often been observed in transfected systems, where it depends on the receptor expression level (for review: (Hermans, 2003)). Transiently expressed α2 adrenergic receptors inhibit adenylate cyclase at low agonist concentrations but activate the enzyme at high agonist concentrations (Fraser et al., 1989). Adenylate cyclase inhibition but not activation is prevented by Gi protein inactivation by pertussis toxin (Fraser et al., 1989): these results indicate that α2 adrenergic receptors are capable of activating both Gi and GS. The equations above predict that the relative activation rate of "Gi" and "GS" is proportional to their relative concentrations. Activation of GS by α2 adrenergic receptors is observed only at very high agonist concentrations: this suggests that, at very high agonist concentrations, Gi becomes unable to compete for receptor activation: in contrast with Gi-GDP, the activated Gi-GTP complex is probably unable to recognize agonist-bound α2 adrenergic receptors (Waelbroeck, 2001).

A few GPCRs are capable of activating several G proteins in physiological settings: the G protein specificity is not always "absolute". Although this is unusual in Family A, some G protein coupled receptors can be expressed as related isoforms due to alternative splicing of RNA expressed from a single gene or to RNA editing: this may lead to receptor isoforms with different abilities to activate G proteins (Hermans, 2003; Bresson-Bepoldin et al., 1998). Alternatively, post-translational modifications such as phosphorylation of the receptor may alter its G protein specificity: β2-adrenergic receptors activate $G_S$ proteins, leading to adenylate cyclase and protein kinase A stimulation, then – after phosphorylation by protein kinase A – activate $G_i$ proteins (Zamah et al., 2002). The TSH receptor is able to activate G proteins from all four families (Allgeier et al., 1997; Laugwitz et al., 1996). Its binding properties are compatible with the hypothesis that it forms a stable dimer, and that occupancy of the dimer by one TSH molecule decreases the affinity of the second binding site ("negative cooperativity" (Urizar et al., 2005)). While signaling through $G_S$ is induced at very low TSH concentrations, low affinity occupancy of two binding sites per dimer appears to be necessary to drive receptor activation of $G_i$ (Allen et al., 2011).

## 8.2 GPCR phosphorylation and desensitization

Activated rhodopsin (metarhodopsin II) activates the rhodopsin kinase (GRK1), which in turn phosphorylates preferentially the activated rhodopsin (Premont and Gainetdinov, 2007). Both activation to metarhodopsin II and phosphorylation synergistically increase the rhodopsin affinity for an adaptor protein, arrestin. This protein competitively inhibits transducin recognition by steric hindrance – resulting in rhodopsin desensitization: light activated rhodopsin becomes unable to activate transducin and signaling is "arrested".

Likewise, ligand-activated GPCRs recognize and activate "GRKs" (G protein coupled Receptor Kinases), that in turn preferentially phosphorylate activated GPCRs (Premont and Gainetdinov, 2007; Huang and Tesmer, 2011). Most GPCRs are, in addition, targets for "second messenger activated kinases": they possess consensus sequences for protein kinase A that is activated in response to the increased cAMP, or for protein kinase C, activated by the phospholipase C signaling pathway (cytosolic Ca2+ and diacylglycerol). Receptor phosphorylation by these kinases will lead to "heterodesensitization", since a given agonist can induce the desensitization of receptors it does not activate.

Mammalian cells express seven GRKs: two of them (GRK1 and 7) are found only in rod and cone cells in the retina; GRK4 is found mainly in the testes and to a lesser extent in some brain regions and in the kidney, and the last four (GRK2, 3, 5 and 6) are ubiquitous (Yang and Xia, 2006). They can be subdivided in three subgroups: GRK1 and 7; GRK2 and 3, and GRK 4, 5 and 6. The C-terminal region of GRK2 and 3 is longer than in other GRKs and possesses a "Pleckstrin Homology" (PH) domain: these two GRKs are cytosolic and recruited by $G\beta\gamma$ in response to G protein activation (Yang and Xia, 2006). In contrast, GRK 1 and 7 are C-terminal farnesylated, GRK4 and 6 are palmitoylated on C-terminal cysteines and GRK4-6 have a highly conserved binding site for phosphatidyl inositol 4-phosphate: the PH domain and post-translational modifications facilitate the permanent localization of these GRKs at the plasma membrane (Yang and Xia, 2006). The N-terminal region of all GRKs is similar and important for receptor recognition – GRKs are highly specific in their receptor preference (Yang and Xia, 2006).

Three dimensional structures for at least one representative of the three GRK families have been determined by X-ray diffraction (Figure 9). In most structures, the N-terminal region (that is essential for receptor recognition) is undefined, and the active cleft is too "open" for substrate recognition, suggesting that the kinases usually crystallize in the resting conformation. Very recently, GRK6 was crystallized in a form very likely resembling its' active conformation, with a relatively "closed" active cleft (Figure 10): this structure probably resembles the active GRK (Boguth et al., 2010). It is characterized by a well defined extended N-terminal α-helix, that could easily be fitted – superimposed on the G protein C-terminal α-helix (Boguth et al., 2010) - in the intracellular pocket formed in the metarhodopsin II structure (Figure 10). This would bring the active cleft in close proximity to the receptors IC3 and Cter – the two regions that are phosphorylated by GRKs.

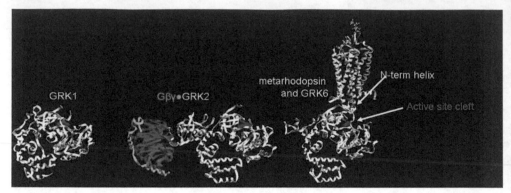

Fig. 9. Representative X-ray structures of GRKs from the three families. Ribbon structure of GRK1 (left: 3C4W), of the GRK2-Gβγ complex (center: 3KRW), and of the presumed "active conformation" of GRK6 (right: 3DQB) superimposed on the transducin C-term peptide in close apposition with the metarhodopsin II structure (3NYN), to form a hypothetical active GRK-receptor complex (according to (Boguth *et al.*, 2010)). The GRKs are shown by a yellow ribbon, co-crystallized ATP or ATP analogues in pink to identify the active site; Gβ and Gγ is shown in blue and green, respectively, and metarhodopsin II, in light grey.

Most GRKs are probably able to regulate GPCR signal transduction by phosphorylation-independent mechanisms. All GRKs have a "Regulator of G protein Signaling (RGS) homology" (RH) domain, and GRK2 and 3 have been shown to specifically interact through this domain with $G\alpha_q$ family members, thereby blocking their interaction with their effector, phospholipase C (see Figure 5). At least some GRKs are able to compete with G protein recognition by the activated receptor and/or compete with effector proteins for Gβγ recognition (Yang and Xia, 2006). By phosphorylating the receptor, they also increase markedly the receptor affinity for "arrestin" molecules that compete with G proteins for receptor recognition, facilitate receptor internalization in endosomes, and may serve as "scaffold", allowing "G protein independent signaling" (see below) (Premont and Gainetdinov, 2007; Huang and Tesmer, 2011).

Since GRKs and G proteins compete for the same (active) receptor conformation, the sequence of receptor recognition is important: GPCRs should recognize first the G proteins, then GRKs. "Sequential" recognition of two ligands is easily explained under the assumption that they have different *dissociation* rate constants (Motulsky and Mahan, 1984): the ligand with the faster dissociation rate constant will occupy the receptor rapidly, then progressively give place to the ligand with the slower dissociation rate constant (see Figure 10).

The most important factor under non equilibrium conditions is the relative **dissociation** (not association) rate constant of the two ligands. This might seem counterintuitive, but can easily be explained. Let us first examine the case of two ligands with different affinities due to different association rate constants. The lower affinity ligand will be needed in larger concentrations to significantly occupy the receptors at equilibrium: its lower association rate constant is then automatically compensated by the larger ligand concentration used. (The association rate is equal to $k_{on}[L]$, where $k_{on}$ is the association rate constant and [L], the ligand concentration). In contrast, if the two ligands have different affinities because of different dissociation rate constant: the larger dissociation rate constant of the low affinity ligand cannot be compensated by the larger ligand concentrations used to occupy the receptor at equilibrium: the dissociation rate, $k_{off}[LR]$ does not depend on the free ligand concentration. In order for the G protein, GRKs (and arrestin) to recognize sequentially the receptors, it is therefore necessary and sufficient that they have a different dissociation rate constants from the receptor. This is not a problem, as a very rapid G protein dissociation from the receptor is also necessary to allow receptor recycling and efficient catalytic activation of the G proteins...

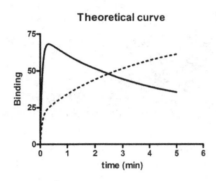

**Theoretical curve**

Fig. 10. Competitive binding of two ligands to the same receptor as a function of time. Ligand A (full line) has a $k_{on}=10^8M^{-1}min^{-1}$, $k_{off} = 1$ min$^{-1}$ and is present at a concentration of 100 nM (10 $K_D$); it will occupy 24% of the receptors at equilibrium in the presence of ligand B. Ligand B (hatched line), has the same $k_{on} = 10^8min^{-1}M^{-1}$, a lower dissociation rate ($k_{off} = .0.1$ min$^{-1}$) and is present at a concentration of 30 nM (30 $K_D$): it will occupy 73% of the receptors at equilibrium, in the presence of ligand A.

## 8.3 Arrestin recognition by GPCRs

All mammalian cells express at least one of the four "arrestins": rod and cone cells from the visual system express arrestins 1 and 4, respectively; arrestins 2 and 3 (also known as β-arrestin 1 and 2) are ubiquitously expressed. These proteins recognize and are activated by multi-phosphorylated, activated GPCRs: arrestin (arrestin 1) is specific for rhodopsin, arrestin 4, for the iodopsins, and arrestins 2 and 3 recognize most if not all GPCRs. Phosphorylation and receptor activation synergistically enhance rhodopsin-arrestin interactions: light activated rhodopsin and resting but phosphorylated rhodopsin have a 10-100 fold lower affinity for arrestin, and rhodopsin does not detectably interact with arrestin 1. β-arrestin binding to "traditional" GPCRs is affected more by phosphorylation than by agonist binding (Gurevich et al., 1995).

All known arrestin 3D structures are rather similar to visual arrestin (Figure 11). They can be subdivided into two concave β-sheet domains held together by a hinge region, an ionic bridge network between two arginine and three aspartate side chains (center of the structure on Figure 11), and by interactions between the C-term tail, the first N-term β strand and the α helix (left of Figure 11).

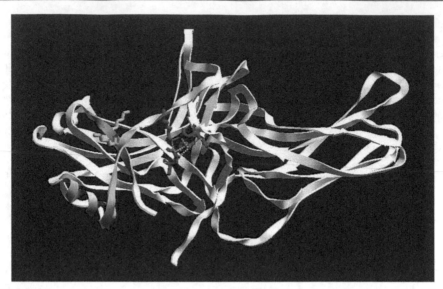

Fig. 11. The visual arrestin 1 crystal structure 1CF1. The N-terminal lobe is shown on the left, and the C-terminal lobe on the right. They are joined by a single random coil stretch (yellow) but held together through ionic interactions between buried arginine (blue) and aspartate (red) side chains, and by close contacts of the first N-terminal β-sheet stretch and α-helix with a C-terminal β-sheet stretch (orange). The two lysine side chains that are important for preferential phosphorylated>non-phosphorylated (light activated) rhodopsin recognition are shown in orange (left).

Each of the two arrestin domains is large enough to interact with a rhodopsin monomer (Figure 12). Even though visual arrestin forms a one to one complex with rhodopsin both in vitro (Bayburt *et al.*, 2011) and in vivo (Hanson *et al.*, 2007), several side chains covering both domains are implicated in rhodopsin recognition or rhodopsin – GPCR discrimination (Bayburt *et al.*, 2011; Vishnivetskiy *et al.*, 2011; Skegro *et al.*, 2007) (Figure 12). This indicates that arrestin undergoes a significant conformation change when it recognizes the phosphorylated receptors. This is confirmed by the observation that the arrestin sensitivity to proteolytic degradation increases upon GPCR recognition, and that the intramolecular BRET between the N- and C-terminal region of a luciferase–arrestin–Yellow Fluorescent Protein (YFP) construct is markedly affected by arrestin recognition of agonist-bound receptors (Shukla *et al.*, 2008).

At least two rhodopsin Ser/Thr must be phosphorylated to allow arrestin interaction with metarhodopsin; three phosphates support stronger arrestin binding, and heavier phosphorylation promotes arrestin binding , in addition, to neighbouring dark (inactive) rhodopsin and to phospho-opsin, two unpreferred rhodopsin forms (Vishnivetskiy *et al.*, 2007). Likewise, β-arrestin recognition increases mainly in response to multi-phosphorylation of the GPCRs C-terminal or IC3 sequence rather than in response to agonist binding (Gurevich *et al.*, 1995; Oakley *et al.*, 2000). "Phosphoserine/phosphothreonine rich" patches are necessary for stable, high affinity arrestin recognition (Oakley *et al.*, 2001). GPCRs that present patches of phosphorylated Ser/Thr residues (angiotensin II type 1A, neurotensin 1, vasopressin V2, thyrotropin-releasing hormone and substance P receptors) have a high affinity and do not discriminate the arrestin 1, 2 and 3 isoforms; they are rapidly internalized and recycle inefficiently or not at all. In contrast, $\beta_2$-adrenergic, μ opioid, endothelin type A, dopamine $D_{1A}$, and $\alpha_{1b}$ adrenergic receptors (with separate phosphorylated Ser/Thr residues) have a low affinity for β-arrestin 2 (arrestin 3), an even lower affinity for β-arrestin 1 (arrestin 2) and do not detectably recruit arrestin 1. Upon internalization, these receptors are rapidly dephosphorylated and recycled to the plasma membrane (Oakley et al., 2000).

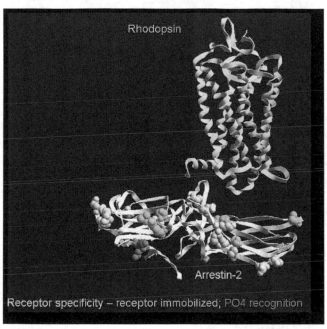

Fig. 12. Arrestin 3 (β-arrestin 2), close to opsin structure. Grey ribbon: arrestin 3 ribbon structure structure (1JSY), showing some of the side chains that are immobilized upon dark phosphorhodopsin recognition (light green), involved in the discrimination of light activated phosphorhodopsin from carbachol-activated phosphorylated M2 muscarinic receptor (yellow) or necessary for recognition of the phosphoserine/threonines (tan) (Vishnivetskiy et al., 2011). Green ribbon: opsin structure 3CAP is shown for size comparison.

Arrestins change conformation upon receptor recognition (Shukla et al., 2008) and behave as receptor-dependent "scaffold proteins" bringing together a number of other proteins (for review: (Premont and Gainetdinov, 2007; DeFea, 2011)). Some of their binding sites are shown in Figure 13.

Several β-arrestin scaffolds have been identified: this protein can recruit either MAP kinase partners, PI3Kinase or Akt, phosphodiesterase of actin assembly proteins scaffolds when bound to activated, phosphorylated receptors (DeFea, 2011). The different binding sites are very close (Figure 13): only some well-defined complexes can be formed or dissociated in response to agonist-receptor recognition by arrestin. The factors determining which complex is formed in response to a given receptor are still elusive.

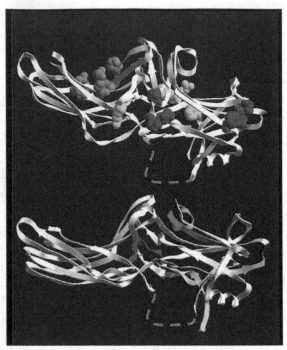

Fig. 13. Scaffolding sites on arrestin. Top: 3 arrestin 3 structure 3P2D The Proline-rich regions that allow SH3 domains recognition are shown in fuchsia, the $PIP_2$ binding site, in blue and side chains essential for β-adaptin binding, in green. The clathrin recognition site LφxφE is in the hatched (unstructured) region. R307, that is essential for cRaf1 recognition, is shown in light blue. Bottom : arrestin 2 structure (1JSY) showing the partially overlapping Ask1 and MEK binding regions (kaki), the MKK4 (dark green), PDED5 (light green), ERK2 (fuchsia) binding sites, and the partially overlapping Akt and clathrin binding sites (light blue) (according to (DeFea, 2011)).

## 9. References

Albery WJ and Knowles JR (1976) Free-Energy Profile of the Reaction Catalyzed by Triosephosphate Isomerase. *Biochemistry* 15: 5627-5631.

Allen MD, Neumann S and Gershengorn MC (2011) Occupancy of Both Sites on the Thyrotropin (TSH) Receptor Dimer Is Necessary for Phosphoinositide Signaling. *FASEB J.*

Allgeier A, Laugwitz KL, Van Sande J, Schultz G and Dumont JE (1997) Multiple G-Protein Coupling of the Dog Thyrotropin Receptor. *Mol Cell Endocrinol* 127: 81-90.

Altenbach C, Kusnetzow AK, Ernst OP, Hofmann KP and Hubbell WL (2008) High-Resolution Distance Mapping in Rhodopsin Reveals the Pattern of Helix Movement Due to Activation. *Proc Natl Acad Sci U S A* 105: 7439-7444.

Bayburt TH, Vishnivetskiy SA, McLean MA, Morizumi T, Huang CC, Tesmer JJ, Ernst OP, Sligar SG and Gurevich VV (2011) Monomeric Rhodopsin Is Sufficient for Normal Rhodopsin Kinase (GRK1) Phosphorylation and Arrestin-1 Binding. *J Biol Chem* 286: 1420-1428.

Berrie CP, Birdsall NJ, Burgen AS and Hulme EC (1979) Guanine Nucleotides Modulate Muscarinic Receptor Binding in the Heart. *Biochem Biophys Res Commun* 87: 1000-1005.

Bhattacharya S and Vaidehi N (2010) Computational Mapping of the Conformational Transitions in Agonist Selective Pathways of a G-Protein Coupled Receptor. *J Am Chem Soc* 132: 5205-5214.

Birdsall NJ (2010) Class A GPCR Heterodimers: Evidence From Binding Studies. *Trends Pharmacol Sci* 31: 499-508.

Birnbaumer L (2007) Expansion of Signal Transduction by G Proteins. The Second 15 Years or So: From 3 to 16 Alpha Subunits Plus Betagamma Dimers. *Biochim Biophys Acta* 1768: 772-793.

Boguth CA, Singh P, Huang CC and Tesmer JJ (2010) Molecular Basis for Activation of G Protein-Coupled Receptor Kinases. *EMBO J* 29: 3249-3259.

Bresson-Bepoldin L, Jacquot MC, Schlegel W and Rawlings SR (1998) Multiple Splice Variants of the Pituitary Adenylate Cyclase-Activating Polypeptide Type 1 Receptor Detected by RT-PCR in Single Rat Pituitary Cells. *J Mol Endocrinol* 21: 109-120.

Choe HW, Kim YJ, Park JH, Morizumi T, Pai EF, Krauss N, Hofmann KP, Scheerer P and Ernst OP (2011) Crystal Structure of Metarhodopsin II. *Nature* 471: 651-655.

De Lean A, Stadel JM and Lefkowitz RJ (1980) A Ternary Complex Model Explains the Agonist-Specific Binding Properties of the Adenylate Cyclase-Coupled Beta-Adrenergic Receptor. *J Biol Chem* 255: 7108-7117.

DeFea KA (2011) Beta-Arrestins As Regulators of Signal Termination and Transduction: How Do They Determine What to Scaffold? *Cell Signal* 23: 621-629.

Ernst OP, Gramse V, Kolbe M, Hofmann KP and Heck M (2007) Monomeric G Protein-Coupled Receptor Rhodopsin in Solution Activates Its G Protein Transducin at the Diffusion Limit. *Proc Natl Acad Sci U S A* 104: 10859-10864.

Fraser CM, Arakawa S, McCombie WR and Venter JC (1989) Cloning, Sequence Analysis, and Permanent Expression of a Human Alpha 2-Adrenergic Receptor in Chinese Hamster Ovary Cells. Evidence for Independent Pathways of Receptor Coupling to Adenylate Cyclase Attenuation and Activation. *J Biol Chem* 264: 11754-11761.

Fredriksson R, Lagerstrom MC, Lundin LG and Schioth HB (2003) The G-Protein-Coupled Receptors in the Human Genome Form Five Main Families. Phylogenetic Analysis, Paralogon Groups, and Fingerprints. *Mol Pharmacol* 63: 1256-1272.

Gurevich VV, Dion SB, Onorato JJ, Ptasienski J, Kim CM, Sterne-Marr R, Hosey MM and Benovic JL (1995) Arrestin Interactions With G Protein-Coupled Receptors. Direct Binding Studies of Wild Type and Mutant Arrestins With Rhodopsin, Beta 2-Adrenergic, and M2 Muscarinic Cholinergic Receptors. *J Biol Chem* 270: 720-731.

Hamm HE (1998) The Many Faces of G Protein Signaling. *J Biol Chem* 273: 669-672.

Hanson SM, Gurevich EV, Vishnivetskiy SA, Ahmed MR, Song X and Gurevich VV (2007) Each Rhodopsin Molecule Binds Its Own Arrestin. *Proc Natl Acad Sci U S A* 104: 3125-3128.

Heck M and Hofmann KP (2001) Maximal Rate and Nucleotide Dependence of Rhodopsin-Catalyzed Transducin Activation: Initial Rate Analysis Based on a Double Displacement Mechanism. *J Biol Chem* 276: 10000-10009.

Heredia VV, Thomson J, Nettleton D and Sun S (2006) Glucose-Induced Conformational Changes in Glucokinase Mediate Allosteric Regulation: Transient Kinetic Analysis. *Biochemistry* 45: 7553-7562.

Hermans E (2003) Biochemical and Pharmacological Control of the Multiplicity of Coupling at G-Protein-Coupled Receptors. *Pharmacology & Therapeutics* 99: 25-44.

Hern JA, Baig AH, Mashanov GI, Birdsall B, Corrie JE, Lazareno S, Molloy JE and Birdsall NJ (2010) Formation and Dissociation of M1 Muscarinic Receptor Dimers Seen by Total Internal Reflection Fluorescence Imaging of Single Molecules. *Proc Natl Acad Sci U S A* 107: 2693-2698.

Hu J, Wang Y, Zhang X, Lloyd JR, Li JH, Karpiak J, Costanzi S and Wess J (2010) Structural Basis of G Protein-Coupled Receptor-G Protein Interactions. *Nat Chem Biol* 6: 541-548.

Huang CC and Tesmer JJ (2011) Recognition in the Face of Diversity: Interactions of Heterotrimeric G Proteins and G Protein-Coupled Receptor (GPCR) Kinases With Activated GPCRs. *J Biol Chem* 286: 7715-7721.

Jensen AA, Greenwood JR and Brauner-Osborne H (2002) The Dance of the Clams: Twists and Turns in the Family C GPCR Homodimer. *Trends Pharmacol Sci* 23: 491-493.

Jian X, Clark WA, Kowalak J, Markey SP, Simonds WF and Northup JK (2001) Gbetagamma Affinity for Bovine Rhodopsin Is Determined by the Carboxyl-Terminal Sequences of the Gamma Subunit. *J Biol Chem* 276: 48518-48525.

Johnston JM, Aburi M, Provasi D, Bortolato A, Urizar E, Lambert NA, Javitch JA and Filizola M (2011) Making Structural Sense of Dimerization Interfaces of Delta Opioid Receptor Homodimers. *Biochemistry* 50: 1682-1690.

Jones KA, Borowsky B, Tamm JA, Craig DA, Durkin MM, Dai M, Yao WJ, Johnson M, Gunwaldsen C, Huang LY, Tang C, Shen Q, Salon JA, Morse K, Laz T, Smith KE, Nagarathnam D, Noble SA, Branchek TA and Gerald C (1998) GABA(B) Receptors Function As a Heteromeric Assembly of the Subunits GABA(B)R1 and GABA(B)R2. *Nature* 396: 674-679.

Katadae M, Hagiwara K, Wada A, Ito M, Umeda M, Casey PJ and Fukada Y (2008) Interacting Targets of the Farnesyl of Transducin Gamma-Subunit. *Biochemistry* 47: 8424-8433.

Kisselev OG and Downs MA (2003) Rhodopsin Controls a Conformational Switch on the Transducin Gamma Subunit. *Structure* 11: 367-373.

Kolakowski LF, Jr. (1994) GCRDb: a G-Protein-Coupled Receptor Database. *Receptors Channels* 2: 1-7.

Kostenis E, Waelbroeck M and Milligan G (2005) Techniques: Promiscuous Galpha Proteins in Basic Research and Drug Discovery. *Trends Pharmacol Sci* 26: 595-602.

Laugwitz KL, Allgeier A, Offermanns S, Spicher K, Van Sande J, Dumont JE and Schultz G (1996) The Human Thyrotropin Receptor: a Heptahelical Receptor Capable of Stimulating Members of All Four G Protein Families. *Proc Natl Acad Sci U S A* 93: 116-120.

Lebon G, Warne T, Edwards PC, Bennett K, Langmead CJ, Leslie AG and Tate CG (2011) Agonist-Bound Adenosine A2A Receptor Structures Reveal Common Features of GPCR Activation. *Nature* 474: 521-525.

Lefkowitz RJ, Mullikin D and Caron MG (1976) Regulation of Beta-Adrenergic Receptors by Guanyl-5'-Yl Imidodiphosphate and Other Purine Nucleotides. *J Biol Chem* 251: 4686-4692.

Ma B and Nussinov R (2010) Enzyme Dynamics Point to Stepwise Conformational Selection in Catalysis. *Curr Opin Chem Biol* 14: 652-659.

Magalhaes AC, Dunn H and Ferguson SS (2011) Regulation of G Protein-Coupled Receptor Activity, Trafficking and Localization by GPCR-Interacting Proteins. *Br J Pharmacol.*

McMillin SM, Heusel M, Liu T, Costanzi S and Wess J (2011) Structural Basis of M3 Muscarinic Receptor Dimer/Oligomer Formation. *J Biol Chem* 286: 28584-28598.

Milligan G (2009) G Protein-Coupled Receptor Hetero-Dimerization: Contribution to Pharmacology and Function. *Br J Pharmacol* 158: 5-14.

Milligan G (2010) The Role of Dimerisation in the Cellular Trafficking of G-Protein-Coupled Receptors. *Curr Opin Pharmacol* 10: 23-29.

Mizuno N and Itoh H (2011) Signal Transduction Mediated Through Adhesion-GPCRs. *Adv Exp Med Biol* 706: 157-166.

Molnes J, Teigen K, Aukrust I, Bjorkhaug L, Sovik O, Flatmark T and Njolstad PR (2011) Binding of ATP at the Active Site of Human Pancreatic Glucokinase--Nucleotide-Induced Conformational Changes With Possible Implications for Its Kinetic Cooperativity. *FEBS J* 278: 2372-2386.

Monod J, Wyman J And Changeux JP (1965) On the Nature of Allosteric Transitions: a Plausible Model. *J Mol Biol* 12: 88-118.

Motulsky HJ and Mahan LC (1984) The Kinetics of Competitive Radioligand Binding Predicted by the Law of Mass Action. *Mol Pharmacol* 25: 1-9.

Moukhametzianov R, Warne T, Edwards PC, Serrano-Vega MJ, Leslie AG, Tate CG and Schertler GF (2011) Two Distinct Conformations of Helix 6 Observed in Antagonist-Bound Structures of a Beta1-Adrenergic Receptor. *Proc Natl Acad Sci U S A* 108: 8228-8232.

Niesen MJ, Bhattacharya S and Vaidehi N (2011) The Role of Conformational Ensembles in Ligand Recognition in G-Protein Coupled Receptors. *J Am Chem Soc.*

Oakley RH, Laporte SA, Holt JA, Barak LS and Caron MG (2001) Molecular Determinants Underlying the Formation of Stable Intracellular G Protein-Coupled Receptor-Beta-Arrestin Complexes After Receptor Endocytosis*. *J Biol Chem* 276: 19452-19460.

Oakley RH, Laporte SA, Holt JA, Caron MG and Barak LS (2000) Differential Affinities of Visual Arrestin, Beta Arrestin1, and Beta Arrestin2 for G Protein-Coupled

Receptors Delineate Two Major Classes of Receptors. *J Biol Chem* 275: 17201-17210.

Oldham WM and Hamm HE (2006) Structural Basis of Function in Heterotrimeric G Proteins. *Q Rev Biophys* 39: 117-166.

Park JH, Scheerer P, Hofmann KP, Choe HW and Ernst OP (2008) Crystal Structure of the Ligand-Free G-Protein-Coupled Receptor Opsin. *Nature* 454: 183-187.

Premont RT and Gainetdinov RR (2007) Physiological Roles of G Protein-Coupled Receptor Kinases and Arrestins. *Annu Rev Physiol* 69: 511-534.

Provasi D and Filizola M (2010) Putative Active States of a Prototypic G-Protein-Coupled Receptor From Biased Molecular Dynamics. *Biophys J* 98: 2347-2355.

Rasmussen SG, Choi HJ, Fung JJ, Pardon E, Casarosa P, Chae PS, Devree BT, Rosenbaum DM, Thian FS, Kobilka TS, Schnapp A, Konetzki I, Sunahara RK, Gellman SH, Pautsch A, Steyaert J, Weis WI and Kobilka BK (2011a) Structure of a Nanobody-Stabilized Active State of the Beta(2) Adrenoceptor. *Nature* 469: 175-180.

Rasmussen SG, Devree BT, Zou Y, Kruse AC, Chung KY, Kobilka TS, Thian FS, Chae PS, Pardon E, Calinski D, Mathiesen JM, Shah ST, Lyons JA, Caffrey M, Gellman SH, Steyaert J, Skiniotis G, Weis WI, Sunahara RK and Kobilka BK (2011b) Crystal Structure of the Beta(2) Adrenergic Receptor-Gs Protein Complex. *Nature*.

Roberts DJ and Waelbroeck M (2004) G Protein Activation by G Protein Coupled Receptors: Ternary Complex Formation or Catalyzed Reaction? *Biochem Pharmacol* 68: 799-806.

Rosenbaum DM, Zhang C, Lyons JA, Holl R, Aragao D, Arlow DH, Rasmussen SG, Choi HJ, Devree BT, Sunahara RK, Chae PS, Gellman SH, Dror RO, Shaw DE, Weis WI, Caffrey M, Gmeiner P and Kobilka BK (2011) Structure and Function of an Irreversible Agonist-Beta(2) Adrenoceptor Complex. *Nature* 469: 236-240.

Scheerer P, Park JH, Hildebrand PW, Kim YJ, Krauss N, Choe HW, Hofmann KP and Ernst OP (2008) Crystal Structure of Opsin in Its G-Protein-Interacting Conformation. *Nature* 455: 497-502.

Shukla AK, Violin JD, Whalen EJ, Gesty-Palmer D, Shenoy SK and Lefkowitz RJ (2008) Distinct Conformational Changes in Beta-Arrestin Report Biased Agonism at Seven-Transmembrane Receptors. *Proc Natl Acad Sci U S A* 105: 9988-9993.

Skegro D, Pulvermuller A, Krafft B, Granzin J, Hofmann KP, Buldt G and Schlesinger R (2007) N-Terminal and C-Terminal Domains of Arrestin Both Contribute in Binding to Rhodopsin. *Photochem Photobiol* 83: 385-392.

Springael JY, Urizar E and Parmentier M (2005) Dimerization of Chemokine Receptors and Its Functional Consequences. *Cytokine Growth Factor Rev* 16: 611-623.

Standfuss J, Edwards PC, D'Antona A, Fransen M, Xie G, Oprian DD and Schertler GF (2011) The Structural Basis of Agonist-Induced Activation in Constitutively Active Rhodopsin. *Nature* 471: 656-660.

Sykes DA, Dowling MR and Charlton SJ (2009) Exploring the Mechanism of Agonist Efficacy: a Relationship Between Efficacy and Agonist Dissociation Rate at the Muscarinic M3 Receptor. *Mol Pharmacol* 76: 543-551.

Temussi PA (2009) Sweet, Bitter and Umami Receptors: a Complex Relationship. *Trends Biochem Sci* 34: 296-302.

Urizar E, Montanelli L, Loy T, Bonomi M, Swillens S, Gales C, Bouvier M, Smits G, Vassart G and Costagliola S (2005) Glycoprotein Hormone Receptors: Link Between Receptor Homodimerization and Negative Cooperativity. *EMBO J* 24: 1954-1964.

Vishnivetskiy SA, Gimenez LE, Francis DJ, Hanson SM, Hubbell WL, Klug CS and Gurevich VV (2011) Few Residues Within an Extensive Binding Interface Drive Receptor Interaction and Determine the Specificity of Arrestin Proteins. *J Biol Chem* 286: 24288-24299.

Vishnivetskiy SA, Raman D, Wei J, Kennedy MJ, Hurley JB and Gurevich VV (2007) Regulation of Arrestin Binding by Rhodopsin Phosphorylation Level. *J Biol Chem* 282: 32075-32083.

Waelbroeck M (1999) Kinetics Versus Equilibrium: the Importance of GTP in GPCR Activation. *Trends Pharmacol Sci* 20: 477-481.

Waelbroeck M (2001) Activation of Guanosine 5'-[Gamma-(35)S]Thio-Triphosphate Binding Through M(1) Muscarinic Receptors in Transfected Chinese Hamster Ovary Cell Membranes; 1. Mathematical Analysis of Catalytic G Protein Activation. *Mol Pharmacol* 59: 875-885.

Waelbroeck M, Boufrahi L and Swillens S (1997) Seven Helix Receptors Are Enzymes Catalysing G Protein Activation. What Is the Agonist Kact? *J Theor Biol* 187: 15-37.

Warne T, Moukhametzianov R, Baker JG, Nehme R, Edwards PC, Leslie AG, Schertler GF and Tate CG (2011) The Structural Basis for Agonist and Partial Agonist Action on a Beta(1)-Adrenergic Receptor. *Nature* 469: 241-244.

Weiss JM, Morgan PH, Lutz MW and Kenakin TP (1996a) The Cubic Ternary Complex Receptor-Occupancy Model. I. Model Description. *J Theor Biol* 178: 151-167.

Weiss JM, Morgan PH, Lutz MW and Kenakin TP (1996b) The Cubic Ternary Complex Receptor-Occupancy Model. II.Understanding Apparent Affinity. *J Theor Biol* 178: 169-182.

Weiss JM, Morgan PH, Lutz MW and Kenakin TP (1996c) The Cubic Ternary Complex Receptor-Occupancy Model. III. Resurrecting Efficacy. *J Theor Biol* 181: 381-397.

Wellendorph P and Brauner-Osborne H (2009) Molecular Basis for Amino Acid Sensing by Family C G-Protein-Coupled Receptors. *Br J Pharmacol* 156: 869-884.

White JH, Wise A and Marshall FH (2002) Heterodimerization of Gamma-Aminobutyric Acid B Receptor Subunits As Revealed by the Yeast Two-Hybrid System. *Methods* 27: 301-310.

Whorton MR, Bokoch MP, Rasmussen SG, Huang B, Zare RN, Kobilka B and Sunahara RK (2007) A Monomeric G Protein-Coupled Receptor Isolated in a High-Density Lipoprotein Particle Efficiently Activates Its G Protein. *Proc Natl Acad Sci U S A* 104: 7682-7687.

Whorton MR, Jastrzebska B, Park PS, Fotiadis D, Engel A, Palczewski K and Sunahara RK (2008) Efficient Coupling of Transducin to Monomeric Rhodopsin in a Phospholipid Bilayer. *J Biol Chem* 283: 4387-4394.

Xu F, Wu H, Katritch V, Han GW, Jacobson KA, Gao ZG, Cherezov V and Stevens RC (2011) Structure of an Agonist-Bound Human A2A Adenosine Receptor. *Science* 332: 322-327.

Yang W and Xia SH (2006) Mechanisms of Regulation and Function of G-Protein-Coupled Receptor Kinases. *World J Gastroenterol* 12: 7753-7757.

Zamah AM, Delahunty M, Luttrell LM and Lefkowitz RJ (2002) Protein Kinase A-Mediated Phosphorylation of the Beta 2-Adrenergic Receptor Regulates Its Coupling to Gs and Gi. Demonstration in a Reconstituted System. *J Biol Chem* 277: 31249-31256.

# 7

# Application of Quantitative Immunogold Electron Microscopy to Determine the Distribution and Relative Expression of Homo- and Heteromeric Purinergic Adenosine A1 and P2Y Receptors

Kazunori Namba

*National Institute of Sensory Organs, National Tokyo Medical Center, Tokyo*
*Japan*

## 1. Introduction

The idea that G-protein-coupled receptors (GPCRs) may generate or modify various functions as dimmers or higher-order oligomers is now generally accepted. Significant numbers of GPCRs exit as heteromeric assemblies (refered to as hetero-oligomerization), generating novel functions for ligand binding and second messengers, and in turn creating unique receptor trafficking systems for pharmacological profiles (Angers *et al.*, 2002, Bulenger *et al.*, 2005). This is also true of the purinergic receptor family. Over recent years, we have explored many biochemical and pharmacological aspects of this particular family via hetero-oligomerization between metabotropic (i.e. G protein-coupled) purinergic receptors (particularly between P1 and P2), in which the agonists are metabolites playing important role in the purinergic signaling cascade.

Purines such as adenosine triphosphate (ATP), via their specific P1 and P2 receptors, mediate a variety of physiological processes including pathophysiology, neurotransmission, neuromodulation, pain, cardiac function, immune responses and almost every aspect of development (Abbracchio *et al.*, 2009; Burnstock, 2007; Burnstock, 2008; Ralevic *et al.*, 1998). P1 receptors are further sub-classified into $A_1$, $A_{2A}$, $A_{2B}$ and $A_3$ sub-types, all of which are G protein-coupled receptors (GPCRs). The adenosine $A_1$ receptor ($A_1R$) is known to regulate $Ca^{2+}/K^+$ channels, adenylate cyclase, and phospholipase C by coupling to $G_{i/o}$ proteins (Ralevic *et al.*, 1998). The P2 receptors can be further sub-classified into ligand-gated ion channel-type $P2X_{(1-7)}$ receptors, and G protein-coupled $P2Y_{(1, 2, 4, 6, 11, 12, 13, 14)}$ receptors. $P2Y_2R$-specific pharmacology (induction of $Ca^{2+}$ release) has been analyzed in detail using CHO-K1 cells (Mehta *et al.*, 2008). In hippocampal astrocytes, $P2Y_1R$- and $P2Y_2R$-mediated $Ca^{2+}$ responses differentially show two forms of activity-dependent negative feedback of synaptic transmission via the phospholipase C beta-$IP^3$ pathway (Fam *et al.*, 2003). $P2Y_2$ R modulation of pain responses has also been reported (Molliver *et al.*, 2002). Today, homo- or hetero-oligomers of many kinds of GPCRs have been reported (Bouvier, 2001) and the hetero-oligomerization of GPCRs affects various aspects of receptor function, including the alteration of ligand-binding specificity and cellular trafficking. We previously demonstrated that $A_1R$ associates with $P2Y_1R$ in co-transfected HEK293T cells and in rat brain homogenates, whereby a $P2Y_1R$ agonist stimulates $A_1R$ signaling via $G_{i/o}$ (Yoshioka et al., 2001, Yoshioka et al., 2002). Furthermore, in co-transfected HEK293T cells, hetero-oligomers display unique pharmacology whereby simultaneous activation of the two receptors attenuates $A_1R$ signaling via $G_{i/o}$, but synergistically enhances $P2Y_2R$ signaling via $G_{q/11}$ (Suzuki et al., 2006). Because $A_1R$ are widely expressed in the brain (Yoshioka et al., 2002), it is likely that these receptors also associate directly in situ; however, direct evidence of their oligomerization or precise co-localization in brain has yet to be demonstrated. In our laboratory, we are developing a new method, immunogold electron microscopic observation using different sized-immunogold particles enable visualize the oligomerization of $A_1R$ and $P2Y_2R$ (Namba et al., 2010). The aim of the study was to determine whether $A_1R$ and $P2Y_2R$ associate with each other in the rat brain by looking for receptor complexes with immunogold electron microscopy (IEM). This method also provides information concerning the localization and density of GPCR monomers and oligomers expressed in transfected cells, that are also applicable to tissues such as brain.

In this chapter, we describe both pre- and post-embedding electronmicroscopic techniques to identify cells or tissues expressing GPCRs utilizing differently-sized immunogold particles, and review IEM quantification as an efficient approach to analyze two specific types of data. One data set represents the classification of receptor formations. $A_1R$ and $P2Y_1R$ ($P2Y_2R$) produce five receptor formations which are made up of monomers ($A_1R$, P2YR), homo-oligomers ($A_1R$- $A_1R$, P2YR- P2YR) and hetero-oligomers ($A_1R$-P2YR). The second dataset describes the estimation of receptor expression levels by counting immunoreactive immunogold particles at the cell surface.

Establishing specific expression patterns of GPCRs at the ultrastructural level, and detecting homo- and hetero-oligomers of GPCRs in both co-transfected cultured cells and tissues, will enable us to visually understand some of the phenomena underlying signal transduction signalling pathways operating via GPCRs in a heteromeric dependent manner. It is widely accepted that drug discovery targets for rapid remedies are likely to be specific receptors expressed upon the cytoplasmic membrane. In order to establish the precise effects of new drugs, the expression patterns and expression level of $A_1R$ and $P2Y_1R$ ($P2Y_2R$) represent significant factors to be considered, especially with regard to their association with cross-talk systems.

## 2. Immunostaining of GPCRs in transfected HEK293T cells and brain sections

Double immunofluororescence microscopic methods is now generally employed for studying the co-localization of GPCRs in transfected cells. Transient transfection using HEK293T cells with epitope tagged-receptors (Hemagglutinin: HA- or Myc-) in expression plasmids has been performed routinely in our laboratory (Yoshioka et al., 2002; Nakata et al., 2006). Before commencing immunogold electron microscopy, we routinely analyse, the subcelluar distribution of HA-$A_1R$ and Myc-$P2Y_1R$ ($P2Y_2R$) in co-transfected cells by immunocytochemistry and confocal laser microscopy (Yoshioka et al., 2001; Namba et al., 2010). Confocal imaging of co-localized GPCRs provides highly detailed information regarding their co-localization upon cellular organelles, an important feature for the subsequent analysis of co-localization in ultrastructural images obtained by transmission electronmicroscopy. If two genes are co-localized at specific cellular organelles, then there is a much higher probability of hetero-oligomerization. Thus, confocal images of co-localization between $A_1R$ and $P2Y_1R$ provide an important opportunity to determine whether immunoelectronmicroscopy is possible.

## 2.1 Results: Co-localization of $A_1R$ and $P2Y_1R$ ($P2Y_2$) in transfected HEK293T cells

Confocal imaging for studying the GPCRs using transfected HEK293T cells is the most common method of co-localization of GPCRs. In our laboratory, the co-localization of $A_1R$ and $P2Y_1R$ ($P2Y_2R$) in co-transfected HEK293T cells has been examined by the double immunostaining of HA-$A_1R$ and Myc-$P2Y_1R$ (or HA-$A_1R$ and Myc-$P2Y_2R$) in order to compare localization pattern. Confocal images of co-transfected HEK293T cells double labeled for HA-$A_1R$ (red) and Myc-$P2Y_2R$ (green) are shown in Fig.1. As co-localization occurred upon the plasma membrane, this data supports the heteromeric association of $A_1R$ and $P2Y_2R$. A similar pattern of co-localization for $A_1R$ and $P2Y_2R$ has been demonstrated in rat brain sections as shown in Fig.2.

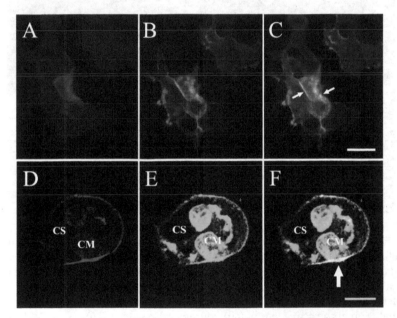

Fig. 1. Co-localization of $A_1R$ and $P2Y_2R$. A-C. Confocal images of double immunostained HA-$A_1R$ (A; red), Myc-$P2Y_2R$ (B; green), and their merged images (C; yellow) in co-transfected HEK293T cells. The co-localization of HA-$A_1R$ and Myc-$P2Y_2R$ is evident at the cell surface membrane (C; small arrow). White bar = 50 μm (A-C). Confocal images of double immunofluorescence for HA-$A_1R$ (D; red), Myc-$P2Y_2R$ (E; green), and their merged images (F; yellow) in co-transfected HEK293T cells are also obtained. Co-localizations of $A_1R$ and $P2Y_1R$ (F) was detected upon the cell surface membrane, but was not as evident upon inner cellular membranes (F; arrow). Cyan bar = 10 μm (D-F). Fluorescent images were obtained via confocal laser scanning microscopy (Zeiss LSM410, Carl Zeiss, Oberkochen, Germany) at two levels: 30-μm(A-C), and 15-μm(D-F). At each level, serial images were collected at 1-μm intervals through a total sectional thickness of 40-μm. Serial optical sections were recorded using an air objective lens of (20 X and 40X, numerical aperture; 0.6).

Both receptors were localized predominantly upon the cell surface and cytosolic membranes (Fig. 1. A,B). Merged images showed co-localization mainly in cell membranes (Fig. 1. C.). Our negative controls showed no positive signals in non-transfected HEK293T cells, indicating that the immunoreactivity observed in Fig. 1 was specific to the expressed receptors (data not shown).

## 2.2 Results: Immunohistochemical studies in rat brain

We examined the expression of $A_1R$ and P2YR in brain using using immunohistochemistry (Yoshioka *et al.*, 2002, Namba *et al.*, 2010). Prominent staining of $A_1R$ and $P2Y_2R$ were observed, particularly in Purkinje cells (Fig. 2A-C). Expression was predominantly restricted to cell bodies and neuronal dendrites. Importantly, co-localization of $A_1R$ and $P2Y_2R$ was observed in cell bodies within the cerebellum, but was detected within the nucleus of Purkinje cells.

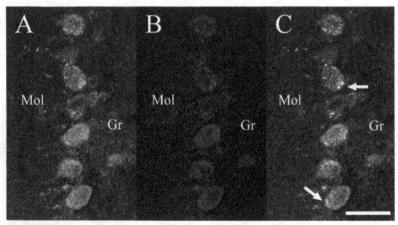

Fig. 2. Confocal images of double immunofluorescence stained $A_1R$ (A; green), $P2Y_2R$ (B; red), and their merged images (C; yellow) in Purkinje cells. Mol: cerebellar molecular layer, Gr: cerebellar granule cell layer. Co-localization of $A_1R$ and $P2Y_2R$ (C; yellow) were detected in the soma of the Purkinje cells (arrows). Bar = 50 μm. Fluorescent images were collected via confocal laser scanning microscopy (Zeiss LSM410, Carl Zeiss) and each 10-μm optical slice consisted of of a stack of of 20 sections (0.5-μm thick). Serial optical sections were recorded using an air objective lens of (40X, numerical aperture; 0.6).

## 3. Pre and post-embedding immunogold electron microscopy of transfected HEK293T cells

The monomeric- or hetero-oligomerization of intrinsic GPCRs cannot be ascertained by immunoelectronmicroscopic examination of brain tissues alone. Data concerning the hetero-oligomerization of GPCRs in brain tissues is typically acquired from three experimental phases. The first phase involves immunoelectronmicroscopic data acquired from pre-embedding methods and gene transfected cells. This provides important information as to whether hetero-expressed GPCRs can oligomerize or not. In other words, this method compares the expression patterns and co-localization of differently-sized immunogold particles using transfected- or non transfected-cells, yielding data that can determine the occurrence of GPCR hetero-oligomerization when using the same antibodies and immunoreactive conditions. Additionally, a particular advantage of using pre-embedding methods is that native antigenicity is maintained. The second phase is to acquire data from gene transfected cells using post-embedding methods. The reason why immunoelectronmicroscopic observation of tissues is applied with post-embedding methods is because it is difficult for a specific epitope anti-body to penetrate into the cytoplasmic region of tissue cells. Before experimenting on tissues, it is important to confirm patterns of immune-reaction with transfected cells using post-embedding methods in order to form positive controls for specific tissues. The last phase is to acquire data from tissues using post-embedding methods. It is suggested that the hetero-oligomerization of GPCRs in tissues would be very precise as an oligomer of different-sized gold particles in a given case of comparative data from single transfected GPCRs and co-transfected GPCRs.

### 3.1 Results: Immunogold electron microscopic observations of HA-A$_1$R and Myc-P2Y$_2$R expressed in transfected cells

In our laboratory, we examined the cellular localization of HA-A$_1$R/Myc-P2Y$_2$R in co-transfected HEK293T cells using post-embedding methods, anti-HA or anti-Myc IEM (Figs. 3A-D) (Namba et al., 2010). Immunogold particles were localized individually or in clusters, indicating that both HA-A$_1$R and Myc-P2Y$_2$R form monomers and homo-oligomers. Specificities of the gold-labeled anti-HA and anti-Myc antibodies were demonstrated by incubating A$_1$R-transfected HEK293T cells with a mixture of both antibodies. Data showed that only A$_1$R-labeled particles were present (Fig. 3A). No significant patterns were detected with either anti-HA and anti-Myc antibodies in mock-transfected HEK293T cells or with only secondary alone (i.e., no primary antibodies) in HA-A$_1$R-transfected HEK293T cells (data not shown). Also, when Myc-P2Y$_2$R-transfected HEK293T cells were incubated with both anti-HA and anti-Myc antibodies, we detected single particles (monomers) scattered all over the cells (Fig.3B). Another control for hetero-oligomerrization, HA-A$_1$R-transfected HEK293T cells incubated with both anti-HA and Myc-P2Y$_2$R, we observed all over the cells (Fig. 3D, inner cellular site). In HEK293T cells co-transfected with both HA-A$_1$R and Myc-P2Y$_2$R, clusters of different-sized particles were observed mainly at the cell surface (Fig. 3C) suggesting the formation of hetero-oligomers.

We would also like to introduce another means of investigating the cellular localization of anti-A$_1$R/anti-P2Y$_1$R in co-transfected HEK293T cells using post-embedding methods. Using mouse anti-A$_1$R or rabbit anti- P2Y$_1$R antibodies, hetero-oligomeric gold particles were clearly observed, predominantly at the cell surface (Fig.3C). This pattern concurred with patterns defined using pre-embedding methods and gold-labeled anti-HA and anti-Myc antibodies. The frequency of A$_1$R and P2Y$_1$R (P2Y$_2$R) hetero-oligomers detected using post- embedding methods was smaller than that detected with pre-embedding methods using fresh specimens. This was likely to be due to polymerization occuring during the embedding process (data not shown). We consider that the native antigenicities of GPCRs in transfected cells may be reduced by polymerization treatment with LR-white, though closely-related patterns of immunoreactivity were obtained in our laboratory across differeing methods (Fig. 4). HA-A$_1$R transfected HEK293T cells incubated with mouse anti-A$_1$R using post-embedding methods indicated patterns (Fig. 3A, large particles) identical to those arising from pre-embedding methods (Data not shown). This data indicates that the immunoreactivety of mouse anti-$A_1R$ antibodies using LR-white post-embedding were effective in HA-$A_1R$ transfected HEK293T cells. Hetero-oligomeric gold particles of the mouse anti-$A_1R$ or rabbit anti-$P2Y_1R$ antibodies were observed at the cell surface (Fig.4B). HA-$A_1R$-transfected HEK293T cells incubated with either mouse anti-$A_1R$ or rabbit anti-$P2Y_1R$ were also seen scattered all over the cells (Fig. 4C, cellular surface).

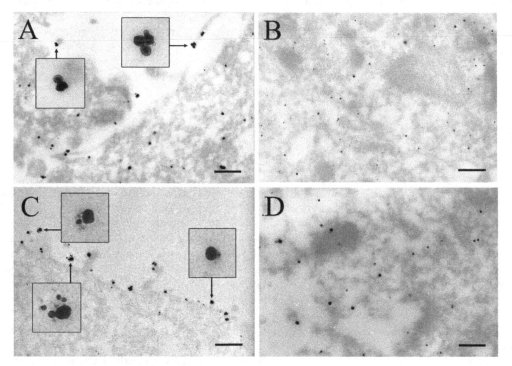

Fig. 3. Immunogold electron microscopy (post-embedding) method to visualise $A_1R$ and $P2Y_2R$ in transfected HEK293T cells using nanogold particles. A: Localization of HA-$A_1R$ (large particles) detected with anti-HA in HA-$A_1R$-transfected HEK293T cells. B: Localization of Myc-$P2Y_2R$ (small particles) detected with anti-Myc in Myc-$P2Y_2R$-transfected HEK293T cells. C: Anti-HA and anti-Myc immuno-localization of anti-$A_1R$ and Myc-$P2Y_2R$ in co-transfected HEK293T cells. D: HA-$A_1R$-transfected HEK293T cells incubated with both anti-HA and anti-Myc. Bars represent 100 nm.

## 4. Post-embedding immunogold electron microscopy of brain tissues

In keeping with observations gained by post-embedding methods for the study of co-transfected HEK293T cells described in Section 3.1. of this chapter, we should highlight that it is also possible to apply immunogold staining using post-embedding methods for the study of brain tissues. Comparing the dose of immunoreactivity from gold particles reflecting hetero-oligomers using co-transfected culture cells and post-embedding methods is essential in acquiring immunogold pattern data from hetero-oligomers *in situ*.

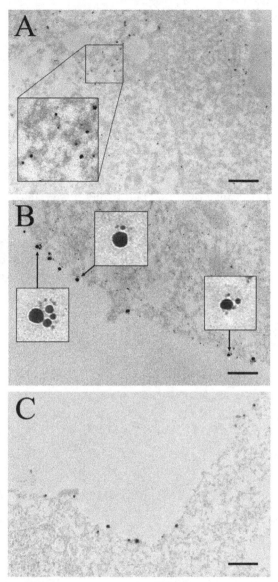

Fig. 4. Immunogold electron microscopy to visualise $A_1R$ and $P2Y_1R$ in transfected HEK293T cells using nanogold particles. A: Localization of HA-$A_1R$ (small particles) detected with mouse anti- $A_1R$ in HA-$A_1R$ -transfected HEK293T cells. B: Mouse anti-$A_1R$ and rabbit anti-$P2Y_1R$ immuno-localization of HA-$A_1R$ and Myc-$P2Y_1R$ in co-transfected HEK293T cells. C: HA-$A_1R$-transfected HEK293T cells incubated with both mouse anti-$A_1R$ and rabbit anti-$P2Y_1R$. Bars represent 100 nm.

There are two reasons for this. Firstly, the antigenicity for the two receptor antibodies must accurately reflect hetero-oligomers or single expression. Secondly, the immunoreactivity of a particular antibody could be variable depending upon the methodology utilized, for example whether post- or pre-embedding methods were deployed. Usually, immunoreactive conditions during pre-embedding methods are much better than during post-embedding methods. However, immunoreactions involving inner tissues are technically difficult to perform. In the following section, we introduce how we can image the hetero-oligomerization of $A_1R$ and $P2Y_1R$ in brain tissues using post-embedding immunogold electron microscopy.

## 4.1 Results: Immunogold electron microscopic observations of $A_1R$ and P2YR expressed in brain tissues

We incubated post-embedded, primary antibody-stained rat brain tissues with two secondary antibodies labeled with gold particles (a 5-nm gold particle-conjugated goat anti-mouse IgG antibody for $A_1R$, and a 10-nm gold particle-conjugated goat anti-rabbit IgG antibody for $P2Y_1R$). As negative controls, brain tissues were stained with only secondary antibodies conjugated with different sized gold particles; no significant immunoreactivity was observed under the experimental conditions (data not shown). As found with transfected HEK293T cells (3.2-3.4), we observed clusters of different-sized gold particles at cytoplasmic membranes in cell bodies, indicating the presence of heteromeric complexes of endogenous $A_1R$ and $P2Y_1R$ in the rat cerebellum (Fig. 5). Significant immunoreactivity was

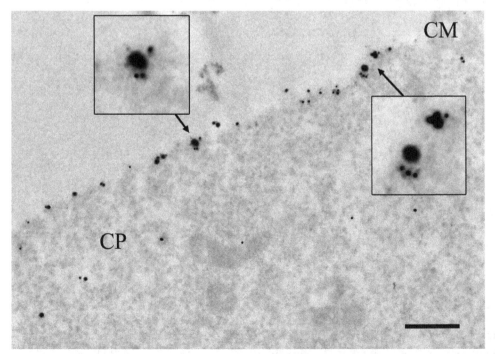

Fig. 5. Immunogold electron microscopy (post-embedding) method for the visualisation of $A_1R$ and $P2Y_1R$ in rat brain using nanogold particles. Localization of $A_1R$ (small particles) and $P2Y_1R$ (large particles) in cell surface of Purkinje cells detected with both anti-$A_1R$ and anti-$P2Y_1R$ Arrows indicate two adjacent receptors on the cell membrane. Bars represent 100 nm. CM, cell membrane; CP, cytoplasm.

detected in the cell surface region (Fig. 3F). In our earlier experiments, oligomerization of $A_1R$ and $P2Y_2R$ in rat brain tissues under the same experimental conditions involved hippocampal pyramidal cells, cerebellum, and pyramidal cells in the forebrain (Namba et al., 2010). Hetero- and homo-oligomers of both $A_1R/P2Y_1R$ and $A_1R/P2Y_2R$ were detected in significant numbers at the cell surface in both transfected HEK293T cells and native brains.

## 5. Data analysis: Comparison of the frequencies of monomers, homo-oligomers, and hetero-oligomers between $P2Y_1R/A_1R$ and $P2Y_2R/A_1R$

Gold-staining was quantified in the following way. Firstly, gene-transfected HEK293T cells exhibiting the highest number of total immuno-reacted gold particles were defined as 100% labeling. Since co-transfected HEK293T cells that displayed unique pharmacology in our previous study (Suzuki et al., 2006) exhibited more than 20% hetero-oligomeric gold particles, we used this number as a threshold in the current study. Thus, cells with more than 20% hetero-oligomeric particles were defined as being "significantly stained", and those with 20% or less were defined as "not significantly stained".

The proportions of relative distributions for $A_1R$ and P2YR between cell surfaces and inner cytoplasmic membranes were clearly different (Fig. 6). The tendency for the proportional distribution of $A_1R$ and P2YR at the surface of HEK293T cells concur with data from brain

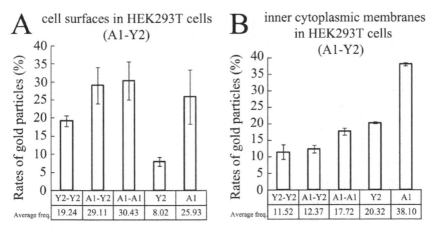

Fig. 6. Comparison of the relative distributions for $A_1R$- and $P2Y_2R$-conjugated gold particles between the cell surface and inner cytoplasmic membranes.

tissues. Based on this, the numbers of immunogold particles at the surface of each cell type in brain tissues were determined. We defined single particles located independently as monomers ($A_1R$ and P2YR in Fig. 7), complexes composed of clusters of the same-sized gold particles as "homo-oligomers" ($A_1R$-$A_1R$ or P2YR-P2YR in Fig. 7), and those of different sized gold particles as "hetero-oligomers" ($A_1R$-P2YR in Fig. 7). Separate calculations were carried out for particles in Purkinje cells (Fig. 7A, C), hippocampal pyramidal neurons (Fig. 7B, D), and cortical neurons (Fig. 7E). To do this, gold particles were counted the number of in three cells for each region. We previously counted immunogold particles in co-transfected HEK293T cells (Namba et al., 2010). The total number of immunoreactive gold particles on

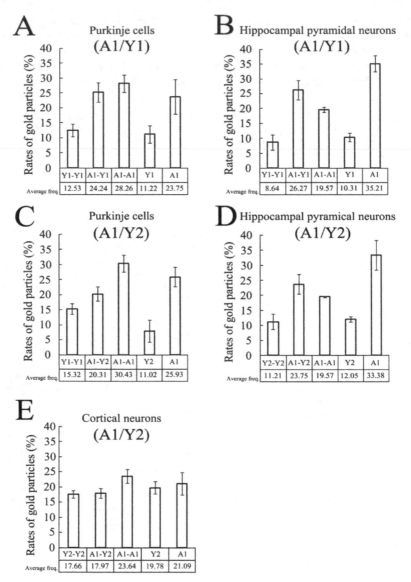

Fig. 7. Bar graphs comparing the relative distributions of $A_1R(A1)/P2Y_1R(Y1)$ immunoreactive elements in Purkinje cells (A) Hippocampal pyramidal cells (B), and $A_1R(A1)/P2Y_2R(Y2)$ immunoreactive elements in Purkinje cells (C), Hippocampal pyramidal cells (D) and Cortical neurons (E). P2YR-P2YR, $A_1R$-$A_1R$ and $A_1R$-P2YR oligomers are indicated by Y1-Y1 (Y2-Y2), A1-A1 and A1-Y1 (A1-Y2), respectively. The total number of immunoreactive gold particles on the cell surface was defined as 100%. Each column represents the average frequency (± SD) from three cells. Data describing the average numbers of gold particles are shown in tables under the graphs. Data represent the mean of three independent experiments.

each cell surface was defined as 100%. From a total of 12 photos from each brain area (i.e., 36 photos) and from transfected cells that were reacted under the same conditions as the brain sections for each immunostaining, we selected three photos of each specimen containing whole cells for comparison.

We then counted gold particles on the surfaces of cells in the cerebellum, hippocampus and cortical neurons, and classified them as monomers, homo-oligomers, or hetero-oligomers. While the homo-oligomerization ratios ($A_1R$-$A_1R$/P2YR-P2YR) displayed different patterns between Purkinje cells and hippocampal pyramidal cells, the rates of hetero-oligomerization were particularly prominent in hippocampal pyramidal cells among them. Curiously, the frequency of $A_1R$ or $P2Y_1R$ hetero-oligomerization was slightly higher than that of $A_1R$ or $P2Y_2R$ in both tissues (Fig. 7A, B). This indicated that the hetero-oligomerization of $A_1R$ or $P2Y_1R$ are the dominant form in both Purkinje cells and hippocampal pyramidal cells.

## 6. Discussion

Previous reports describe electron microscopic studies of plasma membranes for homo-oligomeric $B_1$ Bradykinin receptor complexes (Kang *et al.*, 2005), heteromeric-oligomerization of $GABA_B$ R1 and R2 receptors (Charara *et al.*, 2004), and the localization of $A_1R$ with caveolin-3 in rat ventricular cardiomyocytes (Lasley *et al.*, 2000). An immunological study suggested that $A_1R$ forms oligomers the cortex of the pig brain (Ciruela *et al.*, 1995), and a FRET study demonstrated the oligomerization of $P2Y_2R$ in transfected HEK293 cells (Kotevic *et al.*, 2005). The hetero-oligomerization of $A_1R$-$P2Y_1R$ on postsynaptic neurons was also analyzed by IEM (Tonazzini *et al.*, 2007). The present study provides the first detailed evidence of an interaction between endogenous $A_1R$ and $P2Y_2R$ in brains using IEM.

The homo- oligomerization of $A_1R$ and its structural profile were previously analyzed in our laboratory by computational prediction, co-immunoprecipitation, and BRET analysis with differently tagged $A_1Rs$ (Suzuki *et al.*, 2009); homo-oligomers and monomers were easily distinguished by IEM. This particular study confirmed the existence of homo-oligomers ($A_1R$-$A_1R$ and $P2Y_2R$-$P2Y_2R$) using IEM. Interestingly, the percentage of $A_1R$ homo-oligomers was higher than that of $P2Y_2R$ in both rat brain and transfected HEK293T cells (Namba *et al.*, 2010). By contrast, the ratio of heteromeric gold-particle clusters were different in the cortex, hippocampus, and cerebellum. Importantly, both homo-oligomeric and hetero-oligomeric gold-particles were reduced in number at inner cytoplasmic membranes than at the cell surface (data not shown). In general, most GPCRs oligomers have been observed at the cell surface (Minneman, 2007; Bulenger *et al.*, 2005).

While the frequencies of $A_1R$ and $P2Y_1R$ homo-oligomers and monomers were similar in the cerebellum (Fig. 5) and in transfected HEK293T cells (Fig. 4B), the ratio of the different receptor oligomers occurred in different patterns in each of the three brain areas (Fig. 7). Total numbers of hetero-oligomers observed on the cell surface and in the cytoplasm were clearly different (Fig. 6A, B) and may reflect the process of receptor maturation and association of the $A_1R$-P2YR complex. However, hetero-oligomers were unmistakably detected at the cell surface by IEM (Fig. 3C, Fig. 4B).

As a signaling pathway, $P2Y_1R$ and $P2Y_2R$ display different ligand specificities. As ligands, ATP and UTP fully activate $P2Y_2R$. However, UTP is not an agonist for $P2Y_1R$. In addition,

ADP is a strong agonist for $P2Y_1R$ but not $P2Y_2R$ (Abbracchio *et al.*, 2006). Many previous studies suggest that $A_1R$-$P2Y_1R$ and $A_1R$-$P2Y_2R$ hetero-oligomers exhibit general pharmacological profiles, possibly because of differences in the conformational changes induced by oligomerization (Nakata *et al.*, 2010). The hetero-oligomerization of $A_1R$-$P2Y_1R$ inhibits adenylyl cyclase activity via the $G_{i/o}$ protein linked effector. The hetero-oligomerization of $A_1R$-$P2Y_2R$ resulted in an increase in intracellular $Ca^{2+}$ levels induced by $P2Y_2R$ activation of $G_{q/11}$ which was synergistically enhanced by the simultaneous addition of an $A_1R$ agonist in the co-expressing cells (Suzuki *et al.*, 2006). Differences in the amounts of hetero-oligomerization between $A_1R$-$P2Y_1R$ and $A_1R$-$P2Y_2R$ were observed (Fig. 7). Assuming that the number of hetero-oligomers formed is functionally dominant, the dominancy of the signaling via $A_1R$-$P2Y_1R$ may be generated by competitive antagonism in pharmacology between $P2Y_1R$ and $P2Y_2R$ in order to oligomerize with $A_1R$. This hypothesis, however, requires further investigation.

In our previous study, the hippocampal hetero-oligomerization of $A_1R$ and $P2Y_2R$ was far more pronounced than in other regions of the brain (Namba *et al.*, 2010). Another research group suggested that the hetero-oligomerization, or cross-talk between $A_1R$ and $P2Y_1R$ is involved in regulation of glutamate release in the hippocampus (Tonazzini *et al.*, 2007). The relative distributions of immunoreactivity for $GABA_B$ R2 and $GABA_B$ R1 were also different in the basal ganglia and globus pallidus/substantia nigra, which suggests the possible co-existence and hetero-oligomerization of the two types of receptors at various pre-/postsynaptic sites (Charara *et al.*, 2004). From the present study, it can be speculated that the $A_1R$/$P2Y_2R$ hetero-oligomer might be responsible for down regulation, via hippocampal $Ca^{2+}$ secretion, of synaptic functions (Safiulina *et al.*, 2006). Furthermore, the abundant formation of $A_1R$/$P2Y_1R$ or $A_1R$/$P2Y_2R$ hetero-oligomers in the cerebellum revealed in this present study supports the idea that the unique signal transduction generated by hetero-oligomerization, including the enhancement of $Ca^{2+}$ signaling via $G_{q/11}$, observed in transfected cells, also occurs in the cerebellum.

There are many families of GPCRs expressed in whole brain, most of which remain a mystery. However, it is clear that GPCR hetero-oligomerization is common in the brain and exhibits unique pharmacology in this region, thus implying that associated signal transduction pathways can be anticipated in this region. The methodology described here using immunogolod particles is one of the most influential techniques available to elucidate the ingenious mechanism underlying GPCR hetero-oligomerization.

## 7. Summary

In summary, IEM provided direct evidence for the existence of homo- and hetero-oligomers of $A_1R$ and $P2Y_2R$, not only in co-transfected cultured cells, but also *in situ* on the surface of neurons in various brain regions. The molecular mechanisms responsible for the control of $A_1R$ and P2YR monomer/homo-oligomer/hetero-oligomer ratios remain to be elucidated. Future investigation of GPCR oligomer formation is indispensable for revealing the elaborate mechanisms of cellular function.

The importance of these novel experimental procedures using IEM is to provide information concerning crosstalk between small molecules with high angle views of whole cells, although these methods do require a high level of technical skill. The development of ingenious histochemical and immunoelectronmicroscopic methods has made it possible to

visuallize crosstalk and provide specific insight into the nature of hetero-oligomers of not
only GPCRs, but also various proteins expressed by cells.

## 8. Acknowledgments

Most of the procedures described in this work were performed in Dr. Hiroyasu Nakata's
laboratory at the Tokyo Metropolitan Institute for Neuroscience, Tokyo, Japan. The author
wishes to express sincere appreciation and gratitude to Dr. Nakata and his collaborators.

## 9. References

Abbracchio, M. P., Burnstock, G., Boeynaems, J. M., Barnard, E. A., Boyer, J. L., Kennedy, C.,
Knight, G. E., Fumagalli, M., Gachet, C., Jacobson, K. A., Weisman, G. A. (2006).
International Union of Pharmacology LVIII: update on the P2Y G protein-coupled
nucleotide receptors: from molecular mechanisms and pathophysiology to therapy.
Pharmacol Rev. 58(3): 281–341.

Abbracchio, M. P., Burnstock, G., Verkhratsky, A. & Zimmermann, H. (2009). Purinergic
signalling in the nervous system: an overview. Trends Neurosci. 32(1): 19–29.

Angers, S., Salahpour, A. & Bouvier M. (2002). Dimerization: an emerging concept for G
protein-coupled receptor ontogeny and function. Annu Rev Pharmacol Toxicol. 42:
409-435.

Bouvier, M. (2001). Oligomerization of G-protein-coupled transmitter receptors. Nat Rev
Neurosci. 2(4): 274-286.

Bulenger, S., Marullo, S. & Bouvier M. (2005). Emerging role of homo- and
heterodimerization in G-protein-coupled receptor biosynthesis and maturation.
Trends Pharmacol Sci. 26(3): 131-137.

Burnstock, G. (2007). Physiology and pathophysiology of purinergic neurotransmission.
Physiol Rev. 87(2): 659–797.

Burnstock, G. (2008). Purinergic signalling and disorders of the central nervous system. Nat
Rev Drug Discov. 7(7): 575–590.

Charara, A., Galvan, A., Kuwajima, M., Hall, R. A. & Smith, Y. (2004). An electron
microscope immunocytochemical study of GABA(B) R2 receptors in the monkey
basal ganglia: a comparative analysis with GABA(B) R1 receptor distribution. J
Comp Neurol. 476(1): 65-79.

Ciruela, F., Casado, V., Mallol, J., Canela, E. I., Lluis, C. & Franco, R. (1995). Immunological
identification of $A_1$ adenosine receptors in brain cortex. J Neurosci Res. 42(6): 818-
828.

Fam, S. R., Gallagher, C. J., Kalia, L. V. & Salter, M. W. (2003). Differential frequency
dependence of P2Y1- and P2Y2- mediated Ca 2+ signaling in astrocytes. J Neurosci.
23(11): 4437-4444.

Kang, D. S., Gustafsson, C., Mörgelin, M. & Leeb-Lundberg, L. M. (2005). B1 bradykinin
receptor homo-oligomers in receptor cell surface expression and signaling: effects
of receptor fragments. Mol Pharmacol. 67(1): 309-318.

Kotevic, I., Kirschner, K. M., Porzig, H. & Baltensperger, K. (2005). Constitutive interaction
of the P2Y2 receptor with the hematopoietic cell-specific G protein G(alpha16) and
evidence for receptor oligomers. Cell Signal. 17(7): 869-880.

Lasley, R. D., Narayan, P., Uittenbogaard, A. & Smart, E. J. (2000). Activated cardiac adenosine A(1) receptors translocate out of caveolae. *J Biol Chem*. 275(6): 4417-4421.

Mehta, B., Begum, G., Joshi, N. B. & Joshi, P. G. (2008). Nitric oxide-mediated modulation of synaptic activity by astrocytic P2Y receptors. *J Gen Physiol*. 132(3): 339-349.

Minneman, K. P. (2007). Heterodimerization and surface localization of G protein coupled receptors. *Biochem Pharmacol*. 73(8): 1043-1050.

Molliver, D. C., Cook, S. P., Carlsten, J. A., Wright, D. E. & McCleskey, E. W. (2002). ATP and UTP excite sensory neurons and induce CREB phosphorylation through the metabotropic receptor, P2Y2. *Eur J Neurosci*. 16(10): 1850-1860.

Nakata, H., Suzuki, T., Namba, K. & Oyanagi, K. (2010). Dimerization of G protein-coupled purinergic receptors: increasing the diversity of purinergic receptor signal responses and receptor functions. *J Recept Signal Transduct Res*. 30(5): 337-346.

Nakata, H., Yoshioka, K., Namba, K. & Kamiya, T. (2006). *Oligomerization of G protein-coupled Purinergic Receptors*, InTech, ISBN 978-0-8493-2771-1, Gunma University, Japan.

Namba, K., Suzuki, T. & Nakata, H. (2010). Immunogold electron microscopic evidence of in situ formation of homo- and heteromeric purinergic adenosine A1 and P2Y2 receptors in rat brain. *BMC Res Notes*. 3: 323.

Ochiishi, T., Chen, L., Yukawa, A., Saitoh, Y., Sekino, Y., Arai, T., Nakata, H. & Miyamoto, H. (1999). Cellular localization of adenosine A1 receptors in rat forebrain: immunohistochemical analysis using adenosine A1 receptor-specific monoclonal antibody. *J Comp Neurol*. 411(2): 301-316.

Ralevic, V., Burnstock, G. (1998). Receptors for purines and pyrimidines. *Pharmacol Rev*. 50(3): 413–492.

Safiulina, V. F., Afzalov, R., Khiroug, L., Cherubini, E. & Giniatullin, R. (2006). Reactive oxygen species mediate the potentiating effects of ATP on GABAergic synaptic transmission in the immature hippocampus. *J Biol Chem*. 281(33):23464-23470.

Suzuki, T., Namba, K., Tsuga, H. & Nakata, H. (2006). Regulation of pharmacology by hetero-oligomerization between A1 adenosine receptor and P2Y2 receptor. *Biochem Biophys Res Commun*. 351(2): 559–565.

Suzuki, T., Namba, K., Yamagishi, R., Kaneko, H., Haga, T. & Nakata, H. (2009). A highly conserved tryptophan residue in the fourth transmembrane domain of the A adenosine receptor is essential for ligand binding but not receptor homodimerization. *J Neurochem*. 110(4):1352-1362.

Tonazzini, I., Trincavelli, M. L., Storm-Mathisen, J., Martini, C. & Bergersen, L. H. (2007). Co-localization and functional cross-talk between A1 and P2Y1 purine receptors in rat hippocampus. *Eur J Neurosci*. 26(4): 890-902.

Yoshioka, K., Saitoh O. & Nakata, H. (2001). Heteromeric association creates a P2Y-like adenosine receptor. *Proc Natl Acad Sci USA*. 98(13): 7617-7622.

Yoshioka, K., Hosoda, R., Kuroda, Y. & Nakata, H. (2002). Heterooligomerization of adenosine A1 receptors with P2Y1 receptors in rat brains. *FEBS Lett*. 531(2): 299–303.

Yoshioka, K., Saitoh, O. & Nakata, H. (2002). Agonist-promoted heteromeric oligomerization between adenosine A(1) and P2Y(1) receptors in living cells. *FEBS Lett*. 523(1-3): 147-151.

# Part 2

# Enzymes

# 8

# Carbonic Anhydrase and Heavy Metals

Maria Giulia Lionetto, Roberto Caricato, Maria Elena Giordano,
Elisa Erroi and Trifone Schettino
*University of Salento - Dept. of Biological and Environmental Sciences and Technologies*
*Italy*

## 1. Introduction

Carbonic anhydrase (CA; EC 4.2.1.1) is a zinc metalloenzyme catalysing the reversible hydration of $CO_2$ to produce $H^+$ and $HCO_3^-$. Its activity is virtually ubiquitous in nature. The fundamental role of this biochemical reaction in diverse biological systems has driven the evolution of several distinct and unrelated families of CAs. Five CA families, referred as α-, β-, γ-CA, δ, and ζ-CAs have been identified in animals, plants and bacteria (Hewett-Emmett and Tashian, 1996; Supuran, 2010). These are the α-CAs, present in vertebrates, bacteria, algae and plants; the β-CAs, predominantly in bacteria, algae and plants; the γ-CAs, mainly present in archaea and some bacteria; the δ-CAs and ζ-CAs only found in some marine diatoms (Supuran, 2010).

The monomeric α-carbonic anhydrases are by far the best studied, being found in animals. In mammals at least 16 different CA isoforms were isolated and several novel isozymes have also been identified in non-mammalian vertebrates. The α-CA isoenzymes differ in their kinetic properties, their tissue distribution and subcellular localization, and their susceptibility to various inhibitors. In general, there are three distinct groups of CA isozymes within the α-CA gene family. One of these groups contains the cytoplasmic CAs, which includes mammalian CA I, II, III, V, VII and XIII. These isozymes are found in the cytoplasm of various tissues, with the exception of the mitochondrial confined CA V. Another group of isozymes, termed the membrane-bound CAs, consists of mammalian CA IV, IX, XII, XIV and XV (Esbaugh and Tufts, 2006). These isozymes are associated with the plasma membranes of many different tissue types. The final group contains several very intriguing isozymes, CA VIII, X and XI, which are termed the CA-related proteins (CA-RP; Tashian et al., 2000). These isozymes have lost classical CA activity – the hydration/dehydration of $CO_2$ – and have no known physiological function; however, their highly conserved nature does suggest a very important role in vertebrates (Tashian et al., 2000).

The β-carbonic anhydrases are dimers, tetramers, or octamers and include the majority of the higher plant CA isoforms (Kimber and Pai, 2000). The γ-carbonic anhydrase is a homotrimer that has been reported for the bacterium *Methanosarcina thermophila* (Alber and Ferry, 1994). The δ class has its prototype in the monomeric CA TWCA1 from the marine diatom *Thalassiosira weissflogii* (Roberts et al., 1997; Tripp et al., 2001). The ζ-CAs are probably monomer with three slightly different active sites on the same protein backbone (Xu et al 2008).

All CAs are metalloenzymes but whereas $\alpha$-, $\beta$-, and $\delta$-CAs use Zn(II) ions at the active site, the $\gamma$-CAs are probably Fe(II) enzymes (Ferry et al., 2010), but they are active also with bound Zn(II) or Co(II) ions, and the $\zeta$-class uses also Cd(II) to perform the physiologic reaction catalysis (Lane et al., 2000; Lane et al., 2005).

CA plays key roles in a wide variety of physiological processes involving $CO_2$ and $HCO_3^-$. In animals the various CA isozymes are found in many different tissues and are involved in a number of different physiological processes, including bone resorption, calcification, ion transport, acid–base transport, and a number of different metabolic processes such as biosynthetic reactions (gluconeogenesis, lipogenesis, and ureagenesis). In algae and plants they play an important role in photosynthesis (Ivanov et al, 2007; Zhang et al., 2010; Cannon Gordon et al., 2010).

Considerable advances towards a detailed understanding of the catalytic mechanism of the zinc enzyme carbonic anhydrase have been made during the past years as a result of the application of crystallographic and kinetic methods to wild-type and mutant enzymes. Moreover, a great amount of work has been performed on CA inhibitors, first of all sulfonamides, $RSO_2NH_2$, which represent the classical CA inhibitors (CAIs) and are in clinical use for more than 50 years as diuretics and systemically acting antiglaucoma drugs (Supuran, 2010).

The review focuses on one interesting but less investigated aspect of the biochemistry of this metalloenzyme, encompassing several areas of interest from human health to environmental science: the relationships between carbonic anhydrase and heavy metals. Heavy metals are chemical elements with a density higher than 5.0 g/cm³, characterized by high reactivity, redox behaviour, and complex formation based on the characteristic of the outer $d$ electron shell. In the scientific literature the following elements are normally ascribed to the heavy metal groups: aluminium, iron, silver, barium, beryllium, manganese, mercury, molybdenum, nickel, lead, copper, tin, titanium, tallium, vanadium, zinc. Some metalloids, such as arsenic, bismuthum, and selenium, are also included in the heavy metals groups.

Heavy metals generally regarded as essential for animals in trace amounts include zinc, the known cofactor of CAs, iron, copper, manganese, chromium, molybdenum and selenium. They are essential because they form an integral part of one or more enzymes involved in a metabolic or biochemical process. Besides essential metals, a number of other heavy metals, such as arsenic, lead, cadmium, mercury, have no known function in the body and are referred as toxic metals. However, also essential metals become toxic when their levels in the body exceed the homeostatic capacity of the organism. The intracellular levels of essential metals are regulated by transporters (which translocate metal across the plasma membrane) as well as by metallothionein and other metal binding proteins (Maret and Wolfgang, 2011). The toxicity of heavy metals is generally ascribed to their high affinity for nucleophilic groups like sulfhydryls. In fact they are soft donors and will therefore readily bind to soft acceptors such as sulphydryl groups.

Recently, a number body of evidence has emerged regarding the effect of several heavy metals on carbonic anhydrase catalytic activity and protein expression. These studies encompass a wide area of interest from human health to environmental sciences.

## 2. Heavy metals as carbonic anhydrase cofactors

CAs catalyze the reversible hydration of carbon dioxide to bicarbonate and protons by means of a metal-hydroxide $(Lig^3M^{2+}(OH)-)$ mechanism, although the α-CAs possess other catalytic activities such as esterase, phosphatase, cyanate/cyanamide hydrase, etc. (Supuran et al., 2003; Supuran and Scozzafava 2007; Innocenti et al., 2008). In the α-, γ, and δ-CA classes, $Lig^3$ is always constituted by three His residues. The metal (M) is ZnII for all classes. The zinc atom is in the +2 state and is located in a cleft near the center of the enzyme. The role of zinc in carbonic anhydrase is to facilitate the deprotonization of water with the formation of the nucleophilic hydroxide ion, which can attack carbonyl group of carbon dioxide to convert it into bicarbonate. This is obtained through the +2 charge of the zinc ion which attracts the oxygen of water, deprotonates water, thus converting it into a better nucleophile able to attack the carbon dioxide.

Water naturally deprotonates itself, but it is a rather slow process. Zinc deprotonates water by providing a positive charge for the hydroxide ion. The proton is donated temporarily to the surrounding amino acid residues, and then it is given to the environment, while allowing the reaction to continue. Zinc is able to help the deprotonation of water by lowering the pKa of water. Therefore, more water molecules are now able to deprotonate at a lower pH than normal, increasing the number of hydroxide ions available for the nucleophilic attack to carbon dioxide (Berg, 2007).

The affinity of carbonic anhydrase for zinc is in subpicomolar range, as assessed for studies on the α-class (Tripp et al., 2001). Cox et al. (2000) and Hunt et al. (1999) ascribed a role for hydrophobic core residues in human CA-II that are important for preorienting the histidine ligands in a geometry that favours zinc binding and destabilizes geometries that favour other metals. In particular, mutagenesis experiments demonstrated that substitutions of these amino acids at position 93, 95, and 97 decrease the affinity of zinc, thereby altering the metal binding specificity up to $10^4$-fold. Furthermore, the free energy of the stability of native CAII, determined by solvent-induced denaturation, correlates positively with increased hydrophobicity of the amino acids at positions 93, 95, and 97 as well as with zinc affinity (Hunt et al., 1999).

β-CAs, present in green plants and cyanobacteria, contain also $Zn^{2+}$ in the active site but are differentiated from α-CAs by virtue of the fact that the active site is coordinated by a pair of cysteine residues and a single histidine residue, whereas the fourth ligand may be either a water molecule/hydroxide ion, or a carboxylate from a conserved aspartate residue in some β-CAs (Type II β-CAs) [Trip et al., 2001; Xu et al., 2008]. The metal hydroxide catalytic mechanism seems to be also valid for these enzymes [Supuran, 2008].

Besides zinc, other metals have demonstrated to be physiologically relevant cofactor for some CAs. In fact, in the γ-CAs metal may also be FeII (Ferry et al., 2010). Cam, the prototypic γ-class carbonic anhydrase, from the anaerobic methane producing Archaea species *Methanosarcina thermophila*, contains zinc in the active site when overproduced in *Escherichia coli* and purified aerobically [Alber et al., 1996], while it has 3-fold greater carbonic anhydrase activity and contains $Fe^{2+}$ in the active site (Fe-Cam) when purified anaerobically from *E. coli* or overproduced in the closely related species M. *acetivorans* and purified anaerobically. Soluble $Fe^{2+}$ is abundant in oxygen free environments and available to anaerobic microbes. The different results obtained in aerobic and anaerobic conditions is explained by the fact that in aerobic conditions $Fe^{3+}$ is oxidized and rapidly loss from CAM enzyme, substituted by $Zn^{2+}$ contaminating buffers not treated with chelating agents. These results indicate $Fe^{2+}$ as the physiologically relevant metal [MacAuley et al., 2009; Tripp et al., 2004] in the active site for CAM enzyme. Interestingly, evidence for the role of ferrous ion in CA has been obtained also for the α class. In fact carbonic anhydrase activity from duck erythrocytes is increased in the presence of iron in the incubation medium suggesting a role for iron in the active site (Wu et al., 2007).

The ζ-CA naturally uses $Cd^{2+}$ as its catalytic metal in marine diatoms (Lane and Morel, 2000; Lane et al., 2005; Park et al., 2008). This cdmium-CA (CDCA1) consists of three tandem CA repeats (R1–R3), which share 85% identity in their primary sequences (Lane et al., 2005). Although CDCA1 was initially isolated as a Cd enzyme, it is actually a "cambialistic" enzyme since it can use either Zn or Cd for catalysis—and spontaneously exchanges the two metals (Xu et al., 2008). Kinetic data show that the replacement of Zn by Cd results nonetheless in a decrease in catalytic efficiency (Xu et al., 2008). In the active site, Cd is coordinated by three invariant residues in CDCA of all diatom species (Park et al., 2007): Cys 263, His 315 and Cys 325. The tetrahedral coordination of Cd is completed by a water molecule. The use of Cd in CDCA is thought to explain the nutrient-like concentration profile of Cd in the oceans, where the metal is impoverished at the surface by phytoplankton uptake and regenerated at depth by remineralization of sinking organic matter (Lane and Morel 2000). It is cycled in the water column like an algal nutrient. It is thought that the expression of a CDCA in diatoms, which are responsible for about 40% of net marine primary production, represents an adaptation to life in a medium containing vanishingly small concentrations of essential metals (Xu et al., 2008). As suggested by Xu et al. (2008) the remarkable ability to make use of cadmium, an element known for its toxicity, gave presumably a significant competitive advantage to diatoms in the oceans, poor in metals, with respect to other species, and could have contributed to the global ratiation of diatoms during the Cenozoic Era and to the parallel decrease in atmospheric $CO_2$.

Moreover, Co(II) has been shown to replace Zn(II) in α-, β and γ-CA (Hoffmann et al., 2011). Cobalt ionic radius and polarizability are very similar to those of Zn(II). In contrast to Zn(II) ($d^{10}$), the $d^7$electron configuration of Co(II) is accessible to electronic spectroscopic methods (, yielding information about the interactions protein-metal. As a result, spectroscopy of Co(II) substituted CA isozymes has been used to probe the environment of the metal ions in the active sites and get information on the nature of the first coordination sphere of the metal (Hoffmann et al., 2011). The Co-containing form of the enzyme generally shows a marked decrease in activity compared with the native Zn form (Tu and Silverman, 1985). The demonstration that Zn can be extracted from a protein and replaced with Co *in vitro* does not demonstrate that such metal substitution takes place in vivo. The evidence for in vivo Co substitution in a CA was for the first time provided by Morel et al (1994) and Yee and Morel (1996) in the diatoms *T. weissflogii*, who demonstrated $^{65}$Zn and $^{57}$Co bands to co-migrate with a single band of CA activity on a native gel of diatom proteins.

## 3. Heavy metals as inhibitors of carbonic anhydrase activity

Several heavy metals were demonstrated to *in vitro* inhibit CA activity in a variety of organisms, including fishes, crabs, bovines, and humans.

The early work of Christensen and Tucker (1976) demonstrated carbonic anhydrase inhibition by heavy metals for the first time in fish. The study was carried out on red blood cells CA of the teleost *Oncorhynchus mykiss*. Erythrocyte CA, which represents the most abundant pool of the enzyme in fish, appeared significantly in *vitro* inhibited by several heavy metals cations, such as $Cd^{2+}$, $Cu^{2+}$, $Ag^+$, and $Zn^{2+}$ (Tab1).

In the intestine and gills of the European eel, *Anguilla anguilla*, Lionetto et al. (1998; 2000) found cadmium to significantly inhibit carbonic anhydrase activity. The inhibition appeared tissue specific (Lionetto et al., 1998; Lionetto et al., 2000). The gill CA was much more sensitive to the heavy metal as compared to the enzyme activity in the intestine, as observed by comparing the IC50 values (Tab1). In particular in the intestine the inhibitory effect of cadmium was more pronounced on the cytosolic than the membrane-bound CA, which revealed only a partial inhibition at high concentrations. Moreover CA activity inhibition showed a certain time-dependence, with a delay of at least 10 min and 30 min for the cytosolic isoform and the membrane bound isoform respectively. The authors attributed this behaviour to the time required by cadmium for displacing the metal (zinc) associated with the enzyme, giving an inactive Cd-substituted carbonic anhydrase. Cadmium is a bivalent metal, similar in many respects to zinc: both are in the same group of the periodic table, contain the same common oxidation state (+2), and when ionized have almost the same size. Due to these similarities, cadmium can replace zinc in many biological systems. Moreover, the delayed inhibition of membrane-bound CA with respect to the cytosolic isoform was explained by a more difficult access of cadmium to the active site of the enzyme bound to the membrane. In fact, it has to be considered that the membrane-bound CA is stabilised by disulfide bonds (Whitney and Briggle, 1982) which could contribute to a less sensitivity of the membrane bound CA to cadmium.

As suggested by Lionetto et al (2000), the observed *in vitro* inhibition of cadmium on CA activity could be useful in the understanding of the toxic effects that the heavy metal can elicits on fish physiology *in vivo*. The inhibitory effect on gill CA activity suggests that the heavy metal might interfere with a number of physiological functions in which gill CA is involved as gas exchanges (Randall and Daxbaeck, 1984), acid–base balance (Heisler, 1984), osmoregulation (Henry, 1984) and clearance of the waste products from nitrogenous metabolism (Evans and Cameron, 1986). Morgan et al (2004) directly demonstrated in *in vivo* expoxure experiments on rainbow trout that inhibition of branchial CA was able to induce an early decline in the gill $Na^+$ and $Cl^-$ uptake. With regards to the intestine, the physiological role of the cytosolic CA is that of generating $HCO_3-$ from metabolic $CO_2$ while the role of the CA enzyme associated to the brush-border membrane should be that of mediating the environmental $HCO_3-$ uptake (Maffia et al., 1996). Therefore, the inhibitory effect of cadmium on intestinal CA isoforms should interfere with bicarbonate balance and in turn with systemic acid–base balance and osmoregulation in fish. In fact, as previously shown (Schettino et al., 1992), the $HCO_3^-$ entry via the membrane-bound CA in the cell across the luminal membrane of the enterocytes seems to be essential for maintaining a steady intracellular $HCO_3^-$ concentration and/or pHi; as a consequence the salt transport in eel intestine occurs at a highest rate and the passive water loss is recovered, so solving in part the osmoregulatory problem in marine fish. Therefore, inhibition of CA enzymes by cadmium could alter $[HCO_3-]i$ and/or pHi leading to a reduction of salt absorption and consequently impairing the osmoregulation of marine fish.

More recently, Soyut et al (2008) demonstrated $Co^{2+}$, $Cu^{2+}$, $Zn^{2+}$, $Ag^+$, and $Cd^{2+}$ to be potent inhibitor for brain CA enzyme activity in Rainbow trout (*Oncorhynchus mykiss*), with the following sequence $Co^{2+} > Zn^{2+} > Cu^{2+} > Cd^{2+} > Ag^+$. They also demonstrated that $Co^{2+}$, $Ag^+$, and $Cd^{2+}$ inhibit the enzyme with competitive manner, $Cu^{2+}$ inhibits with noncompetitive manner, and $Zn^{2+}$ with uncompetitive manner.

Ceyhun et al., 2011 *in vitro* demonstrated $Al^{+3}$, $Cu^{+2}$, $Pb^{+2}$, $Co^{+3}$, $Ag^{+1}$, $Zn^{+2}$ and $Hg^{+2}$ to exert inhibitory effects on fish liver CA. Metal ions inhibited the enzyme activity at low concentrations. $Al^{+3}$ and $Cu^{2+}$ resulted the most potent inhibitors of the CA enzyme. All the metals inhibited CA in competitive manner and aluminium showed to be the best inhibitor for fish liver CA. Concerning the mechanism of inhibition, the authors argued a possible interaction of the metal with the histidines exposed on the surface of the molecule and/or other aminoacids around the active site.

In invertebrates Vitale et al (1999)demonstrated cadmium, copper and zinc to *in vitro* inhibit CA activity in the gills of the estuarine crabs *Chasmagnathus granulate* (Tab.1). The inhibitory potentials of the three metals on CA was in the following sequence: $Cu^{2+} > Zn^{2+} > Cd^{2+}$. The observed inhibitory effect *in vitro* was confirmed by a corresponding inhibitory effect *in vivo*.

In the euryhaline crabs *Callinectes sapidus and Carcinus maenas* Skaggs et al (2002) also documented a significant *in vitro* inhibition of gill CA by $Ag^+$, $Cd^{2+}$, $Cu^{2+}$ and $Zn^{2+}$. The binding affinities of the metals were one thousand times weaker for cytoplasmic CA from the gills of *C. maenas* than that from *C. sapidus*. The large differences in Ki values (Tab.1) suggests the presence of two different CA isoforms in the gills of these species, with *Callinectes sapidus* possessing a highly metal-sensitive CA isoform and *Carcinus maenas* having a metal-resistant isoform. Interestingly, heavy metal inhibition of CA from the gills of another euryhaline crab, *Chasmagnathus granulata*, (as reported by Vitale et al., 1999, see above) appears to be intermediate between that found in the other two species. Moreover, in *Callinectes sapidus* CA isolated from the cytoplasmic pool of gill homogenates was much more sensitive to heavy metal inhibition than was CA from the microsomal fraction, which is believed to be anchored to the basolateral membrane, and as such, it exists within a lipid-rich environment. The authors argued that metal could be sequestered in the lipid component of the microsomal fraction and, therefore, higher amounts of metals are required to achieve an effective concentration of free metals available for CA inhibition. However, the authors did not considered the time-dependence of the inhibition, which can be an important aspect to be taken into account (see Lionetto et al., 2000) in the analysis of membrane bound vs cytosolic isoform CA inhibition.

In humans Ekinci et al (2007) demonstrated the inhibition of two human carbonic anhydrase isozymes *in vitro*, the cytosolic HCA I and II by lead, cobalt and mercury. Lead was a noncompetitive inhibitor for HCA-I and competitive for HCA-II, cobalt was competitive for HCA-I and noncompetitive for HCA-II and mercury was uncompetitive for both HCA-I and HCA-II. Lead was the best inhibitor for both HCA-I and HCA-II.

In tab.1 the Ki, IC50 values and the type of inhibition for several heavy metals on CA from different vertebrate and invertebrate species is summarized. A great variability among species, tissues and metals can be observed. This suggests that the inhibitory mechanisms through which heavy metals exert their effect on carbonic anhydrase activity could be different for different isoenzymes and that also small structural differences between CA isoforms could result in different metal binding affinities.

| Metal | Average value of $K_i$ (M) | IC50 (M) | Type of inhibition | Tissue | Species | Ref |
|---|---|---|---|---|---|---|
| $Cd^{2+}$ | n.d. | $9.979\ 10^{-6}$ | n.d. | gills | *Anguilla anguilla* | Lionetto et al 2000 |
| | n.d. | $3.64\ 10^{-5}$ | n.d. | Intestine (cytosolic isoform) | *Anguilla anguilla* | Lionetto et al 2000 |
| | n.d. | $2.15\ 10^{-5}$ | n.d. | gills | *Chasmagnathus granulata* | Vitale et al., 1999 |
| | n.d. | $9.00\ 10^{-4}$ | n.d. | Red blood cells | *Ictalurus punctatus* | Christensen and Tucker, 1976 |
| | $94.16\ 10^{-3}$M | $8.25 \pm 10^{-2}$ | Competitive | brain | *Oncorhynchus mykiss* | Soyut et al., 2008 |
| | $5.0\ 10^{-7}$ | n.d. | n.d. | Gills (cytoplasmic isofom) | *Callinectes sapidus* | Skaggs and Hery, 2002 |
| | $6.0\ \text{-}25.0\ 10^{-4}$ | n.d. | n.d. | Gills (cytoplasmic isofom) | *Carcinus maenas* | Skaggs and Hery, 2002 |
| $Ag^{+}$ | $193.8\ 10^{-3}$M | $1.59\ 10^{-1}$ | Competitive | brain | *Oncorhynchus mykiss* | Soyut et al., 2008 |
| | $6.40\ 10^{-4}$ | $3.79\ 10^{-4}$ | Competitive | liver | *Dicentrarchus labrax* | Ceyhun et al., 2011 |
| | n.d. | $3.50\ 10^{-5}$ | n.d. | Red blood cells | *Ictalurus punctatus* | Christensen and Tucker, 1976 |
| | $5.0\text{--}0.10\ 10^{-8}$ | n.d. | n.d. | Gills (cytoplasmic isofom) | *Callinectes sapidus* | Skaggs and Hery, 2002 |
| | $6.0\ \text{-}25.0\ 10^{-4}$ | n.d. | n.d. | Gills (cytoplasmic isofom) | *Carcinus maenas* | Skaggs and Hery, 2002 |
| $Zn^{2+}$ | $2.15\ 10^{-3}$M | $3.10\ 10^{-4}$ | Uncompetitive | brain | *Oncorhynchus mykiss* | Soyut et al., 2008 |
| | $7.21\ 10^{-4}$ | $3.90\ 10^{-4}$ | Competitive | liver | *Dicentrarchus labrax* | Ceyhun et al., 2011 |
| | n.d. | $7.00\ 10^{-4}$ | n.d. | Red blood cells | *Ictalurus punctatus* | Christensen and Tucker, 1976 |
| | n.d. | $1.62\ 10^{-5}$ | n.d. | gills | *Chasmagnathus granulata* | Vitale et al., 1999 |
| | $6.0\ \text{-}25.0\ 10^{-4}$ | n.d. | n.d. | Gills (cytoplasmic isofom) | *Carcinus maenas* | Skaggs and Hery, 2002 |

| Metal | Average value of $K_i$ (M) | IC50 (M) | Type of inhibition | Tissue | Species | Ref |
|---|---|---|---|---|---|---|
| $Cu^{2+}$ | $27.6\ 10^{-3}M$ | $3.00\ 10^{-2}$ | Non competitive | brain | *Oncorhynchus mykiss* | Soyut et al., 2008 |
| | $1.75\ 10^{-5}$ | $7.15\ 10^{-5}$ | Competitive | liver | *Dicentrarchus labrax* | Ceyhun et al., 2011 |
| | n.d. | $6.50\ 10^{-5}$ | n.d. | Red blood cells | *Ictalurus punctatus* | Christensen and Tucker, 1976 |
| | n.d. | $3.75\ 10^{-6}$ | n.d. | gills | *Chasmagnathus granulata* | Vitale et al., 1999 |
| | $3.60\ 10^{-7}$ | n.d. | n.d. | Gills (cytoplasmic isofom) | *Callinectes sapidus* | Skaggs and Hery, 2002 |
| | $6.0\ -25.0\ 10^{-4}$ | n.d. | n.d. | Gills (cytoplasmic isofom) | *Carcinus maenas* | Skaggs and Hery, 2002 |
| $Co^{2+}$ | $5\ 10^{-5}M$ | $1.40\ 10^{-5}$ | competitive | brain | *Oncorhynchus mykiss* | Soyut et al., 2008 |
| | $5.32\ 10^{-4}$ | $3.16\ 10^{-4}$ | competitive | liver | *Dicentrarchus labrax* | Ceyhun et al., 2011 |
| | $3.91\ 10^{-3}$ | n.d. | competititve | Erytrocytes (CAI) | *Homo sapiens* | Ekinci et al., 2007 |
| | $1.7\ 10^{-3}$ | n.d. | non competitive | Erytrocytes (CAII) | *Homo sapiens* | Ekinci et al., 2007 |
| $Al^{3+}$ | $1.48\ 10^{-4}$ | $6.92\ 10^{-5}$ | competitive | liver | *Dicentrarchus labrax* | Ceyhun et al., 2011 |
| $Pb^{2+}$ | $2.42\ 10^{-4}$ | $1.13\ 10^{-4}$ | competitive | liver | *Dicentrarchus labrax* | Ceyhun et al., 2011 |
| | $9.90\ 10^{-4}$ | n.d. | Non competitive | Erytrocytes (CAI) | *Homo sapiens* | Ekinci et al., 2007 |
| | $5.6\ 10^{-5}$ | n.d. | uncompetitive | Erytrocytes (CAII) | *Homo sapiens* | Ekinci et al., 2007 |
| $Hg^{2+}$ | $7.68\ 10^{-4}$ | $4.48\ 10^{-4}$ | competitive | liver | *Dicentrarchus labrax* | Ceyhun et al., 2011 |
| | $1.42\ 10^{-3}$ | n.d. | uncompetitive | Erytrocytes (CAI) | *Homo sapiens* | Ekinci et al., 2007 |
| | $3.12\ 10^{-4}$ | n.d. | uncompetitive | Erytrocytes (CAII) | *Homo sapiens* | Ekinci et al., 2007 |

Table 1. Ki, IC50 and type of inhibition for several heavy metals in different species and tissues as assessed in *in vitro* studies.

Concerning the mechanisms of inhibition some heavy metals are believed to bind to CA not at the specific catalytic site of $CO_2$ hydration but nearby in a pocket, the so called 'proton shuttle' as demonstrated for human CAII (Tu et al., 1981). His-64 is a proton shuttle in catalysis, where it accepts the proton product (via the bridging solvent molecules) from zinc-bound water as zinc-bound hydroxide is regenerated; subsequently, the proton product is passed along to buffer (Liang et al, 1988; Tu et al., 1989; Vedani et al., 1989). The mechanism of inhibition of heavy metals on proton shuttle has been elucidated for copper on human CA II. $Cu^{2+}$ is believed to competitively inhibit CAII by binding to the imidazole side chain of His-64, blocking its role in proton transfer from the zinc-bound water molecule to buffer molecules located outside of the active site region [Tu et al., 1981]. However, the knowledge of the mechanism of action of other metals on different CA isoforms is lacking. It cannot be excluded the CA binding to other different parts of the protein, possibly cysteine residues, as demonstrated in studies with other enzymes for silver and mercury.

## 4. Heavy metals as modulators of carbonic anhydrase activity and expression

If it has been widely demonstrated *in vitro* that heavy metals are able to inhibit CA activity in a variety of organisms, on the contrary little is known about the *in vivo* effects of trace metals on the activity and the expression of this metalloenzyme. The major information regards $Zn^{2+}$, while very few is known about other metals.

In humans early studies demonstrated that dietary zinc deficiency significantly reduces zinc concentrations of serum and in turn CA activity in erythrocytes (Hove,1940; Rahman et al., 1961; Kirchgessner et al., 1975) suggesting a possible influence of $Zn^{2+}$ on CA protein expression. These early data have been more recently confirmed by Lukaski (2005) who demonstrated zinc concentration of serum and erythrocyte to be positively correlated to CA activity *in vivo*. Low dietary zinc decreases erythrocyte carbonic anhydrase activity and, in turn, impairs cardiorespiratory function in men during exercise (Lukaski et al., 2005). In ducks $Zn^{2+}$ at a low level (up to 1.25 μM Zn) induced the rise of CA activity in erythrocytes (Wu et al., 2007). In parotid saliva of patients with CAVI deficiency $Zn^{2+}$ treatment was able to stimulate synthesis/secretion of CAVI (Henkin et al., 1999), probably through stimulation of CAIV gene. In rats $Zn^{2+}$ deficiency significantly reduced CAII protein expression in the submandibular gland (Goto et al., 2008).

As regards other metals Grimes et al (1997) reported the depression of CAIII mRNA and, in turn, CAIII protein in the mouse mutant 'toxic milk' (tx) liver following copper accumulation, Kuhara et al (2011) found CAIII suppression by copper accumulation during carcinogenesis, while Wu et al (2007) found iron at low levels to induce a rise in CA activity in duck erythrocytes.

Recently, Caricato et al (2010) demonstrated for the first time CA activity and protein expression to be enhanced by the exposure to the trace element cadmium in animals, opening new perspective in the comprehension of the functioning and regulation of this enzyme. Digestive gland CA activity showed a weak sensitivity to *in vitro* cadmium exposure since only high concentrations of $CdCl_2$ (from $10^{-5}$ to $10^{-3}$ M) were able to exert a significant inhibition. On the contrary digestive gland CA activity showed a significant increment in cadmium exposed animals (about 40% after two week of exposure). This was the first time that CA activity appears to be increased by cadmium in animals. Carbonic anhydrases from the microalgae *Chlamydomonas reinhardtii* (Wang at al., 2005) and *Thalassiosira weissflogii* (Morel et al., 1994; Lee et al., 1995) are the only other examples reported in nature of CA activity increase induced by cadmium exposure. Evidence of *in vivo* utilization of Cd in CA has been found in microalgae (Price and Morel, 1990; Morel et al., 1994; Lee et al. 1995, Xu et al., 2008). In these organisms the ability of Cd to substitute for Zn at the active site of the enzyme is reflected in the regulation of the enzyme expression. In *Thalassiosira weissflogii* a cadmium-containing CA was found to be expressed during zinc limitation (Lane and Morel, 2000; Lane et al., 2005). This cadmium CA (CDCA1) which naturally uses Cd as its catalytic metal (Trip et al., 2001; Lane et al., 2005) has been ascribed to a novel $\zeta$-CA class (see above). Genes coding for similar proteins have been identified in other cultured diatoms (Park et al., 2007). In mussel digestive gland western blotting analysis clearly demonstrated the enhancement of CA protein expression following cadmium exposure, according to the enzymatic activity data (Caricato et al., 2010). Laboratory experimental results were confirmed by a field experiment. Mussels exposed for 30 days to an anthropogenic impacted site showed a significant increase in CA activity and protein expression with respect to animals exposed for 30 days in a control site. If the new synthesized enzyme is a Cd-CA is not possible to say at the moment. If it was the case, then the increase in CA would not be a direct adaptive response to Cd pollution; rather, Cd could remove any limitations placed on CA synthesis by the availability of Zn. However, future studies will be needed to clarify this intriguing aspect of the research.

## 5. Carbonic anhydrase and heavy metals interactions: Potential applications

In the last years the interactions between carbonic anhydrase and heavy metals have found a number of applications in environmental and health fields, including the development of biomarkers of pollution exposure, in vitro bioassays, and biosensors.

### 5.1 Carbonic anhydrase sensitivity to heavy metals and development of biomarkers of pollution exposure

Pollution by trace metals is a world-wide problem due to the persistency and continuing accumulation of metals in the environment (de Mora et al. 2004; Hwang et al 2006). Heavy metals may enter the organisms through food, water, air, or absorption through the skin. As a result of mining, waste disposal and fuel combustion the environment is becoming increasingly contaminated with heavy metals.

In recent years the increasing sensibility to pollution problems has promoted the development of environmental "diagnostic" tools for early warning detection of pollution. Pollution monitoring has been increasingly concerned with the use of biological responses to pollutants at molecular and cellular level for evaluating biological hazard of toxic chemicals. Methods based on biological effects and their underlying mechanisms can complement the use of analytical chemistry in environmental monitoring. The major advantages of such biological, mechanism-based methods are their toxicological specificity, rapidity, and low cost. Toxicological specificity refers to the relationship between the assay response and the toxic potential rather than simply the contaminant concentrations (provided by chemical analysis) of the sample being analyzed. Moreover, biological assays provide rapid, sensitive, easily learnt and readily interpretable new useful tools for environmental biomonitoring and risk assessment. They include biomarkers, and *in vivo* and *in vitro* bioassays. It is known that the harmful effects of pollutants are typically first manifested at lower levels of biological organization before disturbances are realized at population, community and ecosystem levels (Adams, 1990). This is the reason why in recent years the study of molecular and cellular effects of pollutants has given important advancement in the developing of biologically-based methodologies useful for environmental biomonitoring and risk assessment. Enzymatic inhibition studies have been a very fruitful field for environmental monitoring application as biomarker of exposure/effect. Biomarkers are defined as pollutant induced variation in cellular or biochemical components occurring in organisms as a result of natural exposure to contaminants in their environment (Depledge, 1994). As reported by several authors, the evaluation of biomarkers in bioindicator organisms sampled in one or more areas suspected of chemical contamination and their comparison with organisms sampled in a control area can allow the evaluation of the potential risk of toxicological exposure of the studied community (Lionetto et al., 2003; Lionetto et al., 2004).

Carbonic anhydrase sensitivity to heavy metal exposure has been recently explored for its possible applications as biomarker of exposure to heavy metal pollution (Lionetto et al. 2006; Caricato et al, 2010b.) in "sentinel" organisms. Lionetto et al., (2006) investigated CA activity inhibition by heavy metals in the filter feeding *Mytilus galloprovincialis*, widely used in pollution monitoring programs as sentinel organism (Jernelov et al., 1996). Following *in vitro* and *in vivo* exposure to cadmium, mantle CA activity was significant inhibited. The inhibitory effect of cadmium on mantle CA activity can explain results previously obtained by Soto et al. (2000), who observed a significant decreased in shell growth in *M. galloprovincialis* exposed to heavy metals. The sensitivity of CA to heavy metals in mussels appears to be tissue-specific. In fact, as reported above, in mussel's digestive gland CA activity and expression was found to increase following Cd exposure (Caricato et al., 2010). Because of the widely application of *M. galloprovincialis* in environmental quality monitoring and assessment, data on tissue specific sensitivity of carbonic anhydrase to heavy metals represent a starting point for future potential application of CA activity changes as biomarker of exposure to heavy metals in the sentinel organism *M. galloprovincialis*.

Other studies carried out on corals have suggested alteration in CA activity as potential biomarker of exposure to environmental chemical stress. CA activity has been demonstrated to be inhibited by heavy metal exposure in anemones and corals (Gilbert and Guzman, 2001), where the enzyme plays a key role in the calcification process. Coral growth has been shown to be an effective indicator of the overall health of a coral reef ecosystem and reduced growth can reflect impaired photosynthetic output of the zooxanthellae and/or changes in enzyme activity (Moya et al., 2008). In an era of climate change and ocean acidification, where factors impacting growth and resilience factors are becoming important, understanding the biological effects of metal exposure to these keystone tropical organisms may be critical (Bielmyer et al., 2010).

## 5.2 Carbonic anhydrase based bioassay

Bioassays use biological systems to detect the presence of toxic chemicals in the environmental matrices (water, sediment, sewage, soil, etc.). In recent years, *in vitro* bioassays, employing cultured cells or cellular extracts, are increasingly being developed and used to detect the presence of contaminants. Examples include assays that measure enzyme inhibition, receptor-binding, or changes in gene expression in *in vitro* systems. Although *in vitro* assay is not a substitute for biomarker approach, it can be used as an adjunct model to whole-animal *in vivo* exposure and to ecotoxicological evaluation of the potential risk of trace pollutants in aquatic environments. They are rapid, low cost and simple tools to be utilized in combination with chemical analysis, for the pre-screening of the environmental samples that should be analyzed. Lionetto et al (2005; 2006) explored the possible application of heavy metal CA inhibition for the development of an *in vitro* bioassay applicable to the determination of the toxicity of environmental aqueous samples. They developed rapid and sensitive chemical hazard detection system for standardizing rapid, sensitive, and low cost CA based *in vitro* bioassay (Schettino et al., 2008).

## 6. Carbonic anhydrase-based biosensing of metal ions

In the last years the affinity of carbonic anhydrase for metal ions has been applied for the development of fluorescence based biosensors for determination of free metal ions in solution using variants of human carbonic anhydrase (apoCA). In particular, $Cu^{2+}$, $Co^{2+}$, $Zn^{2+}$, $Cd^{2+}$, and $Ni^{2+}$ have been determined at concentration down the picomolar range (Fierke and Thompson, 2001; Thompson and Jones, 1993; Mey et al., 2011) by changes in fluorescence emission (Thompson et al., 2000) and excitation wavelength ratios (Thompson et al., 2002), lifetimes (Thompson and Patchan, 1995), and anisotropy (polarization) (Elbaum et al., 1996; Thompson et al., 2000). The sensitivity, selectivity, analyte binding, kinetics and stability of the biosensors have been improved by subtle modification of the protein structure by directed mutagenesis (Kiefer et al., 1995; Hunt et al., 1999; DiTusa et al., 2001; McCall et al., 2004; Burton et al, 2000). These studies have hallowed the development of highly selective and sensitive fluorescence-based biosensors for $Zn^{2+}$ e $Cu^{2+}$, which have been shown to be viable approach in some important applications. In fact, the CA-based $Cu^{2+}$ biosensor has been used to obtain real-time measurement of free Cu(II) at picomolar concentrations in seawater (Zeng et al., 2003), while the CA-base $Zn^{2+}$ biosensor has been used for measurement of free Zn ion at picomolar levels in cultured cells (Bozym et al, 2004).

## 7. Conclusions

Although carbonic anhydrase represents one of the most investigated metalloenzyme in nature, its interaction with heavy metals has been only partially elucidated to date and some issues still remains to be explored. An intriguing aspect that needs more investigation is the *in vivo* effect of heavy metals on CA expression. From the few studies available in literature some metals appear to be important modulator of the expression of this protein. The understanding of the underlying mechanisms could open new perspective in the comprehension of the functioning and regulation of this enzyme. Another intriguing aspect of the biochemistry of CA is the inhibition by heavy metals. It has been documented for some species and some metals, but the mechanisms behind the inhibition, its metal specificity and isoform specificity remains still unknown. These aspects merits in depth examination and open new perspective for drug design and biomarkers development.

## 8. References

Adams, S.M.; Crumby, W.D.; Greeley, M.S.; Ryon, M.G. & Schilling, E.M. (1990). Relationship between Physiological and Fish Population Responses in a Contaminated Stream. Environmental Toxicology and Chemistry, Vol. 11, Issue 11, (November 1992), pp. 1549-1557, ISSN: 0730-7268

Alber, B.E. & Ferry, J.G. (1994). A carbonic anhydrase from the archaeon *Methanosarcina thermophila*. *Proceedings of the National Academy of Sciences of the United States of America*, Vol.91, (July 1994), pp. 6909-6913, ISSN 0027-8424

Alber, B.E. & Ferry, J.G. (1996). Characterization of heterologously produced carbonic anhydrase from *Methanosarcina thermophila*, The Journal of Bacteriology, Vol.178, No.11, (June 1996), pp.3270-3274, ISSN 0021-9193

Alber, B.E.; Colangelo, C.M.; Dong, J.; Staalhandske, C.M.V.; Baird, T.T.; Tu, C.; Fierke, C.A.; Silverman, D.N.; Scott, R.A. & Ferry, J.G. (1999). Kinetic and Spectroscopic Characterization of the Gamma-Carbonic Anhydrase from the Methanoarchaeon *Methanosarcina thermophil. Biochemistry*, Vol.38, Issue 40, (October 1999), pp. 13119-13128, ISSN 0006-2960

Berg, J.M. (2007). Biochemistry, 6th Ed., Sara Tenney. ISBN0-7167-8724-5

Bertini, I. & Luchinat, C. (1984). High spin cobalt(II) as a probe for the investigation of metalloproteins. *Advances in inorganic biochemistry*, Vol. 6, pp. 71-111, ISSN 0190-0218

Bielmyer, G.K.; Grosell, M.; Bhagooli, R.; Baker, A.C.; Langdon, C.; Gillette, P. & Capo, T.R. (2010). Differential effects of copper on three species of scleractinian corals and their algal symbionts (Symbiodinium spp.). *Aquatic Toxicology*, Vol. 97, No. 2, (April 2010), pp. 125-133, ISSN 0166-445X

Bozym, R.A.; Zeng, H.H.; Cramer, M.; Stoddard, A.; Fierke, C.A. & Thompson, R.B. (2004). In vivo and intracellular sensing and imaging of free zinc ion. In: *Proceedings of the SPIE Conference on Advanced Biomedical and Clinical Diagnostic Systems II*, Cohn, G.E., Grundfest, W.S., Benaron, D.A.; and Vo-Dinh, T., eds. Bellingham, WA: SPIE, 2004.

Burton, R.E.; Hunt, J.A.; Fierke, C.A. & Oas, T.G. (2000). Novel disulfide engineering in human carbonic anhydrase II using the PAIRWISE side-chain geometry database. *Protein Science*, Vol. 9, Issue 4, (April 2000), pp. 776-785, ISSN 0961-8368

Cannon, G.C.; Heinhorst, S. & Kerfeld C.A. (2010). Carboxysomal carbonic anhydrases: Structure and role in microbial CO(2) fixation. *Biochimica et Biophysica Acta-Proteins and Proteomics*, Vol. 1804, Issue 2, (February 2010), pp. 382-392, ISSN: 1570-9639

Caricato, R.; Lionetto, M.G.; Dondero, F.; Viarengo, A. & Schettino, T. (2010a) Carbonic anhydrase activity in *Mytilus galloprovincialis* digestive gland: sensitivity to heavy metal exposure. *Comparative Biochemistry and Phyiology Part C: Toxicology & Pharmacology*, Vol.152C, Issue 3, (September 2010), pp.241-247, ISSN 1532-0456

Caricato, R.; Lionetto, M.G. & Schettino, T. (2010b). Seasonal variation of biomarkers in *Mytilus galloprovincialis* sampled inside and outside Mar Piccolo of Taranto (Italy). *Chemistry and Ecology*, Vol.26, supplement 1, (June 2010), pp.143-153, ISSN 0275-7540

Ceyhun, S.B.; Şentürk, M.; Yerlikaya, E.; Erdoğan, O.; Küfrevioğlu, Ö.I. & Ekinci, D. (2011). Purification and characterization of carbonic anhydrase from the teleost fish *Dicentrarchus labrax* (European Seabass) liver and toxicological effects of metals on enzyme activity. *Environmental Toxicology and Pharmacology*. Vol.32, Issue 1, (July 2011), pp. 69-74, ISSN 1382-6689

Christensen, G.M. & Tucker, J.H. (1976). Effects of selected water toxicants on the in vitro activity of fish carbonic anhydrase. *Chemico Biological Interactions*, Vol.13, Issue 2, (May 1976), pp.181-92, ISSN 0009-2797

Cox, J.D.; Hunt, J.A.; Compher, K.M.; Fierke, C.A. & Christianson, D.W. (2000). Structural Influence of Hydrophobic Core Residues on Metal Binding and Specificity in Carbonic Anhydrase II. *Biochemistry*, Vol.39, No.45, (November 2000), pp.13687-13694, ISSN 0006-2960

De Mora, S.; Fowler, S.W.; Wyse, E. & Azemard, S. (2004). Distribution of heavy metals in marine bivalves, fish and coastal sediments in the Gulf and Gulf of Oman, *Marine Pollution Bulletin*, Vol. 49, Issue 5-6, (September 2004), pp.410-424, ISSN 0025-326X

Depledge, M.H. (1994). The rational basis for the use of biomarkers as ecotoxicological tools. In: *Nondestructive Biomarkers in Vertebrates*, M.C., Fossi; C., Leonzio (Eds.), pp. 271-295, Lewis Publisher, ISBN 978-0873716482, Boca Raton, USA

DiTusa, C.A.; McCall, K.A.; Chritensen, T.; Mahapatro, M.; Fierke, C.A. & Toone, E.J. (2001). Thermodynamics of metal ion binding. 2. Metal ion binding by carbonic anhydrase variants. *Biochemistry*, Vol.40, Issue 18, (May 2001), pp.5345-5351, ISSN 0006-2960

Ekinci, D.; Beydemir, Ş. & Küfrevioğlu Ö.İ. (2007). In vitro inhibitory effects of some heavy metals on human erytrocyte carbonic anhydrases. *Journal of Enzyme Inhibition and Medicinal Chemistry*, Vol.22, Issue 6, pp.745-750, ISSN 1475-6366

Elbaum, D.; Nair, S.K.; Patchan, M.W.; Thompson, R.B. & Christianson, D.W. (1996). Structure-based design of a sulfonamide probe for fluorescence anisotropy detection of zinc with a carbonic anhydrase-based biosensor. *Journal of the American Chemical Society*, Vol. 118, Issue 35, (September 1996), pp. 8381-8387, ISSN 0002-7863

Esbaugh, A.J. & Tufts, B.L. (2006). The structure and function of carbonic anhydrase isozymes in the respiratory system of vertebrates. *Respiratory Physiology & Neurobiology*, Vol.154, Issue 1-2, (November 2006), pp. 185-198, ISSN 1569-9048

Evans, D.H. & Cameron, J.N. (1986). Gill ammonia transport. *Journal of Experimental Zoology*, Vol.239, Issue 1, (July 1986), pp. 17-23, ISSN 0022-104X

Ferry, J.F. (2010). The gamma class of carbonic anhydrases. *Biochimica et Biophysica Acta (BBA) - Proteins & Proteomics*, Vol.1804, Issue 2, (February 2010), pp. 374-38, ISSN 1570-9639

Fierke, C.A. & Thompson, R.B. (2001). Fluorescence-based biosensing of zinc using carbonic anhydrase. *Biometals*, Vol. 14, Issue 3-4, (September 2001), pp. 205-222, ISSN 0966-0844

Gilbert, A.L. & Guzman, H.M. (2001). Bioindication potential of carbonic anhydrase activity in anemones and corals. *Marine Pollution Bulletin*, Vol. 42, Issue 9, (September 2001), pp. 742-744, ISSN 0025-326X

Goto, T.; Shirakawa, H.; Furukawa Y. & Komai, M. (2008). Decreased expression of carbonic anhydrase isozyme II, rather than of isozyme VI, in submandibular glands in long-term zinc-deficient rats. British Journal of Nutrition, Vol.99, Issue 2, (February 2008), pp. 248-53, ISSN 0007-1145

Grimes, A.; Paynter, J.; Walker, I.D.; Bhave, M. & Mercer, J.F.B. (1997). Decreased carbonic anhydrase III levels in the liver of the mouse mutant "toxic milk" (tx) due to copper accumulation, *Biochemical Journal*, Vol. 321, Part. 2, (January 1997), pp. 341-346, ISSN 0264-6021

Heisler, N. (1984). Acid-base regulation in fishes. In: *Hoar*, W.S., Randall, D.J. (Eds.), Fish Physiology, vol. 10A. Academic Press, New York, pp. 315-401.

Henkin, R.; Martin, B.M. & Agarwal, R. (1999). Efficacy of exogenous oral zinc in treatment of patients with carbonic anhydrase VI deficiency. *American Journal of the Medical Science, Vol.* 318, Issue 6, (December 1999), pp. 392-405, ISSN 0002-9629

Henry, R.P. (1984). The role of carbonic anhydrase in blood ion and acid–base regulation. *American Zoologist.* Vol. 24(1), pp. 241–253, 0003-1569.

Hewett-Emmett, D. & Tashian, R.E. (1996). Functional diversity, conservation, and convergence in the evolution of the α-, ß-, and γ-carbonic anhydrase gene families. *Molecular Phylogenetics and Evolution, Vol.* 5, Issue 1, (February 1996), pp. 50–77, ISSN 1055-7903

Hoffmann, K.M.; Samardzic, D.; van den Heever, K. & Rowlett, R.S. (2011). Co(II)-substituted Haemophilus influenzae b-carbonic anhydrase: Spectral evidence for allosteric regulation by pH and bicarbonate ion. *Archives of Biochemistry and Biophysics*, Vol. 511, Issue 1-2, (July 2011), pp. 80–87, ISSN 0003-9861

Hove, C.; Elvehjem, C.A. & Hart, E.B. (1940). The relation of zinc to carbonic anhydrase. J Biol Chem 136:425–434

Hunt, J.A. & Fierke, C.A. (1997). Selection of carbonic anhydrase variants displayed on phage. Aromatic residues in zinc binding site enhance metal affinity and equilibration kinetics. *Journal of Biological Chemistry,* Vol.272, Issue 33, (August 1997), pp. 20364−20372, ISSN 0021-9258

Hunt, J.A.; Ahmed, M. & Fierke, C.A. (1999). Metal Binding Specificity in Carbonic Anhydrase is Influenced by Conserved Hydrophobic Amino Acids. *Biochemistry*, Vol. 38, Issue 28, (July 1999), pp. 9054-9060, ISSN 0006-2960

Hwang, H.; Green, P.G.; Higashi, R.M. & Young, T.M. (2006). Tidal salt marsh sediment in California, USA. Part 2: Occurrence and anthropogenic input of trace metals, *Chemosphere*, Vol. 64, Issue 11, (September 2006), pp.1899-1909, ISSN: 0045-6535

Innocenti, A.; Scozzafava, S.; Parkkila, L.; Puccetti, G.; De Simone, G. & Supuran, C.T. (2008). Investigations of the esterase, phosphatase, and sulfatase activities of the cytosolic mammalian carbonic anhydrase isoforms I, II, and XIII with 4-nitrophenyl esters as substrates. *Bioorganic & Medicinal Chemistry Letters*, Vol. 18, Issue 7, (April), pp.2267–2271, ISSN 0960-894X

Ivanov, B.N.; Ignatova, L.K. & Romanova, A.K. (2007). Diversity in Forms and Functions of Carbonic Anhydrase in Terrestrial Higher Plants. *Russian Journal of Plant Physiology*, Vol. 54, No. 2, (March-April 2007), pp. 143–162, ISSN 1021-4437

Jernelov, A. (1996). The international mussel watch: a global assessment of environmental levels of chemical contaminats. *Science of the Total Environment,* Vol.188, Supplement 1, (September 1996), pp. 37-44, ISSN: 0048-9697

Kiefer L.L.; Paterno S.A. & Fierke, C.A. (1995). Hydrogen-Bond Network in the Metal-Binding Site of Carbonic Anhydrase Enhances Zinc Affinity and Catalytic Efficency. *Journal of the American Chemical Society*, Vol. 117, Issue 26, (July 1995), pp. 6831-6837, ISSN 0002-7863

Kimber, M.S. & Pai, E.F. (2000). The active site architecture of Pisum sativum β-carbonic anhydrase is a mirror image of that of α-carbonic anhydrases. *Embo Journal, Vol.* 19, Issue 7, (April 2000), pp.1407–1418, ISSN 0261-4189

Kirchgessner M, Stadler AE, Roth HP (1975) Carbonic anhydrase activity and erythrocyte count in the blood of zinc-deficient rats. *Bioinorganic chemistry*, Vol. 5, Issue 1, pp. 33-38, ISSN: 0006-3061

Kuhara, M.; Wang, J.; Flores, M.J.; Qiao, Z.; Koizumi, Y.; Koyota, S.; Taniguchi, M. & Sugiyama, T. (2011). Sexual dimorphism in LEC rat liver: suppression of carbonic anhydrase III, by copper accumulation during hepatocarcinogenesis. Biomedical Research, Vol. 32 (2), (April 2011), pp. 111-117, ISSN 0388-6107

Lane, T.W. & Morel, F.M.M. (2000). A biological function for cadmium in marine diatoms. Proceedings of the National Academy of Sciences of the United States of America, Vol.97, Issue 9, (April 2000), pp. 4627–4631, ISSN 0027-8424

Lane, T.W.; Saito, M.A.; George, G.N.; Pickering, I.J.; Prince, R.C. & Morel, F.M.M. (2005). A cadmium enzyme from a marine diatom. Nature, Vol. 435, Issue 7038, (May 2005), pp.42–42, - ISSN 0028-0836

Lee, J.G.; Roberts, S.B. & Morel, F.M.M. (1995). Cadmium a nutrient for the marine diatom. Limnology and Oceanography, Vol. 40, Issue 6, (September 1995), pp. 1056-1063, ISSN 0024-3590

Liang, J.Y. & Lipscomb, W. N (1988). Hydration of $CO_2$ By Carbonic-Anhydrase – Intramolecular Proton Transfer between $Zn^{2+}$-Bound $H_2O$ and Histidine-64 in Human Carbonic Anhydrase-II. Biochemistry, Vol. 27, Issue 23, (November 1988), pp. 8676-8682, ISSN 0006-2960

Lionetto, M.G.; Maffia, M.; Cappello, M.S.; Giordano, M.E.; Storelli, C. & Schettino, T. (1998). Effect of cadmium on carbonic anhydrase and Na+-K+-ATPase in eel, Anguilla anguilla, intestine and gills. Comparative Biochemistry and Physiology A- Molecular and Integrative Physiology, Vol.120, Issue 1, (May 1998), pp.89-91, ISSN 1095-6433

Lionetto, M.G.; Giordano, M.E.; Vilella, S. & Schettino, T. (2000). Inhibition of eel enzymatic activities by cadmium. Aquatic Toxicology, Vol. 48,Issue 4, (April 2000), pp. 561-571, ISSN: 0166-445X

Lionetto, M.G.; Caricato, R.; Giordano, M.E.; Pascariello, M.F.; Marinosci, L. & Schettino, T. (2003). Integrated use of biomarkers (acetylcholinesterase and antioxidant enzymatic activities) in Mytilus galloprovincialis and Mullus barbatus in an Italian coastal marine area. Marine Pollution Bulletin, Vol. 46, Issue 3, (March 2003), pp. 324-330, ISSN 0025-326X

Lionetto, M.G.; Caricato, R.; Erroi, E.; Giordano, M.E. & Schettino, T. (2005). Carbonic anhydrase based environmental bioassay. International Journal of Environmental Analytical Chemistry, Vol.85, Issue 12-13, (October-November 2005), pp. 895-903, ISSN: 0306-7319

Lionetto, M.G.; Caricato, R.; Erroi, E.; Giordano, M.E. & Schettino, T. (2006). Potential application of carbonic anhydrase activity in bioassay and biomarker studies. Chemistry and Ecology, Vol. 22, Supplement 1, pp. 119-125, ISSN: 0275-7540

Lukaski, H.C. (2005). Low dietary zinc decreases erythrocyte carbonic anhydrase activities and impairs cardiorespiratory function in men during exercise. American Journal of Clinical Nutrition, Vol 81, Issue 5, (May 2005), pp.1045–1051, ISSN 0002-9165

MacAuley, S.R.; Zimmerman, S.A.; Apolinario, E.E.; Evilia, C.; Hou, Y.; Ferry, J.G. & Sowers, K.R. (2009). The archetype γ-class carbonic anhydrase (Cam) contains iron when synthesized in vivo, Biochemistry, Vol. 48, Issue 5, (February 2009) pp 817–819, ISSN: 0006-2960

Maffia, M.; Trischitta, F.; Lionetto, M.G.; Storelli, C. & Schettino, T. (1996). Bicarbonate absorption in eel intestine: Evidence for the presence of membrane-bound carbonic anhydrase on the brush border membranes of the enterocyte. *Journal of Experimental Zoology*, Vol: 275, Issue 5, (August 1996), pp. 365-373, ISSN 0022-104X

Maret, W. (2011). Metals on the move: zinc ions in cellular regulation and in the coordination dynamics of zinc proteins. *Biometals*,Vol. 24, Issue 3, (June 2011), pp. 411-418, ISSN: 0966-0844

Marouan, R.; Cecchi, A; Montero, J.L.; Innocenti, A.; Vullo, D; Scozzafava, A.; Winum, J.Y. &. Supuran, C.T. (2008) Carbonic Anhydrase Inhibitors: Design of Membrane-Impermeant Copper(II) Complexes of DTPA-, DOTA-, and TETA-Tailed Sulfonamides Targeting the Tumor-Associated Transmembrane Isoform IX. ChemMedChem, Vol. 3, Issue 11, (November 2008), pp. 1780 – 1788, ISSN 1860-7179

McCall, K.A. & Fierke, C.A. (2004). Probing Determinants of the Metal Ion Selectivity in Carbonic Anhydrase Using Mutagenesis. Biochemistry, Vol. 43, Issue 13, (April 2004), pp. 3979-3986, ISSN 0006-2960

Mei, Y.J.; Frederickson, C.J.; Giblin, L.J.; Weiss, J.H.; Medvedeva, Y. & Bentley, PA (2011). Sensitive and selective detection of zinc ions in neuronal vesicles using PYDPY1, a simple turn-on dipyrrin. Chemical Communications, Vol. 47, Issue 25, pp. 7107-7109, ISSN 1359-7345

Morel, F.M.M.; Reinfelder, J.R.; Roberts, S.B.; Chamberlain, C.P.; Lee, J.G. & Yee, D. (1994). Zinc and carbon co-limitation of marine phytoplankton. *Nature*, Vol. 369, Issue 6483, (June 1994), pp. 740–742, ISSN: 0028-0836

Morgan, I.J.; Henry, R.P. & Wood, C.M. (1997). The mechanism of acute silver nitrate toxicity in freshwater rainbow trout (Oncorhynchus mykiss) is inhibition of gill Na+ and Cl- transport. *Aquatic Toxicology*,Vol.38, Issue 1-3, (May 1997), pp. 145–63, ISSN: 0166-445X

Morgan, T.P.; Grosell, M.; Gilmour, K.M.; Playle, R.C. & Wood, C.M. (2004). Time course analysis of the mechanism by which silver inhibits active Na+ and Cl- uptake in gills of rainbow trout. *American Journal of Physiology-Regulatory Integrative and Comparative Physiology*, Vol. 287, Issue 1, (July 2004), ISSN 0363-6119

Moya, A.; Ferrier-Pages, C.; Furla, P.; Richier, S.; Tambutte, E.; Allemand, D. & Tambutte, S. (2008). Calcification and associated physiological parameters during a stress event in the scleractinian coral Stylophora pistillata. *Comparative Biochemistry and Physiology A – Molecular & Integrative Physiology*, Vol. 151, Issue 1, (September 2008), pp. 29–36, ISSN 1095-6433

Park, H.; McGinn, P.J. & Morel, F.M.M. (2008). Expression of cadmium carbonic anhydrase of diatoms in seawater. Aquatic Microbial. Ecology, Vol.51, Issue2, (May 2008), pp. 183–193, ISSN: 0948-3055

Park, H.; Song, B. & Morel, F.M.M. (2007). Diversity of the cadmium-containing carbonic anhydrase in marine diatoms and natural waters. *Environmental Microbioogyl*, Vol. 9, Issue 2, (Feb 2007), pp. 403–413, ISSN: 1462-2912

Price, N.M. & Morel, F.M.M. (1990). Cadmium and cobalt substitution for zinc in a marine diatom. *Nature*, Vol. 344, Issue 6267, (April 1990), pp. 658–660, ISSN 0028-0836

Rahman, M.W.; Davies, R.E.; Deyoe, C.W.; Reid, B.L. &. Couch, J.R. (1961). Role of zinc in the nutrition of growing pullets. *Poultry Science*, Vol. 40, pp. 195-200, ISSN 0032-5791.

Randall, D.J. & Daxbaeck, C. (1984). Oxygen and carbon dioxide transfer across fish gills. In: Hoar, W.S., Randall, D.J. (Eds.), Fish Physiology, vol. 10A. Academic Press, New York, pp. 263–314, ISBN 0-12-350430-9

Roberts, S.B.; Lane, T.W. & Morel, F.M.M. (1997). Carbonic anhydrase in the marine diatom Thalassiosira weissflogii (Bacillariophyceae). Journal of Phycology, Vol. 33, Issue 5, (October 1997), pp. 845–850, ISSN 0022-3646

Rowlett, R.S. (2010). Structure and catalytic mechanism of the $\beta$-carbonic anhydrases. Biochimica et Biophysica Acta (BBA) - Proteins & Proteomics, Vol.1804, Issue 2, (February 2010), pp. 362-373, ISSN: 1570-9639

Schettino, T.; Trischitta, F; Denaro, M.G.; Faggio, C. & Fucile, I. (1992). Requirement of $HCO_3^-$ For $Cl^-$-Absorption in Seawater-Adapted Eel Intestine. Pflugers Archiv-European Journal of Physiology, Vol. 421, Issue 2-3, (June 1992), pp. 146-154, ISSN 0031-6768

Schettino, T.; Lionetto, M.G. & Erroi, E. (2008). Enzymatic method for the dectetion of the tozicity of aqueous environmental matrices. Patent n. MI2008A008813, PCT/EP2008/064703

Skaggs, H.S. & Henry, R.P. (2002). Inhibition of carbonic anhydrase in the gills of two euryhaline crabs, Callinectes sapidus and Carcinus maenas, by heavy metals. Comparative Biochemistry and Physiology C-Toxicology & Pharmacology, Vol. 133, Issue 4, (December 2002), pp. 605-612, ISSN 1532-0456

Soto, M.; Ireland, M.P. & Marigómez, I. (2000). Changes in mussel biometry on exposure to metals: implications in estimation of metal bioavailability in "Mussel-Watch" programmes. The Sciences of the Total Environment, Vol. 247, Issue 2-3, (March 2000), pp. 175-187, ISSN 0048-9697

Soyut, H.; Beydemir, Ş. & Hisar, O. (2008). Effects of Some Metals on Carbonic Anhydrase from Brains of Rainbow Trout. Biological Trace Element Research, Vol.123, Issue 1-3, (June 2008), pp.179–190, ISSN 0163-4984

Supuran, C.T.; Scozzafava, A. & Casini, A. (2003). Carbonic anhydrase inhibitors. Medicinal Research Reviews, Vol.23, Issue 2, (March 2003), pp. 146– 189, ISSN: 0198-6325

Supuran, C.T. & Scozzafava, A. (2007). Carbonic anhydrases as targets for medicinal chemistry. Bioorganic & Medicinal Chemistry, Vol. 15, Issue 13, (July 2007), pp. 4336–4350, ISSN: 0968-0896

Supuran C.T., (2008). Carbonic anhydrases: novel therapeutic applications for inhibitors and activators. Nature Reviews Drug Discovery, Vol.7, Issue 2, (February), pp.168–181, ISSN 1474-1776

Supuran, C.T. (2010). Carbonic anhydrase inhibitor. Bioorganic & Medicinal Chemistry Letters, Vol. 20, Issue 12, (June), pp. 3467-3474, ISSN: 0960-894X

Tashian, R.E.; Hewett-Emmett, D.; Carter, N.D. & Bergenhem, N.C.H. (2000). Carbonic anhydrase (CA)-related proteins (CA-RPs) and transmembrane proteins with CA or CA-RP domains. In: Chegwidden, W.R.; Carter, N.D. & Edwards, Y.H. (Eds.), Carbonic anhydrase (CA)-related proteins (CA-RPs) and transmembrane proteins with CA or CA-RP domains. The Carbonic Anhydases: New Horizons. Birkhauser, Basel, pp. 105–120, ISBN 3-7643-5670-7

Thompson, R.B. & Jones, E.R. (1993). Enzyme-based fiber optic zinc biosensor. Analitical Chemistry, Vol.65, Issue 6, March 1993), pp. 730-734, ISSN 0003-2700

Thompson, R.B. & Patchan, M.W. (1995). Lifetime-Based Fluorescence Energy-Transfer Biosensing of Zinc. *Analytical Biochemistry*, Vol. 227, Issue 1, (May 1995), pp. 123-128, ISSN 0003-2697

Thompson, R.B.; Maliwal, B.P. & Zeng H.H. (2000). Zinc biosensing with multiphoton excitation using carbonic anhydrase and impoved fluorofores. Journal of Biomedical Optics, Vol. 5, Issue 1, (January 2000), pp. 17-22, ISSN 1083-3668

Thompson, R.B.; Thompson, R.B. & Meisinger, J.J. (2002). Fluorescent zinc indicators for neurobiology. Journal of Neuroscience Methods, Vol. 118, Issue 1, (July 2002), pp. 63-75, ISSN 0047-2425

Tripp, B.C.; Bell, C.B.; Cruz, F.; Krebs, C.& Ferry, J.G. (2004). A role for iron in an ancient carbonic anhydrase, Journal of Biological Chemistry, Vol. 279, Issue 20, (May), pp. 21677-21677, ISSN 0021-9258

Tripp, B.C.; Smith K.S. & Ferry J.G. (2001)., Journal of Biological Chemistry, Vol.276, Issue 52, (December), pp. 48615– 48618, ISSN: 0021-9258

Tu, C.; Silverman, D. N.; Forsman, C.; Jonsson, B.-H. & Lindskog, S. (1989). Role of Histidine-64 in the Catalytic Mechanism of Human Carbonic Anhydrase-II Studied with a Site-Specific Mutation. Biochemistry, Vol.28, Issue 19, (September 1989), pp. 7913-7918, ISSN 0006-2960

Tu, C.K. & Silverman, D.N. (1985). Catalysis by cobalt(II)- substituted carbonic anhydrase II of the exchange of oxygen-18 between CO2 and H2O. *Biochemistry*, Vol.24, Issue 21, pp. 5881–5887, ISSN: 0006-2960

Tu, C; Wynns, G.C. & Silverman, D.N. (1981). Inhibition by cupric ions of $^{18}$O exchange catalyzed by human carbonic anhydrase II. Relation to the Interaction Between Carbonic Anhydrase and Hemoglobin. *The Journal of Biological Chemistry*, Vol. 256, No.18, (September 1981), pp. 9466-9470, ISSN 0021-9258

Vedani, A.; Huhta, D. W. & Jacober, S. P. J. (1989). Metal Coordination, H-Bound Network Formation, and Protein-Solvent Interactions in Native and Complexed Human Carbonic Anhydrase-I - a Molecular Mechanism Study. *Journal of the American Chemical Society*, Vol.111, Issue 11, (May 1989), pp. 4075-4081, ISSN: 0002-7863

Viarengo, A.; Ponzano, E.; Dondero, F. & Fabbri, R. (1997). A simple spectrophotometric method for metallothionein evaluation in marine organisms: an application to Mediterranean and Antartic molluscs. *Marine Environmental Research*, Vol. 44, Issue 1, (July 1997), pp. 69–84, ISSN 0141-1136

Vitale, A.M.; Monserrat, J.M.; Casthilo, P. & Rodriguez, E.M. (1999). Inhibitory effects of cadmium on carbonic anhydrase activity and ionic regulation of the estuarine crab, Chasmagnathus granulata (Decapoda, Grapsidae). *Comparative Biochemistry and physiology C-Toxicology & Pharmacology*, Vol. 122, Issue 1, (January 1999), pp. 121-129, ISSN 1532-0456

Wang, B.; Liu, C.Q. & Wu, Y. (2005). Effect of Heavy Metals on the Activity of External Carbonic Anhydrase of Microalga Chlamydomonas reinhardtii and Microalgae from Karst Lakes. *Bulletin of Environmental Contamination and Toxicology*, Vol. 74, Issue 2, (February 2005), pp. 227-233, ISSN 0007-4861

Whitney, P.L. & Briggle, T.V. (1982). Membrane-associated carbonic anhydrase purified from bovine lung. *The Journal of Biological Chemistry*, Vol.257, (October 1982), pp. 12056-12059, ISSN 0021-9258

Wu, Y.; Zhao, X.; Li, P. & Huang, H. (2007). Impact of Zn, Cu, and Fe on the Activity of Carbonic Anhydrase of Erythrocytes in Ducks. *Biological Trace Element Research*, Vol. 118, No. 3, (September 2007), pp. 227-232, ISSN 0163-4984

Xu, Y.; Feng, L.; Jeffrey, P.D.; Shi, Y. & Morel, F.M. (2008). Structure and metal exchange in the cadmium carbonic anhydrase of marine diatoms. *Nature*, Vol. 452, Issue 7183, (March 2008), pp.56-U3, ISSN 0028-0836

Yee, D. & Morel, F.M.M. (1996). In vivo substitution of zinc by cobalt in carbonic anhydrase of a marine diatom. *Limnology and Oceanography*, Vol.41, Issue 3, (May 1996), pp. 573-577, ISSN 0024-3590

Zeng, H.H.; Thompson, R.B.; Maliwal, B.P.; Fones, G.R.; Moffet, J.W. & Fierke, C.A. (2003). Real-time determination of picomolar free Cu(II) in seawater using a fluorescence-based fiber optic biosensor. *Analytical Chemitry*, Vol.75, Issue 24, (December 2003), pp. 6807-6812, ISSN 0003-2700

Zhang, B.Y.; Yang, F.; Wang, G.C. & Peng, G. (2010) Cloning and Quantitative Analysis of the Carbonic Anhydrase Gene from Porphyra yezoensis. *Journal of Phycology*, Vol. 46, Issue 2, (April 2010), pp.290-296, ISSN 0022-3646

# Permissions

The contributors of this book come from diverse backgrounds, making this book a truly international effort. This book will bring forth new frontiers with its revolutionizing research information and detailed analysis of the nascent developments around the world.

We would like to thank Dr. Deniz Ekinci, for lending his expertise to make the book truly unique. He has played a crucial role in the development of this book. Without his invaluable contribution this book wouldn't have been possible. He has made vital efforts to compile up to date information on the varied aspects of this subject to make this book a valuable addition to the collection of many professionals and students.

This book was conceptualized with the vision of imparting up-to-date information and advanced data in this field. To ensure the same, a matchless editorial board was set up. Every individual on the board went through rigorous rounds of assessment to prove their worth. After which they invested a large part of their time researching and compiling the most relevant data for our readers. Conferences and sessions were held from time to time between the editorial board and the contributing authors to present the data in the most comprehensible form. The editorial team has worked tirelessly to provide valuable and valid information to help people across the globe.

Every chapter published in this book has been scrutinized by our experts. Their significance has been extensively debated. The topics covered herein carry significant findings which will fuel the growth of the discipline. They may even be implemented as practical applications or may be referred to as a beginning point for another development. Chapters in this book were first published by InTech; hereby published with permission under the Creative Commons Attribution License or equivalent.

The editorial board has been involved in producing this book since its inception. They have spent rigorous hours researching and exploring the diverse topics which have resulted in the successful publishing of this book. They have passed on their knowledge of decades through this book. To expedite this challenging task, the publisher supported the team at every step. A small team of assistant editors was also appointed to further simplify the editing procedure and attain best results for the readers.

Our editorial team has been hand-picked from every corner of the world. Their multi-ethnicity adds dynamic inputs to the discussions which result in innovative outcomes. These outcomes are then further discussed with the researchers and contributors who give their valuable feedback and opinion regarding the same. The feedback is then collaborated with the researches and they are edited in a comprehensive manner to aid the understanding of the subject.

Apart from the editorial board, the designing team has also invested a significant amount of their time in understanding the subject and creating the most relevant covers. They scrutinized every image to scout for the most suitable representation of the subject and create an appropriate cover for the book.

The publishing team has been involved in this book since its early stages. They were actively engaged in every process, be it collecting the data, connecting with the contributors or procuring relevant information. The team has been an ardent support to the editorial, designing and production team. Their endless efforts to recruit the best for this project, has resulted in the accomplishment of this book. They are a veteran in the field of academics and their pool of knowledge is as vast as their experience in printing. Their expertise and guidance has proved useful at every step. Their uncompromising quality standards have made this book an exceptional effort. Their encouragement from time to time has been an inspiration for everyone.

The publisher and the editorial board hope that this book will prove to be a valuable piece of knowledge for researchers, students, practitioners and scholars across the globe.

# List of Contributors

**Lifeng Cai, Weiguo Shi and Keliang Liu**
Beijing Institute of Pharmacology & Toxicology, Beijing, China

**Mohammad T. Elnakish and Hamdy H. Hassanain**
Department of Anesthesiology, Dorothy M. Davis Heart & Lung Research Institute; Molecular, Cellular and Developmental Biology Program, the Ohio State University, Columbus, OH, USA

**Spyridoula N. Charova, Anastasia D. Gazi, Panagiotis F. Sarris, Nickolas J. Panopoulos, and Michael Kokkinidis**
Institute of Molecular Biology & Biotechnology, Foundation of Research & Technology, Greece
Department of Biology, University of Crete, Vasilika Vouton, Heraklion, Crete, Greece

**Vassiliki E. Fadouloglou**
Department of Biology, University of Crete, Vasilika Vouton, Heraklion, Crete, Greece
Department of Molecular Biology and Biotechnology, Democritus University of Thrace, Alexandroupolis, Greece

**Marianna Kotzabasaki**
Department of Biology, University of Crete, Vasilika Vouton, Heraklion, Crete, Greece

**Ivonne M.C.M. Rietjens**
Toxicology section, Wageningen University, Wageningen, The Netherlands

**Jacques Vervoort**
Laboratory of Biochemistry, Wageningen University, Wageningen, The Netherlands

**Jan-Åke Gustafsson**
Department of Biology and Biochemistry, Center for Nuclear Receptors and Cell Signaling, University of Houston, Science & Engineering Research Center, Houston, USA
Department of Biosciences and Nutrition, Karolinska Institutet, Novum, Huddinge, Sweden

**Ana M. Sotoca**
Toxicology section, Wageningen University, Wageningen, the Netherlands
Laboratory of Biochemistry, Wageningen University, Wageningen, The Netherlands

**Bakás Laura, Maté Sabina, Vazquez Romina and Herlax Vanesa**
Instituto de Investigaciones Bioquímicas La Plata (INIBIOLP), CCT- La Plata, CONICET, Facultad de Ciencias Médicas, Universidad Nacional de La Plata. La Plata, Buenos Aires, Argentina

**Waelbroeck Magali**
Université Libre de Bruxelles, Belgium

**Kazunori Namba**
National Institute of Sensory Organs, National Tokyo Medical Center, Tokyo, Japan

**Maria Giulia Lionetto, Roberto Caricato, Maria Elena Giordano, Elisa Erroi and Trifone Schettino**
University of Salento - Dept. of Biological and Environmental Sciences and Technologies, Italy

Printed in the USA
CPSIA information can be obtained
at www.ICGtesting.com
JSHW011421221024
72173JS00004B/628